Enzymes: Novel Biotechnological Approaches

Enzymes: Novel Biotechnological Approaches

Editor: Caden Hamilton

Callisto Reference
www.callistoreference.com

Callisto Reference,
118-35 Queens Blvd., Suite 400,
Forest Hills, NY 11375, USA

Visit us on the World Wide Web at:
www.callistoreference.com

ISBN: 978-1-64116-796-3 (Hardback)

Cataloging-in-Publication Data

Enzymes : novel biotechnological approaches / edited by Caden Hamilton.
p. cm.
Includes bibliographical references and index.
ISBN 978-1-64116-796-3
1. Enzymes--Textbooks. 2. Enzymes--Biotechnology--Textbooks. 3. Enzymology--Textbooks.
4. Biochemistry--Textbooks. 5. Biotechnology--Textbooks. I. Hamilton, Caden.
QP601 .E59 2023
572.7--dc23

Table of Contents

Preface

This book has been an outcome of determined endeavour from a group of educationists in the field. The primary objective was to involve a broad spectrum of professionals from diverse cultural background involved in the field for developing new researches. The book not only targets students but also scholars pursuing higher research for further enhancement of the theoretical and practical applications of the subject.

Enzymes are proteins that act as biological catalysts, which accelerate the rate of biochemical reactions taking place inside a living organism. Biological processes such as respiration, digestion, excretion, and metabolic processes that take place within the cells are performed with the help of enzymes necessary for sustaining life. Enzymes are affected by various environmental factors such as pH level and temperature. There are six types of enzymes, namely, oxidoreductases, hydrolases, lyases, transferases, isomeras, and ligases. Enzymes are used in various fields. They are specifically used in the field of biotechnology. Enzymes are used as industrial catalysts, therapeutic agents, manipulative tools, and analytic reagents. In medicine, they are used in making cultures and developing medicines. This book provides significant information on enzymes and their biotechnological applications. A number of latest researches have been included to keep the readers updated with the global concepts in this area of study. Researchers and students in the field of biotechnology will be assisted by this book.

It was an honour to edit such a profound book and also a challenging task to compile and examine all the relevant data for accuracy and originality. I wish to acknowledge the efforts of the contributors for submitting such brilliant and diverse chapters in the field and for endlessly working for the completion of the book. Last, but not the least; I thank my family for being a constant source of support in all my research endeavours.

Editor

1

Enzymatic Polymerization on DNA Modified Gold Nanowire for Label-Free Detection of Pathogen DNA

Jaepil Jeong, Hyejin Kim and Jong Bum Lee *

Department of Chemical Engineering, University of Seoul, Seoul 130-743, Korea;
E-Mails: jpjeong@uos.ac.kr (J.J.); ajin2010@uos.ac.kr (H.K.)

* Author to whom correspondence should be addressed; E-Mail: jblee@uos.ac.kr

Academic Editor: Vladimír Křen

Abstract: This paper presents a label-free biosensor for the detection of single-stranded pathogen DNA through the target-enhanced gelation between gold nanowires (AuNW) and the primer DNAs branched on AuNW. The target DNA enables circularization of the linear DNA template, and the primer DNA is elongated continuously via rolling circle amplification. As a result, in the presence of the target DNA, a macroscopic hydrogel was fabricated by the entanglement of the elongated DNA with AuNWs as a scaffold fiber for effective gelation. In contrast, very small separate particles were generated in the absence of the target DNA. This label-free biosensor might be a promising tool for the detection of pathogen DNAs without any devices for further analysis. Moreover, the biosensor based on the weaving of AuNW and DNAs suggests a novel direction for the applications of AuNWs in biological engineering.

Keywords: biosensor; label-free; pathogen DNA detection; rolling circle amplification (RCA); gold nanowire

---

1. Introduction

Deoxyribonucleic acid (DNA) has significant potential as a multifunctional material, and its programmability through base-paring of DNA [1,2] has enabled the use of DNA as a building block. By taking advantage of DNA as a versatile material, a wide range of applications have been introduced, such as drug delivery [3,4], pathologic diagnosis [5,6], microscale patterning [7,8], arraying nanomaterials [9,10], and construction of structures in various shapes [11–13]. Furthermore, the conjugation of DNA with various materials, such as gold [14,15], silver [16,17], and magnetic core [18], have endowed additional functions to the DNA nanostructures. Among them, gold has been adopted widely for its optical and photothermal properties induced by its localized surface plasmon resonance in a variety of shapes, such as nanoparticles [19,20], nanorods [21], nanoplate films [22], and nanoclusters [23]. To date, however, there are few reports on the use of gold nanowires (AuNWs) in biological engineering. In addition, AuNWs in the majority of the studies focused on gold nanoparticles assembled in a line rather than gold nanowires per se [24–26].

This paper proposes a label-free biosensor for single-stranded DNA detection induced by the interlacing of AuNW and primer DNA branched on the AuNW. The rolling circle amplification (RCA) technique was used on the branched primer DNA. RCA is an isothermal and enzymatic process, which enables the continuous amplification of DNA sequences [27–29]. In the present study, macroscopic hydrogel was produced via the entanglement of elongated primer DNA and AuNWs in the presence of the target pathogen DNA. To the best of the authors' knowledge, this is the first approach of the use of AuNWs as a scaffold fiber for effective gelation. Using this biosensor, the existence of pathogen DNA was distinguishable by naked eye, which may offer a novel way for point-of-care diagnosis.

2. Results and Discussion

2.1. Design of AuNW: DNA Based Biosensor

To synthesize the AuNW: DNA based biosensor, the target DNA was first designed from influenza A virus DNA. Subsequently, 13 bases from the 5′ end and 14 bases from the 3′ end of the linear DNA were designed to be complementary to the 27 bases of the target DNA, leaving a few bases at both ends. By rational design, the linear DNA can be circularized in the presence of the target DNA by temperature annealing, as shown in Figure 1A. The discontinuous region between the 5′ and 3′ ends of the linear DNA was connected enzymatically by ligation, and closed circular DNA was formed. On the other hand, in the absence of the target DNA, the complete circularization of linear DNA is impossible, even after ligation.

The 5′ end of primer DNA was modified chemically with an alkane thiol group to immobilize the primer DNA on AuNW, as shown in Figure 1B. After removing the excess primer DNA, the primer DNA immobilized on AuNW was allowed to hybridize with the linear or circular DNA for the initiation of RCA. Moreover, 12 bases (denoted as A) and the other 12 bases (denoted as A′) of the linear DNA were designed complementary to each other, which enabled efficient cross-linking.

Because the surface of AuNW is covered with cationic cetyltrimethylammonium bromide (CTAB) molecules, it is essential to determine if the primer DNAs are functionalized on AuNWs by sulfur-gold bonding or electrostatic interactions due to the negatively charged nature of DNA. To examine the binding characteristic between the primer DNA and AuNW, the melting temperature between the primer DNAs on the AuNW and the linear DNAs were investigated using the DNA-intercalating fluorescence dye. A sudden decrease in the fluorescence intensity on the melt curve shown in Figure 2 suggests that the primer DNAs is immobilized on the AuNW through a sulfur-gold interaction. As the charge-induced binding between DNA and AuNW is irregular, programmed assembly between the primer DNA and linear DNA would be inhibited, and randomly denatured DNA would result in a gradual decrease in the fluorescence intensity on the melt curve rather than a sudden decline [30].

To demonstrate the target-enhanced gelation via rolling circle amplification, the previously circularized DNA template was allowed to hybridize with the primer DNA modified on AuNW. As the 3′ end of the target DNA is designed not to hybridize with linear DNA, the 3′ end of the primer DNA is the only initiation site for phi29 DNA polymerase. As shown in Figure 1B, the DNA strands were elongated continuously via RCA from the previously synthesized circular DNA in the presence of the target DNA. The AuNW is believed to work as a scaffold and amplified DNA strands on AuNW function as the side chains that facilitate expeditious gelation. In absence of the target DNA, however, RCA was not performed with the unligated linear DNA. Therefore, the primer DNA could not be elongated. As a result, macroscopic hydrogel is synthesized only in the presence of the target DNA. Also, same results were obtained when reacson was performed with library of non-complementary DNA, in the presence (see Figure S1) or absence of target DNA (see Figure S2), which implies potential for selective detection.

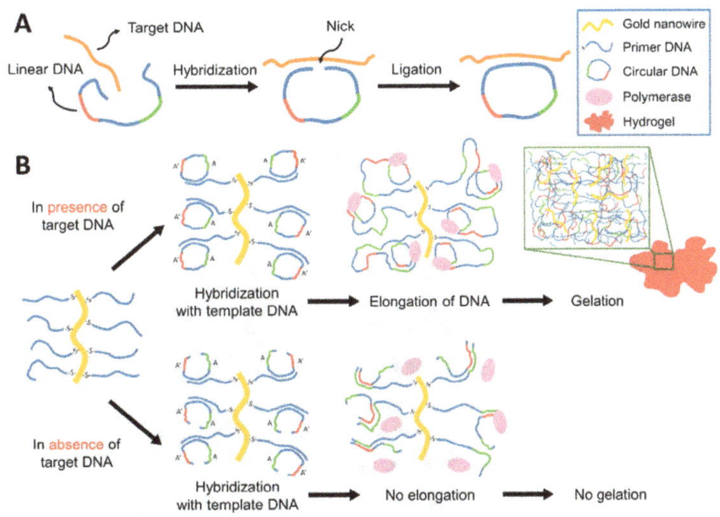

Figure 1. Schematic illustration of the label-free detection of single-stranded DNA. (**A**) Scheme of circular DNA preparation in the presence of target DNA; and (**B**) scheme of hydrogel synthesis with the prepared circular DNA. S indicates thiol group. Sequence of A and A′ are complementary.

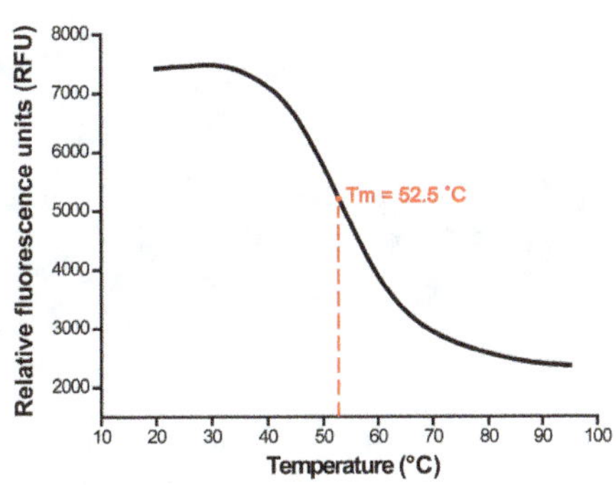

Figure 2. Melt curve of the primer DNA modified on gold nanowires. The temperature was increased gradually by 0.5 °C from 20 to 95 °C.

2.2. Characterization of AuNW: DNA Based Biosensor

After the RCA process was conducted on the AuNW: DNA based biosensor, the products were analyzed to confirm the potential utilization of the AuNW: DNA based biosensor for pathogen DNA detection. The digital camera image of the RCA product shown in Figure 3A showed that the hydrogel formed in the presence of the target DNA is apparently visible to the naked eye. In contrast, no macroscopic structures were observed in the absence of the target DNA (Figure 3C). After staining DNA hydrogel with DNA specific dye (GelRed), strong fluorescence from hydrogel also confirmed that the resulting product was made of DNA (Figure 3B). Moreover, the fluorescence microscopic image of the stained hydrogel showed that many thin DNA film-like structures were closely packed and interlaced with each other in the presence of the target DNA (Figure 3B). However, only small features were observed in the absence of the target DNA (Figure 3D). The microscopic structures of the RCA products in the presence of target DNA were investigated further by scanning electron microscopy (SEM) and transmission electron microscopy (TEM). The surficial structure of the hydrogel was decorated with porous ball-like particles (Figure 3E). The TEM image demonstrated in Figure 3F showed that DNAs were entangled with the gold nanowires as a result of the crosslinking of elongated DNA.

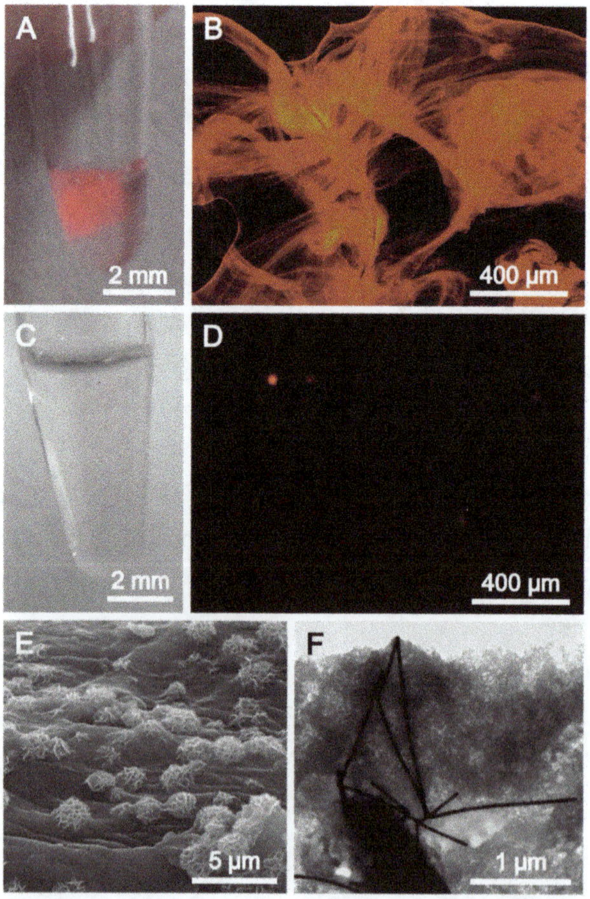

Figure 3. (**A–D**) Digital camera images and fluorescence microscopy images of the products of rolling circle amplification in the presence (**A,B**) and absence (**C,D**) of the target DNA; (**E**) SEM and (**F**) TEM images of the AuNW-DNA hydrogel after rolling circle amplification with the target DNA.

3. Experimental Section

3.1. Materials

All DNA oligonucleotides (Primer DNA: 5′ SH-TTT TTT TTT TTT TTT TTT TTA CGA CGT GTG ACC ATG CA-3′; Linear DNA: 5′-ACT TGC GGC AAT ACA AGT CGT CTC GTC GCA CTC TTT TTG CAT GGT CAC ACG TCG TTC TAT TGT GCG ACG AGA CCG TTT CAA GAT CCC AAT GAT-3′; Target DNA: 5′-TGT TAT TGC CGC AAG TAT CAT TGG GAT CTT GCA CTT-3′) were purchased from IDT. The target DNA used in this study was formulated from the influenza A virus sequences shown in Genbank database (http://www.ncbi.nlm.nih.gov/nuccore/401716582, accessed 20 March 2014). Dithiothreitol (DTT) and gold nanowires (AuNW) were purchased from Sigma-Aldrich (Saint Louis, MO, USA). NAP-5 column (Sephadex G-25 DNA grade) was purchased from G.E. Healthcare (Amersham, UK). T4 DNA Ligase and 10× ligase buffer were purchased from Promega (Madison, WI, USA). Phi29 DNA polymerase, 10× phi29 buffer and deoxyribonucleotide triphosphate mixture (dNTPs) were purchased from Epicentre (Madison, WI, USA). Evagreen was purchased from Biotium (Carlsbad, CA, USA).

3.2. Preparation of Thiolated Oligonucleotide-Modified Gold Nanowires

The thiolated primer DNAs were deprotected by addition of 0.2 M DTT, 0.18 M phosphate buffer (pH 8.0) and incubated at room temperature for 2 h. The primer DNAs were then washed over a NAP-5 desalting column to remove DTT. Purified primer DNAs (final concentration of 1.875 μM) were then mixed with AuNW, phosphate (final concentration of 10 mM) and NaCl (1 M). After 20 h incubation at room temperature, the solution was centrifuged at 13,000 rpm for 20 min, and the supernatant was discarded for the removal of unconjugated primer DNAs. This washing step was repeated 4 times. Finally, the AuNW-Primer DNA conjugates (AuNW:Primer) were redispersed in nuclease-free water.

3.3. Procedure for the Label-Free Detection of Target DNA

To produce the circularized DNA, the linear DNA and target DNA were mixed in nuclease-free water at a final concentration of 7.5 μM each. Temperature annealing was performed for the efficient hybridization between the linear DNA and target DNA. The mixture was heated to 95 °C for 2 min, then cooled gradually to 25 °C over a 60 min period. After 20 min incubation at 25 °C, the annealed DNA was mixed with T4 DNA ligase (3 U·μL^{-1}) and ligase buffer (300 mM Tris-HCl (pH 7.8), 100 mM $MgCl_2$, 100 mM DTT and 10 mM ATP) and incubated overnight at room temperature to close the nick.

Rolling circle amplification (RCA) was then performed as follows. First, 20 μL of the closed circular DNA (final concentration of 3 μM) and AuNW:Primer (30 μL) were combined and incubated for 2 h at room temperature for hybridization between the circular DNA and the primer DNA on AuNW. After incubation, unhybridized circular DNA was removed by discarding the supernatant after centrifugation at 13,000 rpm for 20 min. After washing twice, the resulting product was mixed with phi29 DNA polymerase (10 U·μL^{-1}), phi29 reaction buffer (80 mM Tris-HCl (pH 7.5), 100 mM KCl, 20 mM $MgCl_2$, 10 mM $(NH_4)_2SO_4$ and 8 mM DTT) and dNTPs (2 mM). Finally, the solution was incubated at 30 °C for 20 h.

3.4. Analysis of Characteristics of the Target-Enhanced Gelation

The melt curve was acquired using a CFX96 Touch Real-Time PCR Detection System (Bio-Rad, Foster, CA, USA). The linear DNA (final concentration of 2.5 μM) and AuNW:Primer (10 μL) was combined and incubated for 2 h at room temperature for hybridization. After washing twice, as mentioned above, NaCl (final concentration of 50 mM) and Evagreen were added. After 20 min incubation at 20 °C, the solution was heated gradually to 95 °C at a rate of 0.5 °C/min.

To obtain the fluorescence images of the hydrogel, the hydrogel was stained with GelRed and incubated overnight. After washing thoroughly, the fluorescence images were acquired by Eclipse Ti-U (Nikon, Tokyo, Japan) inverted fluorescent microscopy. A scanning electron microscopy (SEM) image was acquired using FE-SEM (S-4200, Hitachi, Tokyo, Japan) and the sample was coated with Pt. Cryo TEM (Cryo Tecnai F20 G2, FEI, Hillsboro, OR, USA) was used to obtain a TEM image, and the grid was treated with oxygen plasma.

4. Conclusions

A biosensor based on interlacing of DNA modified AuNW was developed for the detection of single-stranded pathogen DNA. Primer DNA coated AuNW formed a macroscopic hydrogel by simple hybridization and RCA. After the RCA process performed on the biosensor, enhanced gelation was observed in the presence of pathogen DNA. Although more studies on detecting various targets like protein or cell using target-binding molecules like aptamer are needed, it is believed that the proposed biosensor can enable label-free and easily recognizable detection without complicated processes or equipment.

Acknowledgments

This research was supported by the Global Innovative Research Center (GiRC) Program through the National Research Foundation of Korea (NRF) funded by the Ministry of Science, ICT and Future Planning (NRF-2012K1A1A2A01056093) and was also supported by the National Research Foundation of Korea (NRF) grant funded by the Korea government (MSIP) (No.2013R1A2A2A04015829) and by Basic Science Research Program through the National Research Foundation of Korea (NRF) funded by the Ministry of Science, ICT & Future Planning (NRF-2015R1A1A1A05001174).

Author Contributions

Jaepil Jeong, Hyejin Kim and Jong Bum Lee designed and analyzed the data. Jaepil Jeong and Hyejin Kim performed the experiments. Jaepil Jeong, Hyejin Kim and Jong Bum Lee wrote the manuscript.

References

1. Roh, Y.H.; Ruiz, R.C.; Peng, S.; Lee, J.B.; Luo, D. Engineering DNA-based functional materials. *Chem. Soc. Rev.* **2011**, *40*, 5730–5744.
2. Feldkamp, U.; Niemeyer, C.M. Rational design of DNA nanoarchitectures. *Angew. Chem. Int. Ed.* **2006**, *45*, 1856–1876.

3. Cao, Z.; Tong, R.; Mishra, A.; Xu, W.; Wong, G.C.; Cheng, J.; Lu, Y. Reversible cell-specific drug delivery with aptamer-functionalized liposomes. *Angew. Chem. Int. Ed.* **2009**, *48*, 6494–6498.
4. De Vries, J.W.; Zhang, F.; Herrmann, A. Drug delivery systems based on nucleic acid nanostructures. *J. Control. Release* **2013**, *172*, 467–483.
5. Kato, D.; Oishi, M. Ultrasensitive detection of DNA and RNA based on enzyme-free click chemical ligation chain reaction on dispersed gold nanoparticles. *ACS Nano* **2014**, *8*, 9988–9997.
6. Storhoff, J.J.; Elghanian, R.; Mucic, R.C.; Mirkin, C.A.; Letsinger, R.L. One-pot colorimetric differentiation of polynucleotides with single base imperfections using gold nanoparticle probes. *J. Am. Chem. Soc.* **1998**, *120*, 1959–1964.
7. Huang, F.; Xu, H.; Tan, W.; Liang, H. Multicolor and erasable DNA photolithography. *ACS Nano* **2014**, *8*, 6849–6855.
8. Jin, Z.; Sun, W.; Ke, Y.; Shih, C.-J.; Paulus, G.L.; Wang, Q.H.; Mu, B.; Yin, P.; Strano, M.S. Metallized DNA nanolithography for encoding and transferring spatial information for graphene patterning. *Nat. Commun.* **2013**, *4*, 1663.
9. Ding, B.; Liu, Y.; Rinker, S.; Yan, H. DNA-templated self-assembly of protein arrays and highly conductive nanowires. In *Encyclopedia of Complexity and Systems Science*; Meryes, R.A., Ed; Springer: New York, NY, USA, 2009; pp. 2104–2113.
10. Mayer-Enthart, E.; Wagenknecht, H.A. Structure-sensitive and self-assembled helical pyrene array based on DNA architecture. *Angew. Chem. Int. Ed.* **2006**, *45*, 3372–3375.
11. Dave, N.; Liu, J. Programmable assembly of DNA-functionalized liposomes by DNA. *ACS Nano* **2011**, *5*, 1304–1312.
12. Hu, R.; Zhang, X.; Zhao, Z.; Zhu, G.; Chen, T.; Fu, T.; Tan, W. DNA nanoflowers for multiplexed cellular imaging and traceable targeted drug delivery. *Angew. Chem. Int. Ed.* **2014**, *126*, 5931–5936.
13. Lee, J.B.; Peng, S.; Yang, D.; Roh, Y.H.; Funabashi, H.; Park, N.; Rice, E.J.; Chen, L.; Long, R.; Wu, M. A mechanical metamaterial made from a DNA hydrogel. *Nat. Nanotechnol.* **2012**, *7*, 816–820.
14. Huschka, R.; Zuloaga, J.; Knight, M.W.; Brown, L.V.; Nordlander, P.; Halas, N.J. Light-induced release of DNA from gold nanoparticles: Nanoshells and nanorods. *J. Am. Chem. Soc.* **2011**, *133*, 12247–12255.
15. Chen, C.-C.; Lin, Y.-P.; Wang, C.-W.; Tzeng, H.-C.; Wu, C.-H.; Chen, Y.-C.; Chen, C.-P.; Chen, L.-C.; Wu, Y.-C. DNA-gold nanorod conjugates for remote control of localized gene expression by near infrared irradiation. *J. Am. Chem. Soc.* **2006**, *128*, 3709–3715.
16. Yeh, H.-C.; Sharma, J.; Han, J.J.; Martinez, J.S.; Werner, J.H. A DNA-silver nanocluster probe that fluoresces upon hybridization. *Nano Lett.* **2010**, *10*, 3106–3110.
17. Thompson, D.G.; Enright, A.; Faulds, K.; Smith, W.E.; Graham, D. Ultrasensitive DNA detection using oligonucleotide-silver nanoparticle conjugates. *Anal. Chem.* **2008**, *80*, 2805–2810.
18. Bamrungsap, S.; Chen, T.; Shukoor, M.I.; Chen, Z.; Sefah, K.; Chen, Y.; Tan, W. Pattern recognition of cancer cells using aptamer-conjugated magnetic nanoparticles. *ACS Nano* **2012**, *6*, 3974–3981.
19. Ding, B.; Deng, Z.; Yan, H.; Cabrini, S.; Zuckermann, R.N.; Bokor, J. Gold nanoparticle self-similar chain structure organized by DNA origami. *J. Am. Chem. Soc.* **2010**, *132*, 3248–3249.

20. Demers, L.M.; Mirkin, C.A.; Mucic, R.C.; Reynolds, R.A.; Letsinger, R.L.; Elghanian, R.; Viswanadham, G. A fluorescence-based method for determining the surface coverage and hybridization efficiency of thiol-capped oligonucleotides bound to gold thin films and nanoparticles. *Anal. Chem.* **2000**, *72*, 5535–5541.
21. Huang, X.; Neretina, S.; El-Sayed, M.A. Gold nanorods: From synthesis and properties to biological and biomedical applications. *Adv. Mater.* **2009**, *21*, 4880–4910.
22. Bi, L.; Rao, Y.; Tao, Q.; Dong, J.; Su, T.; Liu, F.; Qian, W. Fabrication of large-scale gold nanoplate films as highly active sers substrates for label-free DNA detection. *Biosens. Bioelectron.* **2013**, *43*, 193–199.
23. Yao, H.; Miki, K.; Nishida, N.; Sasaki, A.; Kimura, K. Large optical activity of gold nanocluster enantiomers induced by a pair of optically active penicillamines. *J. Am. Chem. Soc.* **2005**, *127*, 15536–15543.
24. Ongaro, A.; Griffin, F.; Beecher, P.; Nagle, L.; Iacopino, D.; Quinn, A.; Redmond, G.; Fitzmaurice, D. DNA-templated assembly of conducting gold nanowires between gold electrodes on a silicon oxide substrate. *Chem. Mater.* **2005**, *17*, 1959–1964.
25. Russell, C.; Welch, K.; Jarvius, J.; Cai, Y.; Brucas, R.; Nikolajeff, F.; Svedlindh, P.; Nilsson, M. Gold nanowire based electrical DNA detection using rolling circle amplification. *ACS Nano* **2014**, *8*, 1147–1153.
26. Mbindyo, J.K.; Reiss, B.D.; Martin, B.R.; Keating, C.D.; Natan, M.J.; Mallouk, T.E. DNA-directed assembly of gold nanowires on complementary surfaces. *Adv. Mater.* **2001**, *13*, 249–254.
27. Blanco, L.; Bernad, A.; Lázaro, J.M.; Martin, G.; Garmendia, C.; Salas, M. Highly efficient DNA synthesis by the phage phi 29 DNA polymerase. Symmetrical mode of DNA replication. *J. Biol. Chem.* **1989**, *264*, 8935–8940.
28. Fire, A.; Xu, S.-Q. Rolling replication of short DNA circles. *Proc. Natl. Acad. Sci. USA.* **1995**, *92*, 4641–4645.
29. Zhao, W.; Ali, M.M.; Brook, M.A.; Li, Y. Rolling circle amplification: Applications in nanotechnology and biodetection with functional nucleic acids. *Angew. Chem. Int. Ed.* **2008**, *47*, 6330–6337.
30. Joo, J.H.; Lee, J.S. Library approach for reliable synthesis and properties of DNA-gold nanorod conjugates. *Anal. Chem.* **2013**, *85*, 6580–6586.

2

Biocatalytic Synthesis of Fluorescent Conjugated Indole Oligomers

Ryan M. Bouldin [1], Ankita Singh [2,3], Michael Magaletta [2], Sean Connor [2], Jayant Kumar [3,4] and Ramaswamy Nagarajan [2,3,*]

[1] Department of Natural and Applied Sciences, Bentley University, Waltham, MA 02452, USA; E-Mail: rbouldin@bentley.edu

[2] Department of Plastics Engineering, University of Massachusetts, Lowell, MA 01854, USA; E-Mails: ankita_singh@student.uml.edu (A.S.); michael_magaletta@student.uml.edu (M.M.); sean_connor@student.uml.edu (S.C.)

[3] Center for Advanced Materials, University of Massachusetts, Lowell, MA 01854, USA; E-Mail: jayant_kumar@uml.edu

[4] Department of Physics, University of Massachusetts, Lowell, MA 01854, USA

* Author to whom correspondence should be addressed; E-Mail: Ramaswamy_Nagarajan@uml.edu

External Editor: Anthony Guiseppi-Elie

Abstract: Fluorescent conjugated materials exhibiting reasonable biocompatibility that are capable of interacting with biological molecules are of interest for bio-sensing and imaging applications. Traditional approaches do not allow for the synthesis of conjugated materials in the presence of biologically relevant substrates. Further conjugated polymers synthesized using conventional methods are doped and not fluorescent. Here we explore the possibility of synthesizing fluorescent oligomers of indole using enzymes as catalyst under mild conditions. The peroxidase catalyzed coupling reaction presented here creates a photoluminescent material that allows for direct utilization (without purification and separation of the dopant) in biosensing applications. The polymerization reaction proceeds smoothly in just deionized water and ethanol. Monitoring of the absorption and fluorescence spectra over one hour shows that the concentration of both absorbing and emitting species grows steadily over time. The presence of anionic buffers and templates is shown to effectively retard the development of light emitting species and instead leads to the formation of an electrically doped conjugated polymer. Structural characterization through FTIR and ^{1}H-NMR analysis suggests that the oligomer is coupled through the 2 and 3 positions on the indole ring.

Keywords: polyindole; indole oligomer; biocatalyic; oxidoreductase; enzymatic polymerization; soybean peroxidase

---

1. Introduction

Conjugated polymers (CP) have attracted considerable attention from materials scientists, chemists and biologists, due to their tunable optical and electrical properties, rendering them suitable for applications in organic electronics and biosensing [1]. Optical sensors based on fluorescence quenching or 'turn-on' of luminescence have been most popular in biosensing applications [2]. The detection schemes for these sensors employ the light harvesting or conformation sensitive optical properties of the CP. The presence/binding of biomolecules or other relevant biological events is detected through Förster resonance energy transfer (FRET) between the CP and an acceptor molecule, or these biomolecules affect the conformation of the conjugated backbone leading to a detectable change in optical properties of the CP [3]. CP have evolved to become novel molecular tools for detection and imaging of biological processes in a complementary fashion to conventional techniques [4].

Key considerations in the design of fluorophores have often included high quantum yield of emission, solubility, ability to interact with biomolecules and a reasonable level of biocompatibility (allowing for detection or imaging without denaturing the biological entity). Solubility of fluorophores (such as conjugated polymers) in predominantly aqueous media is essential for interfacing of the fluorophore with biomacromolecules, such as proteins or nucleic acids [5]. It is becomingly increasingly clear that even for fluorescence-based *in vivo* imaging applications water solubility and reasonable biocompatibility of the fluorophores are crucial [6].

However, most conjugated polymers are intractable and not inherently water-soluble. Incorporation of pendant water-soluble moieties on conjugated polymers is one of the common strategies adopted to improve water solubility. Water-soluble moieties allow for tuning the interactions of conjugated polymers with biomolecules. Super-quenching of water-soluble poly(phenylene vinylene)s and poly(phenylene ethynylene)s has been reported in fluorescence turn-on and turn-off protein assays [7]. In these cases, the energy transfer between poly(fluorene-co-phenylene) derivatives and an acceptor dye is often utilized for fluorescence turn-on protein assays with amplified signal output.

For improved interaction between CP and biological polyelectrolytes such as DNA, cationic or anionic groups have been incorporated in the side chain of CP such as polythiophene. Cationic polythiophenes have been used for colorimetric detection of oligonucleotide hybridization based on conformational changes of the polymer upon interaction with DNA [3]. Signal transduction is typically controlled by electrostatic interactions between the charged species on the side chain of CP and the biological entities.

There is obviously a need for new non-toxic fluorescent conjugated materials that can be synthesized in predominantly aqueous media and near neutral pH conditions. The possibility of synthesizing CP under these conditions can also provide the opportunity to improve the interaction of the CP with biomolecules. In order to explore this aspect, the polymerization of indole monomers is investigated in this paper. Indole monomers and its derivatives are known to be fluorescent in their native state [8]. The fluorescence of indole has been utilized in the quantification of indole-conjugated proteins. Indoles have also been chemically or electrochemically polymerized to yield doped polyindoles with semi-conducting properties. The polymerization of indole has been shown to proceed through different routes yielding at least three distinct types of non-planar polymers [9,10]. Polyindoles have been reported to be reasonably stable, with good redox activity and electrical conductivity close to 0.1 S/cm in the doped state [11]. Figure 1 represents the different possible coupling sites and chemical structures of polyindole.

Figure 1. Possible coupling sites for the polymerization of indole.

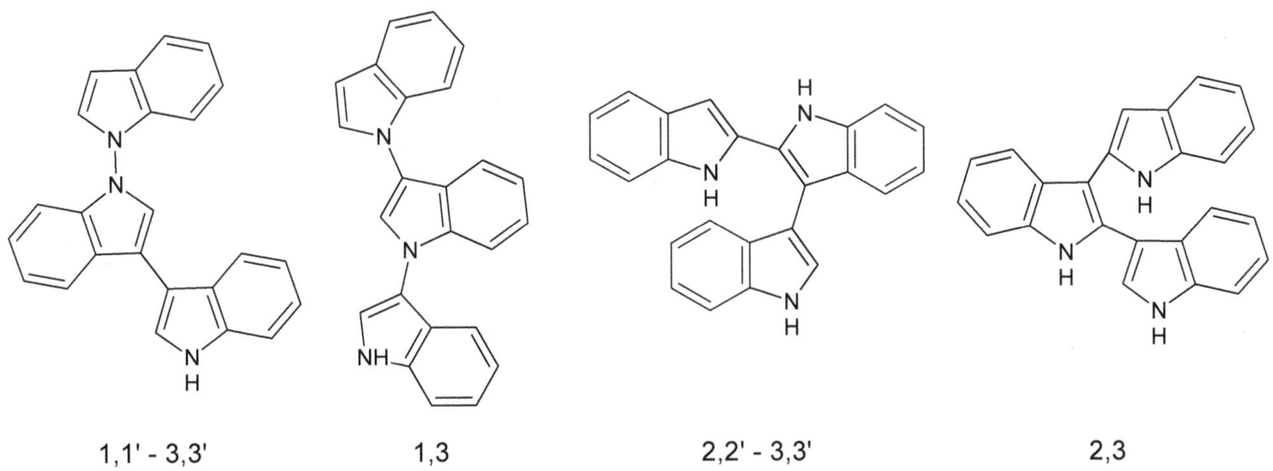

Most approaches reported to date for the polymerization of indole involve electrochemical oxidation or metal-salt catalyzed chemical synthesis. In all the cases the polymer produced is in the doped form and therefore does not exhibit significant photoluminescence. Moreover the use of harsh and hazardous conditions for the polymerization does not allow the direct utilization (without purification and separation of the dopant) of polyindole in biosensing applications. Enzymatic catalysis is an environmentally benign route for the synthesis of conjugated polymers in predominantly aqueous environment. Polypyrrole, polyaniline, and poly(3,4-dioxyethylene thiophene) (PEDOT) have been successfully synthesized with improved yield and greater reaction control by using enzymes, such as horseradish peroxidase (HRP) or soybean peroxidase (SBP), as catalysts [12–15]. The goal of this work was to explore the possibility of using an enzyme catalyst to create fluorescent indole species suitable for biosensor applications.

2. Experimental Section

2.1. Chemical Reagents

Indole (purity > 99%), poly(styrene sulfonic acid)-sodium salt [PSS] (MW = 70,000 Daltons), anhydrous citric acid (purity 99%) and sodium citrate monohydrate (purity 99%), absolute ethanol, hydrogen peroxide (30% wt) were purchased from Sigma-Aldrich Co., St. Louis, MO, USA. 3% (w/v) stock solutions of H_2O_2 were prepared by diluting the 30% solution using DI water and stored in the refrigerator before and after use.

Soybean Peroxidase (SBP) (EC 1.11.1.7) was purchased from Bioresearch Products Inc. North Liberty, IA and obtained as a reddish brown powder. SBP with an RZ (Reinheitszhal value) of 2.8 and a specific activity of 1840 IU per mg dry weight was used for all reported experiments. Prior to use SBP was dissolved in deionized water to a stock solution concentration of 3 mg/mL. Specific activity was assessed through an established guaiacol oxidation assay [16].

DI water was obtained from a Millipore Elix 3 system equipped with Proguard 2 filters. The system produces DI water with a resistance greater than 15 Mohm·cm.

2.2. Procedure

An amount of 28 mg of indole was dissolved in 3 mL of 30 % (v/v) ethanol in deionized water for a monomer concentration of 80 mM. If the reaction was run in a buffered solution, a pH 4.0, 100 mM citrate buffer was used instead of deionized water. If a surfactant or polyelectrolytes was used for the reaction, 150 mg of Triton X-100 (80 mM) or 55 mg of PSS (80 mM) was then added. To the indole solution, 300 μL of enzyme (SBP) stock solution was then added. Immediately, after the addition of SBP, an initial aliquot of 22.6 μL of 3% H_2O_2 was added. Subsequent additions of equal volume and concentration were made at an interval of 5 min for one hour. The reaction was allowed to stir at room temperature (~21 °C) for an additional hour. The final product was isolated by diluting the reaction mixture with deionized water and dialyzing the resulting compound in a dialysis bag (Molecular weight cut-off 500 Dalton) against deionized water.

2.3. Characterization

The absorption spectra of the monomer and all polymeric products were obtained on an Agilent 8453 photodiode array UV-visible spectrometer in a 1 cm quartz cuvette. For absorption measurements, a 100 μL aliquot of the reaction mixture was added to 3 mL solution of 30 % ethanol in DI water. The fluorescence spectra of the monomer and polymeric products were obtained using Agilent Cary Eclipse Fluorescence Spectrophotometer in a 1 cm quartz cuvette. Samples were prepared by dissolving 30 μL of the reaction solution in 3 mL of absolute ethanol. Dilute concentrations were purposely used to limit self-quenching due to aggregation. Fourier Transform Infrared (FTIR) Spectroscopy measurements were carried out on a Thermo Scientific Nicolet 4700 FTIR with a Smart Orbit Attenuated total reflectance (ATR) accessory. ^{1}H-NMR measurements were recorded on a Bruker Avance DRX 500 MHz NMR in deuterated dimethyl sulfoxide (DMSO).

3. Results and Discussion

The oxidative bulk polymerization of indole reported in the literature is most commonly carried out in organic solvents and is catalyzed by strong oxidizing agents [17]. Electrochemical polymerization of indole in acetonitrile in the presence of electrolytes such as $LiClO_4$ or $LiBF_4$, has also been reported [18]. All of these methods of polymerization involve non-biocompatible organic solvents and/or metal catalysts. Moreover, significant purification is essential to obtain the polymer in the fluorescent form and render it useful for biosensing or imaging applications.

3.1. Enzyme Catalysis

Peroxidases are known to catalyze the polymerization of certain types of aromatic amines, thiophenes and pyrroles in mildly acidic conditions (pH 3.5–5) to form conjugated polymers [12,19,20]. In these reactions anionic polyelectrolytes such as PSS are used as templates to (1) improve solubility of the monomer and the polymer (2) electrostatically interact with the radical cationic species of the oxidized monomer and the growing polymer chain, and (3) serve as a charge balancing dopant for the polymer. While the template facilitates the reaction, the products formed are often doped and non-fluorescent.

3.2. Enzymatic Polymerization of Indole

In an attempt to develop a mild and possibly biocompatible method, here we extend the use of peroxidases for the polymerization of undoped and fluorescent indoles.

Our initial investigations on the enzymatic polymerization of indole indicated that electrically conducting polymers of indole can be synthesized in mildly-acidic buffered solutions in the presence of an anionic dopant/template, such as PSS [21]. However, biosensors that are based on measurement of electronic properties of conjugated electroactive polymers have already been reported. However, from these initial reactions and through the use of FTIR spectroscopy, we observed that polyindoles synthesized enzymatically exhibit a characteristic absorption of the N–H bond at ~3398 cm^{-1}. This band together with the band at 1570 cm^{-1} can be ascribed to be the stretching and deformation vibrations of the N–H bond, respectively [22]. The full spectrum is provided in the Appendix as Figure A1. This supports the argument corroborated by other independent studies that the 2–3 coupling (Figure 2) is the most probable route among the four possible coupling mechanisms shown in Figure 1 [23].

Figure 2. Reaction scheme and proposed coupling route for the biocatalytic polymerization of indole with SBP.

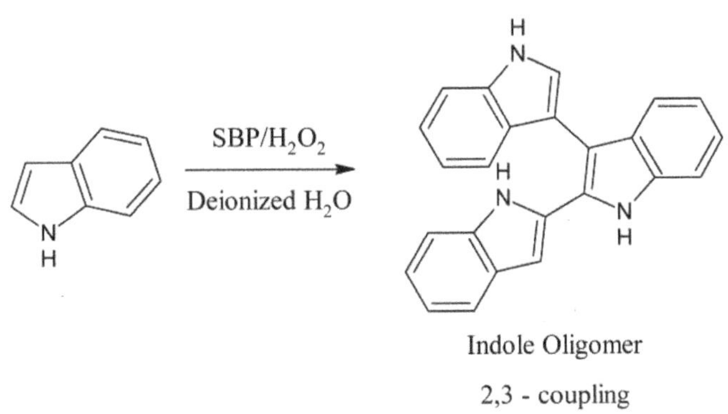

During the characterization of the polyindole-PSS reaction mixture, a faint green fluorescence emission was observed. Upon further purification by precipitating the polyindole-PSS complex, the emission was determined to be from soluble co-products in the reaction mixture and not from the PSS complexed polyindole. This is also supported by our earlier work that has showed that the enzyme-catalyzed polymerization of pyrrole progressed more efficiently to yield an electrically conducting product in the presence of buffered solutions and PSS [18]. However, the generation of an electrically conducting species typically leads a quenched fluorescence emission. Based on previous reports, this luminescence was most likely from soluble undoped oligomeric products [24]. Given this prior work and our recent work on pyrrole oligomer sensors [25], we designed a series of experiments to create fluorescent indoles for biosensor applications. To validate the role of SBP and hydrogen peroxide in the oligomerization reaction, control experiments were performed without SBP but with hydrogen peroxide in the reaction mixture. Similarly another reaction was performed with SBP alone but without the addition of hydrogen peroxide. The UV-Vis spectra from these experiments, provided in the Appendix as Figure A2, clearly indicate that the enzyme catalyzes the oligomerization and the reaction does not proceed in the absence of either SBP or hydrogen peroxide.

In an effort to observe the evolution of the fluorescent conjugated species of indole a series of experiments were carried out. Buffers salts and charged templates are known to interact with or dope growing conjugated polymer chains through electrostatic interactions with radical cations along the polymer backbone. To confirm these deleterious effects some initial experiments were performed on the polymerization of indole catalyzed by SBP in a buffered solution, a buffered solution with PSS, and as a control in deionized water without a buffer or PSS. Additionally, all of the reactions contained ethanol as a cosolvent to help solubilize the indole monomer. Using ethanol as a cosolvent has previously been reported to have little effect on the activity of SBP [26]. In the presence of 30% ethanol, SBP maintains roughly 80% of its original activity, making it a viable catalyst for these investigations [27]. When observed under UV light (365 nm), it is clear that a strongly fluorescent species is formed in the absence of a buffer and PSS, Figure 3. Within the first minute, the reaction containing DI water emits a strong blue/green fluorescence. Reactions containing buffer or buffer and PSS exhibit an emission that is either dramatically diminished or not present. A less intense, but still prevalent emission was also observed when the reaction was run in pure DI water without an ethanol cosolvent.

Figure 3. Image of the initial stages of the enzymatic polymerization of indole under 365 nm blacklight in the presence of DI water, citrate buffer, and buffer and PSS. All solutions contain 30% ethanol as cosolvent.

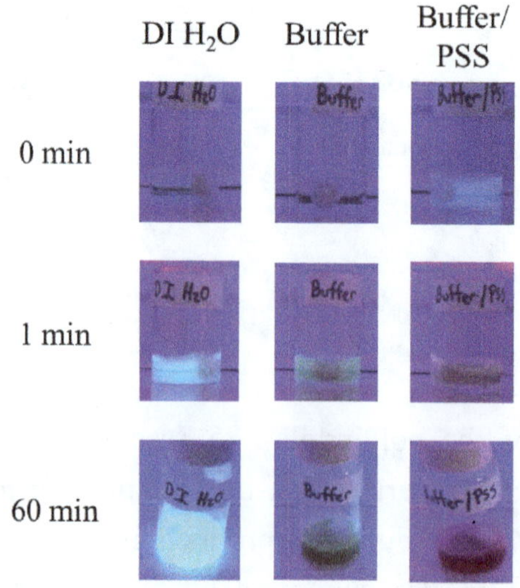

Upon isolation of the product synthesized in DI water by dialysis, further structural characterization was performed. ^{1}H-NMR analysis in deuterated DMSO revealed the final product is likely an oligomer of indole coupled between the 2 and 3 positions. No evidence of any ring opened products was observed as previously reported for the oxidation of indole by horseradish peroxidase [24]. The ^{1}H-NMR spectrum is provided in the Appendix as Figure A3.

3.3. Reaction Monitoring with Absorption and Fluorescence Spectroscopy

To further investigate the emergence of fluorescence during the enzymatic polymerization, absorption and fluorescence spectra of the reaction were monitored over the course of one hour. Figure 4a presents the absorption spectra of the reaction in a DI water/ethanol mixture, as well as in the presence of Triton X-100, a nonionic surfactant. Triton X-100 was selected to improve the solubility of the growing polymer chain without doping the backbone. The activity of SBP is not drastically influenced by the presence of Triton X-100. As reported earlier, SBP is active in the presence of Triton X-100 and has been used for treatment of phenol in wastewater [28].

The initial reaction solution containing only solvent, monomer, and enzyme was colorless. Upon the addition of SBP and the first aliquot of H_2O_2, the reaction solution turns a pale orange. The reaction continues to deepen in color with additional aliquots of hydrogen peroxide. After an hour, a small amount of precipitate is observed in the reaction solution. A similar progression is observed for the reaction containing Triton X-100, however the reaction solution always remained less colored compared to reactions without surfactant. For comparison, the post-reaction absorption spectra for the polymerization in the presence PSS and buffered solutions are included as Figure 4b.

Figure 4. (**a**) Absorption spectra development during unbuffered polymerization reaction. (**b**) Comparison of absorption spectra of indole monomer and final products synthesized in a buffer in the presence of PSS and Triton X-100.

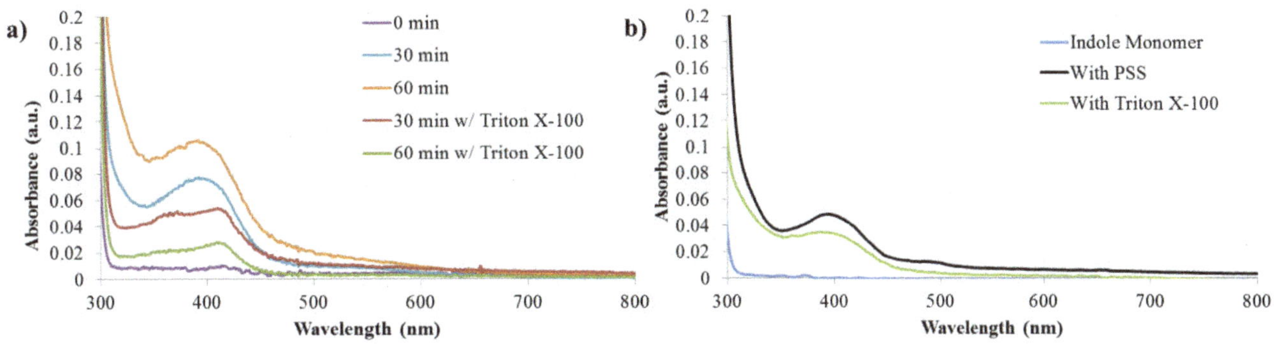

For all reactions, absorption from the indole monomer is observed at 300 nm, Appendix Figure A4. As the reactions proceed, new peaks are generated at 390 nm and 410 nm for solutions without and with Triton X-100, respectively. In both cases, the newly formed peak near 400 nm is believed to be due to the absorption of an indole species with extended conjugation. Similar spectra have previously been reported for polyindoles complexed with DNA when synthesized using the chemical oxidant ferric chloride [29]. As expected, the absorption peak for the reaction products in DI Water/ethanol increases over time as the concentration of absorbing species increases. In contrast, in the presence of Triton X-100, the absorption peak initially increases, but decreases significantly between 30 min and 60 min. It is unclear why this occurs, however, it could be due to the formation of a less soluble species.

Additionally based on the strength of the absorption peak, the concentration of absorbing species in solution is always higher for the reaction run without Triton X-100. This could be the result of a higher concentration of indole monomer interacting with the enzyme when Triton X-100 is not present in the reaction. The enzyme is likely to remain within the bulk phase of the solution and not inside the surfactant molecules. Therefore, in reactions containing Triton X-100, only indole that is not partitioned inside the micelle is free to react with the enzyme. This could ultimately leads to a lower concentration of products formed.

Fluorescence emission (λ_{ex} = 350 nm) during the polymerization reactions with and without Triton X-100 increases steadily over the course of one hour, Figure 5. This is expected as the number of emitters generated by the polymerization reaction increases over time. The emission maximum for both reactions is approximately 465 nm and does not shift upon excitation at higher wavelengths. The fluorescence emissions observed here are similar to those previous reported [26,30]. The reaction run without a surfactant/template has emission intensity almost five times greater than the reaction with Triton X-100. This corresponds well to the absorption data that suggest there is a higher concentration of absorbing/emitting products in solution for the reaction without a surfactant. Emission spectra monitoring at excitation wavelengths of 300 (monomer excitation) and 420 nm are included as Figures A5 and A6 in the Appendix.

Figure 5. Fluorescence emission (λ_{ex} = 350 nm) as a function of reaction time.

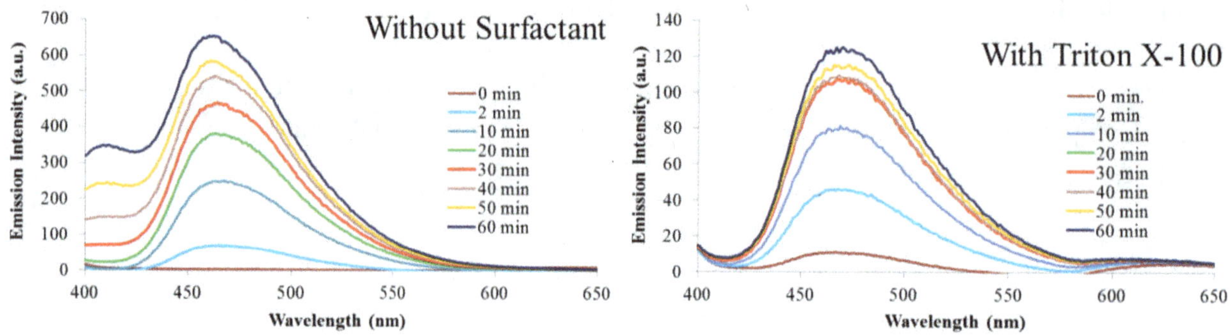

Figure 6. Rate of Normalized Emission Growth at 465 nm during Polymerization (λ_{ex} = 350 nm).

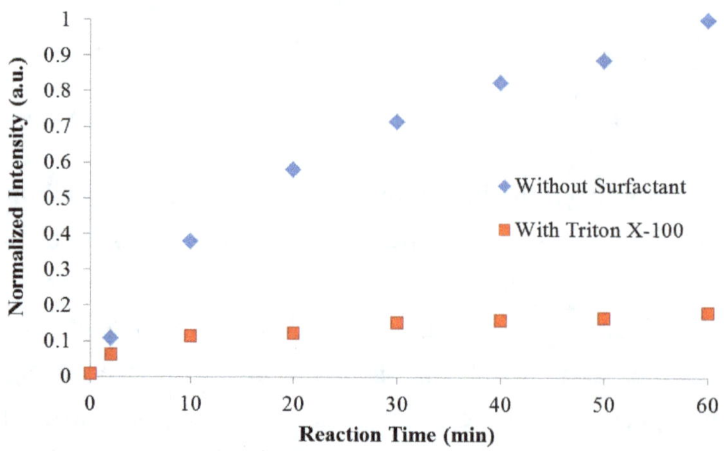

Figure 6 presents a comparison of the rate of normalized emission intensity growth. The spectra were normalized to the emission intensity present 60 min after the start of the reaction without a surfactant. In both reactions there is a rapid increase in fluorescence emission within the first two minutes. This is due to an extended conjugated length created from the initial coupling of indole monomers along the growing chain. When Triton X-100 is present, the rate of emission growth begins to level off after 30 min; when no surfactant was present the emission intensity continues to increase steadily over the course of the first hour. Fluorescence monitoring was stopped at one hour when a precipitate was observed in the reaction without Triton X-100.

4. Conclusions

In conclusion, an improved biocatalytic route is presented here that allows for the easy creation of fluorescent conjugated materials under near neutral pH conditions in biocompatible solvents such as water and ethanol. The enzyme-catalyzed technique is simple and offers a direct route to a conjugated indole oligomer with photoluminescent properties without exhaustive purification steps. The polymerization is sensitive to the presence of anionic buffers and templates that may dope the newly formed conjugated chain and quench the photoluminescence. *In situ* monitoring of absorption and emission spectra shows that when the reaction is conducted in just deionized water with an ethanol cosolvent, a more photoluminescent material is created. The presence of the non-ionic surfactant, Triton X-100 during the reaction results in decreased photoluminescence intensity. ^{1}H-NMR and FTIR analysis suggests that the biocatalytically coupled indoles are linked at the 2 and 3 positions on the indole ring. This improved synthesis technique offers the possibility for enhancing the interactions of fluorescent conjugated materials with biomolecules, rendering them suitable for biosensing and imaging applications.

Acknowledgments

The authors thank Zarif Farhana Mohd Aris and Wendy Gavin for their help obtaining ^{1}H-NMR spectra. The authors also thank Prof. Daniel Sandman for his suggestions and useful discussions.

Author Contributions

Ryan M. Bouldin and Ankita Singh conducted laboratory synthesis and characterization of the polyindole as well as writing significant portions of the manuscript. These two authors contributed equally to the paper. Michael Magaletta and Sean Connor assisted in conducting experiments for the paper in the laboratory. Jayant Kumar and Ramaswamy Nagarajan supervised, reviewed the work and helped in the design of the laboratory experiments and the interpretation of the results.

References

1. Guiseppi-Elie, A.; Wallace, G.G.; Matsue, T. Chemical and Biological Sensors Based on Electrically Conducting Polymers. In *Handbook of Conducting Polymers*, 2nd ed.; Skotheim, T.A., Elsenbaumer, R.L., Reynolds, J.R., Eds.; Marcel Dekker Inc.: New York, NY, USA, 1997; pp. 963–988.
2. Feng, X.; Liu, L.; Wang, S.; Zhu, D. Water-soluble fluorescent conjugated polymers and their interactions with biomacromolecules for sensitive biosensors. *Chem. Soc. Rev.* **2010**, *39*, 2411–2419.

3. Liu, B.; Bazan, G.C. Homogeneous fluorescence-based DNA detection with water-soluble conjugated polymers. *Chem. Mater*. **2004**, *16*, 4467–4476.
4. Chan, Y.-H.; Wu, P.-J. Semiconducting polymer nanoparticles as fluorescent probes for biological imaging and sensing. *Part. Part. Syst. Char.* **2014**, doi:10.1002/ppsc.201400123.
5. Faïd, K.; Leclerc, M. Responsive supramolecular polythiophene assemblies. *J. Am. Chem. Soc.* **1998**, *120*, 5274–5278.
6. Hussain, M.; Wackerlig, J.; Lieberzeit, P.A. Biomimetic strategies for sensing biological species. *Biosensors* **2013**, *3*, 89–107.
7. Chen, L.; McBrach, G.C.; Wang, H.-L.; Helgeson, R.; Wudl, R.; Whitten, D.G. Highly sensitive biological and chemical sensors based on reversible fluorescence quenching in a conjugated polymer. *Proc. Natl. Acad. Sci. USA* **1996**, *96*, 12287–12292.
8. Bridges, J.W.; Williams, R.T. The fluorescence of indoles and aniline derivatives. *Biochem. J.* **1968**, *107*, 225–237.
9. John, A.; Palaniappan, S. Synthesis and characterization of soluble poly(N-heptyl indole). *Polymer* **2005**, *46*, 12037–12039.
10. Zotti, G.; Zecchin, S.; Schiavon, G.; Seraglia, R.; Berlin, A.; Canavesi, A. Structure of polyindoles from anodic coupling of indoles: an electrochemical approach. *Chem. Mater.* **1994**, *6*, 1742–1748.
11. Cai, Z.; Zhang, R.; Shi, X. Preparation and characterization of polyindole nanofibers by electrospinning method. *Synthetic Met.* **2012**, *162*, 2069–2074.
12. Liu, W.; Kumar, J.; Tripathy, S.; Senecal, K.J.; Samuelson, L. Enzymatically synthesized conducting polyaniline. *J. Am. Chem. Soc.* **1999**, *1*, 71–78.
13. Liu, W.; Cholli, A.L.; Nagarajan, R.; Kumar, J.; Tripathy, S.; Bruno, F.F.; Samuelson, L. The role of template in the enzymatic synthesis of conducting polyaniline. *J. Am. Chem. Soc.* **1999**, *121*, 11345–11355.
14. Bruno, F.F.; Fossey, S.A.; Nagarajan, S.; Nagarajan, R.; Kumar, J.; Samuelson, L.A. Biomimetic synthesis of water-soluble conducting copolymers/homopolymers of pyrrole and 3,4-ethylenedioxythiophene. *Biomacromolecules* **2006**, *7*, 586–589.
15. Ravichandran, S; Nagarajan, S.; Kokil, A.; Ponrathnam, T.; Bouldin, R.M.; Bruno, F.F.; Samuelson, L.; Kumar, J.; Nagarajan, R. Micellar nanoreactors for hematin catalyzed synthesis of electrically conducting polypyrrole. *Langmuir* **2012**, *28*, 13380–13386.
16. Sessa, D.J.; Anderson, R.L. Soybean peroxidases: Purification and some properties. *J. Agric. Food Chem*. **1981**, *29*, 960–965.
17. Billaud, D.; Maarouf, E.B.; Hannecart, E. Chemical oxidation and polymerization of indole. *Synthetic Met.* **1995**, *69*, 571–572.
18. Billaud, D.; Maarouf, E.B.; Hannecart, E. Electrochemical polymerization of indole. *Polymer* **1994**, *35*, 2010–2011.
19. Tewari, A.I.; Kokil, A.; Ravichandran, S.; Nagarajan, S.; Bouldin, R.; Bruno, F.F.; Samuelson, L.A.; Nagarajan, R.; Kumar, J. Soybean peroxidase catalyzed enzymatic synthesis of pyrrole/EDOT copolymers. *Macromol. Chem. Phys*. **2010**, *211*, 1610–1617.
20. Bouldin, R.; Ravichandran, S.; Kokil, A.; Garhwal, R.; Nagarajan, S.; Kumar, J.; Bruno, F.F.; Samuelson, L.A.; Nagarajan, R. Synthesis of polypyrrole with fewer structural defects using enzyme catalysis. *Synthetic Met.* **2011**, *161*, 1611–1617.

21. Nagarajan, S.; Jayaraman, S.; Ravichandran, S.; Bouldin, R.; Nagarajan, R.; Kumar, J. Enzyme Mediated Synthesis as an Eco-Friendly Tool for Chemical Education. In Proceedings of the American Chemical Society National Meetings, San Diego, CA, USA, 25–29 March 2012.
22. An, S.; Abdiryim, T.; Ding, Y.; Nurulla, I. A comparative study of the microemulsion and interfacial polymerization for polyindole. *Mater. Lett.* **2008**, *62*, 935–938.
23. Talbi, H.; Ghanbaja, J.; Billaud, D.; Humbert, B. Vibrational properties and structural studies of doped and dedoped polyindole by FTIR, Raman and EEL spectroscopies. *Polymer* **1997**, *38*, 2099–2106.
24. Ximenes, V.F.; Campa, A.; Catalani, L.H. The oxidation of indole derivatives catalyzed by horseradish peroxidase is highly chemiluminescent. *Arch Biochem Biophys*. **2001**, *387*, 173–179.
25. Bouldin, R.M.; Kokil, A.; Ponrathnam, T.; Urban, N.; Kumar, J.; Samuelson, L.A.; Nagarajan, R. Biocatalyic synthesis of unusually photoluminescent oligomers and electrically conducting polymers of 4-(3-pyrrolyl)butyric acid. *J. Appl. Poly. Sci.* **2014**, *131*, doi:10.1002/app.41035.
26. Kim, Y.H.; An, E.S.; Song, B.K.; Kim, D.S.; Chelikani, R. Polymerization of cardanol using soybean peroxidase and its potential application as anti-biofilm coating material. *Biotechnol. Lett.* **2003**, *25*, 1521–1524.
27. Ghaemmaghami, F.I.; Alemzadeh, I.; Motamed, S. Seed coating soybean peroxidase: Extraction and biocatalytic properties determination. *Iranian J. Chem. Eng.* **2010**, *7*, 28–38.
28. Steevensz, A.; Madur, S.; Feng, W.; Taylor, K.E.; Bewtra, J.K.; Biswas, N. Crude soybean hull peroxidase treatment of phenol in synthetic and real wastewater: Enzyme economy enhanced by Triton X-100. *Enzym. Microb. Tech.* **2014**, *55*, 65–71.
29. Hassanien, R.; Al-Hinai, M.; Al-Said, S.A.F.; Little, R.; Siller, L.; Wright, N.G.; Houlton, A.; Horrocks, B.R. Preparation and characterization of conductive and photoluminescent DNA-templated polyindole nanowires. *ACS Nano* **2010**, *4*, 2149–2159.
30. Xu, J.; Nie, G.; Zhang, S.; Han, X.; Hou, J.; Pu, S. Electrosynthesis of free-standing polyindole films in borontrifluoride diethyl etherate. *J. Polymer Sci. Polymer Chem.* **2005**, *43*, 1444–1453.

Appendix

FTIR of Indole Monomer and Enzyme Catalyzed Polyindole

Figure A1. FTIR-ATR spectra of indole and polyindole.

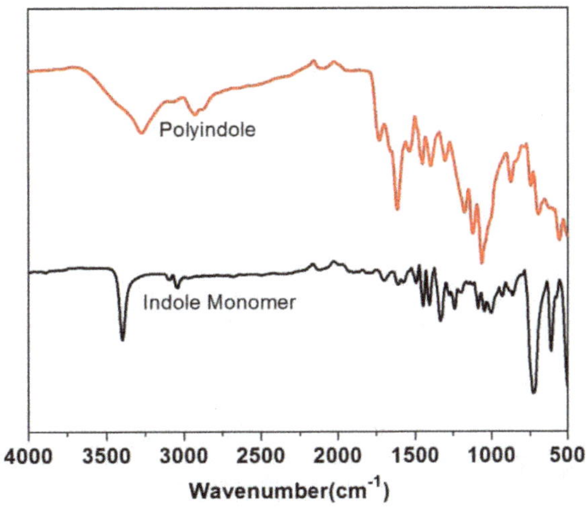

Absorbance Spectra Monitoring

Figure A2. Absorbance spectra monitoring of control reactions with and without SBP and hydrogen peroxide. Measurement recorded from 10 μL reaction sample in 290 μL of solvent system.

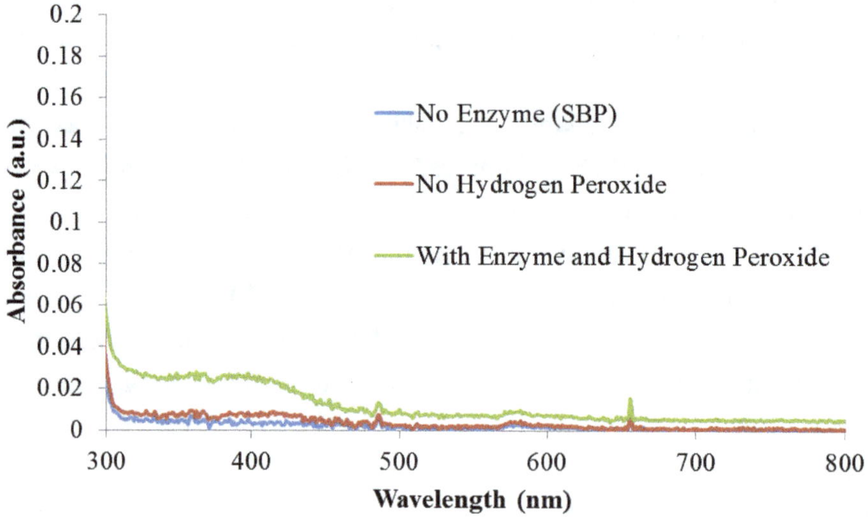

^{1}H-NMR Analysis of Indole Oligomer in d-DMSO

Figure A3. ^{1}H-NMR of SBP catalyzed polymerization product of indole in deionized water and ethanol.

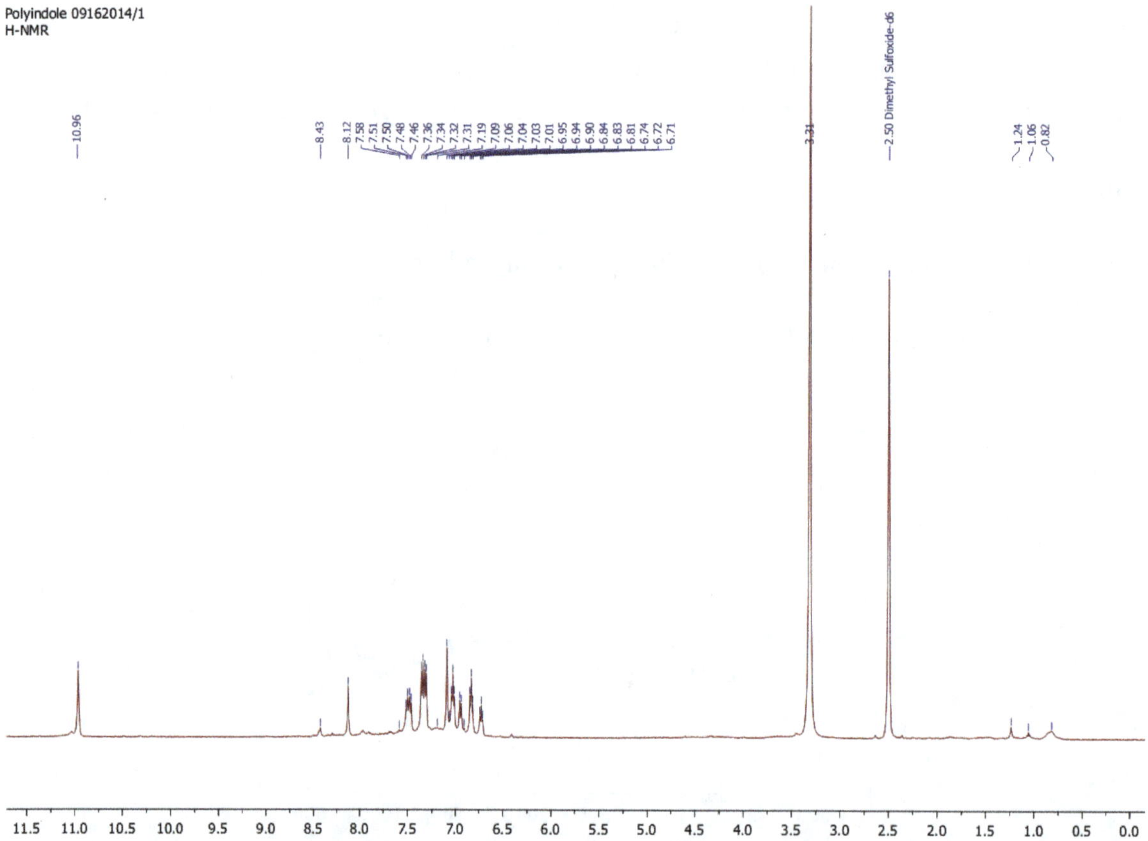

Absorbance Spectra Monitoring

Figure A4. Full absorbance spectra monitoring of reaction with and without Triton X-100.

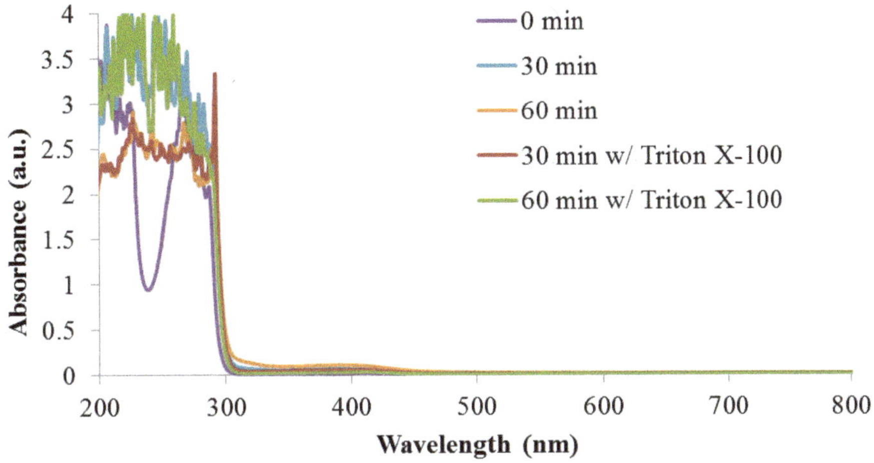

Photoluminescence Monitoring at Various Excitation Wavelengths

Figure A5. Fluorescence emission (λ_{ex} = 300 nm) as a function of reaction time with and without Triton X-100. Excitation at 300 nm corresponds to excitation of the indole monomer.

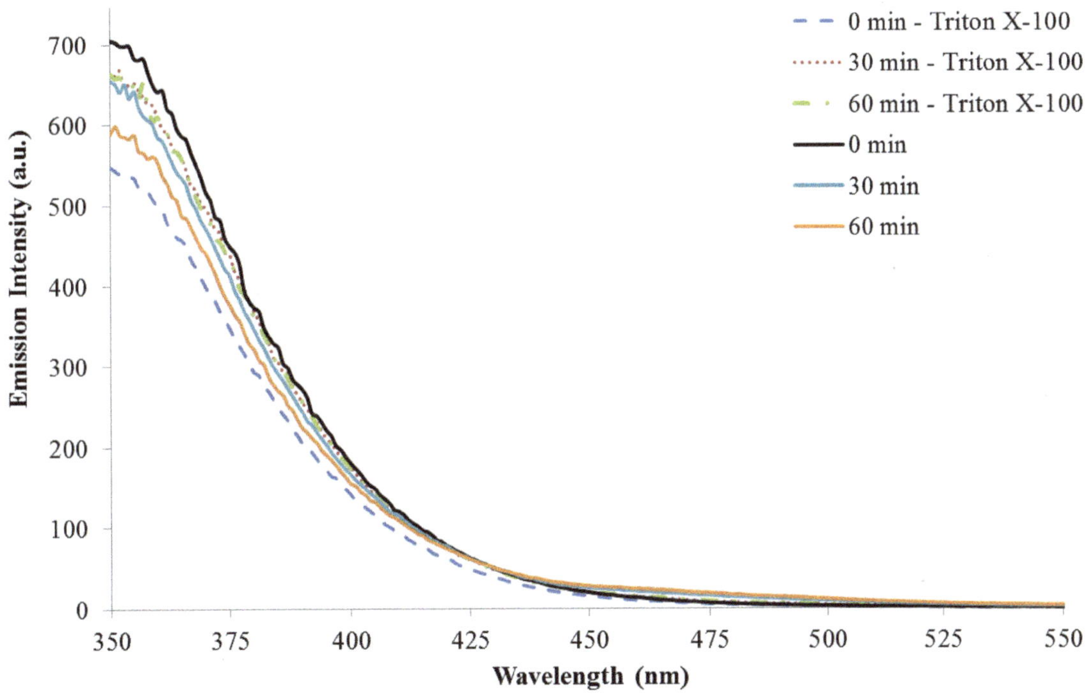

Figure A6. Fluorescence emission (λ_{ex} = 420 nm) as a function of reaction time with and without Triton X-100.

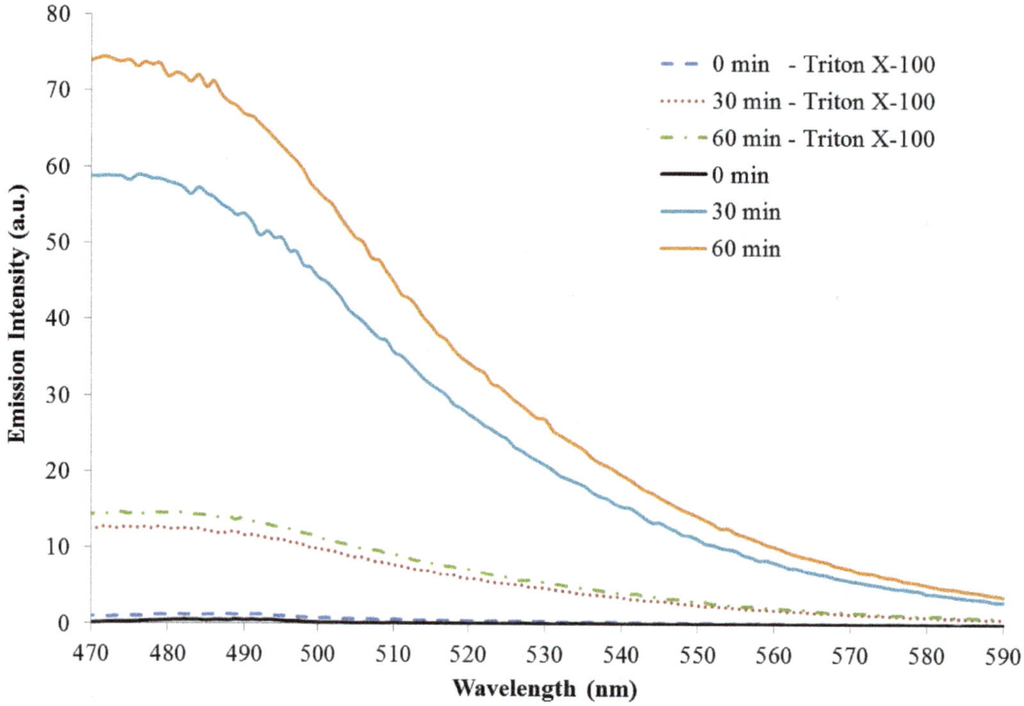

3

Enzyme-Assisted Extraction of Oil from Wet Microalgae *Scenedesmus* sp. G4

Shuhao Huo [1,2], Zhongming Wang [2], Fengjie Cui [1], Bin Zou [1], Pengxiang Zhao [3] and Zhenhong Yuan [2,*]

[1] School of Food and Biological Engineering, Jiangsu University, Zhenjiang 212013, Jiangsu, China; E-Mails: huoshuhao@yeah.net (S.H.); fengjiecui@163.com (F.C.); binzou2009@163.com (B.Z.)

[2] Key Laboratory of Renewable Energy, Chinese Academy of Sciences, Guangzhou 510640, Guangdong, China; E-Mail: wangzm@ms.giec.ac.cn

[3] Beijing Biomass Energy Technology Center, State Grid Energy Conservation Service Ltd., Beijing 100053, China; E-Mail: pengxiang-zhao@sgxy.sgcc.com.cn

* Author to whom correspondence should be addressed; E-Mail: yuanzh@ms.giec.ac.cn

Academic Editor: Thomas E. Amidon

Abstract: The enzyme-assisted aqueous extraction of oil from wet microalgae was employed to avoid the energy consumption of a dewatering process. In this paper, oil-rich microalgae *Scenedesmus* sp. G4 was hydrolyzed by enzyme mixtures for oil extraction. The results showed that the algae concentration had the greatest influence on yield of extracted oil, and the temperature and the ratio of enzyme mixtures affected the results as follows: the maximum yield of oil extracted from *Scenedesmus* sp. G4 reached 86.1% under optimal conditions with an algae concentration at 2.5 g/L, temperature 30/50 °C, pH 3.5/4.5 and a cellulase:pectinase:hemicellulase ratio of 1:1:1 or 1:2:1 (w/w/w). The results also indicated that the enzyme mixtures had a significant impact on the integrity of microalgae cells and the crystallization index increased from 30.7% to around 36.0% after enzymatic hydrolysis treatment. The cell wall of *Scenedesmus* sp. G4 has a high content of cellulose Iα and low crystallization, which is beneficial to the oil extraction by the enzyme-assisted hydrolysis method.

Keywords: *Scenedesmus* sp.; wet microalgae; oil extraction; enzyme-assisted hydrolysis

---

1. Introduction

Oil-rich microalgae are fast-growing organisms, able to fix large amounts of CO_2 and adapt to various environments. Their development and utilization are regarded as an important measure to contain global warming and guarantee national energy security [1,2]. The cost of large-scale utilization of microalgae energy, however, remains high [3]. Conventional oil extraction methods require dry microorganisms, and one of the major challenges is that the energy costs of a dehydration process before oil extraction are relatively high. Moreover, the conventional methods use chemical solvents that display insufficient recovery of extracts and toxicity and their unrenewable petroleum origin is undesirable. Nowadays, green extraction has become part of the sustainable development and industrial strategy [4,5]. Besides green extraction methods like ultrasound or microwaves [6], as shown in Table 1, enzyme-assisted aqueous extraction does not require a drying process and reduces the use of organic reagents in the extraction process, thus improving the economic efficiency and environmental benefits of the utilization of microalgae energy [7]. Compared with traditional chemical processes, extraction of energy substances using enzymatics method will not affect the quality of value-added biomass, and the high value-added coproduction of bioactivators greatly improves its competitiveness to rival fossil fuel [8].

Table 1. Comparison with conventional extraction methods.

Item	Conventional extraction	Enzyme-assisted extraction
Dehydration process	Needed	No
Organic reagent	Excess needed	Small
Quality of value-added biomass	Affected	Unaffected

The structure of the cell wall of microalgae is closely related to the effect of enzymolysis. The characteristic specificity of enzymes causes great limitations in single enzyme enzymolysis processes, and the cell walls of microalgae could be more thoroughly and effectively degraded by combining several kinds of enzymes. However mixed enzymatic hydrolysis studies always focus on large terrestrial plants, and hardly ever on methods for treating microscopic plants like microalgae. Also few reports have described the coordinative mechanism of enzymolysis in enzyme-assisted aqueous extraction of oil-rich microalgae.

In order to obtain a higher oil extraction yield, in this paper the oil-rich wet microalgae *Scenedesmus* sp. G4 was used to extract oil by enzyme-assisted hydrolysis. According to the orthogonal experiments, different extraction conditions were compared such as the ratio of cellulase, pectinase and hemicellulase, and the effects of operational parameters (*i.e.*, pH, temperature and microalgae concentration) on the yield of oil extraction were also investigated. The microalgae enzyme-assisted extraction mechanism was also initially analyzed by using modern methods such as scanning electron microscope (SEM), Fourier transform infrared spectroscopy (FT-IR), and X-ray diffraction (XRD).

2. Materials and Methods

2.1. Microalgal Strain and Culture Conditions

The freshwater green algae *Scenedesmus* sp. G4 was preserved at the Guangzhou Institute of Energy Conversion, Chinese Academy of Sciences (Guangzhou, Guangdong, China). For the cultivation of this strain, BG11 medium [9] was used. About 100 mL of the pre-culture broths mentioned above were inoculated into a vertical tubular photobioreactor containing 1.0 L medium. The vertical tubular photobioreactor consisted of glass tubes of 70.0 cm height and 5.0 cm outside diameter. Light was supplied by white cool fluorescent lamps (Hengxing-T4, 24W) on one side of the photobioreactor (light intensity: 200 ± 50 μE/(m^2 s)). Aeration and mixing were achieved by compressed air enriched with 6.0% CO_2 through a glass-filter, which was inserted in the bottom of the reactor and the flow rate of the gas was 0.5 vvm regulated by a gas flow meter (Model G, Aalborg Instruments & Controls, Inc., Orangeburg, NY, USA). The temperature of the culture media was 25 °C ± 1 °C regulated by the room air conditioner (Gree Electric Appliances Inc., Zhuhai, Guangdong, China). A photo of the vertical tubular photobioreactors is shown in Figure 1. After 7 days of cultivation, the cultures were used for oil extraction by enzyme mixture hydrolysis.

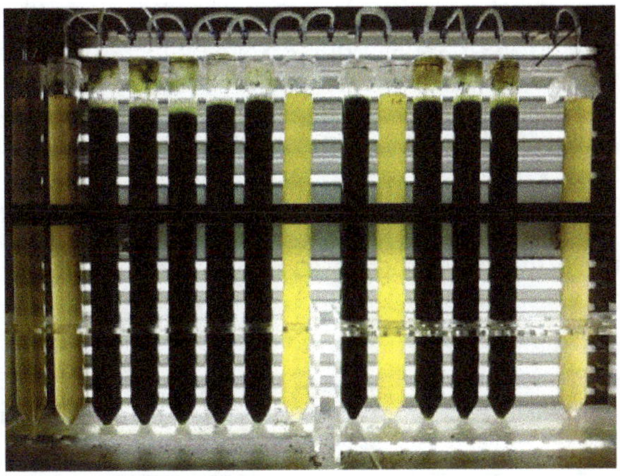

Figure 1. Photo of the vertical tubular photobioreactors.

2.2. Measurement Methods of Oil, Protein and Sugar Content

Total oil, protein and sugar content were measured according to Bigogno *et al.*'s method [10], Chu *et al.*'s method [11] and Yemm and Willis's method [12], respectively. The contents of oil, protein and sugar in dry weight (g/g dry weight) of *Scenedesmus* sp. G4 were 48.8%, 21.8% and 24.1%, respectively.

2.3. Enzyme Mixtures

Cellulase was purchased from Heshibi Biotechnology Co., Ltd. (Yinchuan, Ningxia, China) which was extracted from the fermentation liquor of *Penicillium* sp. cultures. Enzyme protein accounted for 10.08% (g/g dry weight). Pectinase and hemicellulase, both obtained from Yuanye Biotechnology Co., Ltd. (Shanghai, China) were extracted from the fermention of *Aspergillus niger* cultures. Protein accounted for 8.70% and 4.17% (g/g dry weight), respectively, in these substances.

2.4. Enzyme Assays

The filter paper enzyme activity of cellulase was measured according to method of the Commission on Biotechnology of the International Union of Pure and Applied Chemistry (IUPAC) [13]. Pectinase activity was evaluated by the method described by Meneghel *et al.* [14]. Hemicellulase activity was analyzed by xylanase activity measured according to Diogo *et al.*'s method [15]. The filter paper enzyme activity of cellulase was 7.2 U/mg protein. The enzyme activities of pectinase and hemicellulase were 500 U/mg protein and 1.5 U/mg protein, respectively.

2.5. Measurement of Oil Extraction

The enzymatic hydrolysis was performed in a 100 mL conical flask containing 30 mL of buffer solution. The enzymatic treatment was incubated in a water-bath shaker at 150 rpm for 72 h. The enzyme concentration was 2.0%. After enzymatic hydrolysis treatment, 20 mL hexane was added to each sample. The suspension obtained was mixed by a vortex mixer. Thereafter, the suspension was centrifuged at 5000 rpm for 5 min at 25°C. The supernatant was transferred to an evaporation flask. The solvent of the supernatant was eliminated by a vacuum rotary evaporator set at 30°C. The residue was re-extracted twice within 30 min. The flask was cooled and weighed to determine the oil content [7]. During the oil extraction process, 0.01% butylated hydroxytoluene (BHT) was added as antioxidant.

2.6. Orthogonal Array Testing of Enzyme Mixtures Hydrolysis

The orthogonal parameters were designed as shown in Table 2. Sodium azide, accounting for 0.03%, was added. In order to find the optimum condition of oil extraction, intuitive analysis was used.

Table 2. Levels and factors of orthogonal experiment for enzymatic hydrolysis.

Factors		Algae concentration (g/L)	Temperature (°C)	pH	Cellulase:pectinase:hemicellulase (dry weight ratio of protein)
Level	1	2.5	30	3.5	1:1:1
	2	5.0	40	4.0	2:1:1
	3	7.5	50	4.5	1:2:1
	4	10.0	60	5.0	1:1:2

2.7. Scanning Electron Microscopy

In order to observe the variations on the surface morphologies of microalgae after alkaline flocculation, these samples were examined using a high-resolution SEM (Model No.: S-3700N; Hitachi Co., Tokyo, Japan), which was operated at 10–20 kV accelerating potential. Prior to the observation, the surface of the sample was coated with a thin, electric conductive gold film.

2.8. Fourier Transform Infrared Spectroscopy Analysis

Infra-red spectra (IR) of the freeze-dried microalgae were measured on a KBr disk with a Tensor 27 FT-IR (Bruker, Billerica, MA, USA) to determine the functional groups of the microalgae cell surface. Absorbance spectra were collected at wavenumbers between 4500 cm^{-1} and 400 cm^{-1} with a nominal resolution of 2 cm^{-1} with 64 scans collected for data acquisition. An empty well served as the background reference for each microplate.

2.9. X-Ray Diffraction Analysis

XRD measurements of the freeze-dried algal powder before and after alkaline flocculation were carried out on X'Pert Pro-MPD diffractometer (Panalytical, Almelo, The Netherlands) using the planar monochromator method and Cu Kα radiation (wavelength of 0.15406 nm). The samples were placed on a silicon holder. XRD data were collected in the range of 5°–80°, a step length of 0.017°, the radioactive rays under a voltage of 40 kV and an electric current of 40 mA. The crystallization index (*CrI*, %) will be calculated with equation [16,17]:

$$CrI = \frac{I_{main} - I_{am}}{I_{main}} \times 100$$

I_{main} is the diffraction intensity of the main peak ($2\theta = 20.1°$) and I_{am} is the diffraction intensity of base line ($2\theta = 15.3°$).

3. Results and Discussion

3.1. Results of Orthogonal Array Testing on the Enzymatic Hydrolysis

Pectinase, cellulase and hemicellulase were chosen as components of the enzyme mixtures. Compared with single enzymes, enzyme mixtures could further enlarge the channels of algae cell walls, to easily separate the oil molecules from the cells. Table 3 shows that different parameters had very different effects on the oil extraction yield. The algae concentration had the greatest influence on the yield of oil extraction, followed by the temperature and the ratio of the enzyme mixtures. The pH had the least influence on the oil extraction process. According to the test results shown in Table 4, the maximum yield of extraction oil reached 86.1% under the optimal conditions of 2.5 g/L algae concentration, 30/50 °C temperature, pH 3.5/4.5, and a cellulase:pectinase:hemicellulase ratio of 1:1:1 (w/w/w) or 1:2:1 (w/w/w).

Table 3. Analysis of variance (ANOVA) for the orthogonal test.

Factor	Sum of squares	Degree of freedom	*F*-value	Significance
Algae concentration (g/L)	7907.023	3	11.336	*
Temperature (°C)	619.913	3	0.889	-
pH	2.693	3	0.004	-
Cellulase:pectinase:hemicellulase (dry weight ratio of protein)	200.723	3	0.288	-
Pure error	697.53	3	-	-

Table 4. Results of orthogonal tests on the yield of oil extraction from *Scenedesmus* sp. G4 by enzyme-assisted hydrolysis.

No.	Algae concentration (g/L)	Temperature (°C)	pH	Cellulase:pectinase:hemicellulase (dry weight ratio of protein)	Yield (%)
1	2.5	30	3.5	1:1:1	86.1
2	2.5	40	4.0	2:1:1	61.5
3	2.5	50	4.5	1:2:1	86.1
4	2.5	60	5.0	1:1:2	69.7
5	5.0	30	4.0	1:2:1	41.0
6	5.0	40	3.5	1:1:2	30.7
7	5.0	50	5.0	1:1:1	51.2
8	5.0	60	4.5	2:1:1	34.8
9	7.5	30	4.5	1:1:2	23.2
10	7.5	40	5.0	1:2:1	20.5
11	7.5	50	3.5	2:1:1	27.3
12	7.5	60	4.0	1:1:1	28.7
13	10	30	5.0	2:1:1	16.4
14	10	40	4.5	1:1:1	12.3
15	10	50	4.0	1:1:2	28.7
16	10	60	3.5	1:2:1	16.4
I	75.850	41.670	40.125	44.575	-
II	39.425	31.250	39.975	35.000	-
III	24.925	48.325	39.100	41.000	-
IV	18.450	37.400	39.450	38.075	-
R	57.400	17.075	1.025	9.575	-

I, II, III, IV—sum of coverage rate of Levels 1–4, respectively. R: the difference between the largest sum value and the smallest sum value of each factor.

3.2. Scanning Electron Microscope Analysis of the Enzymatic Hydrolysis Products

As Figure 2 shows, before enzymatic hydrolysis *Scenedesmus* sp. G4 cells presented a smooth surface, the cellular morphology was clear and the cell walls complete.

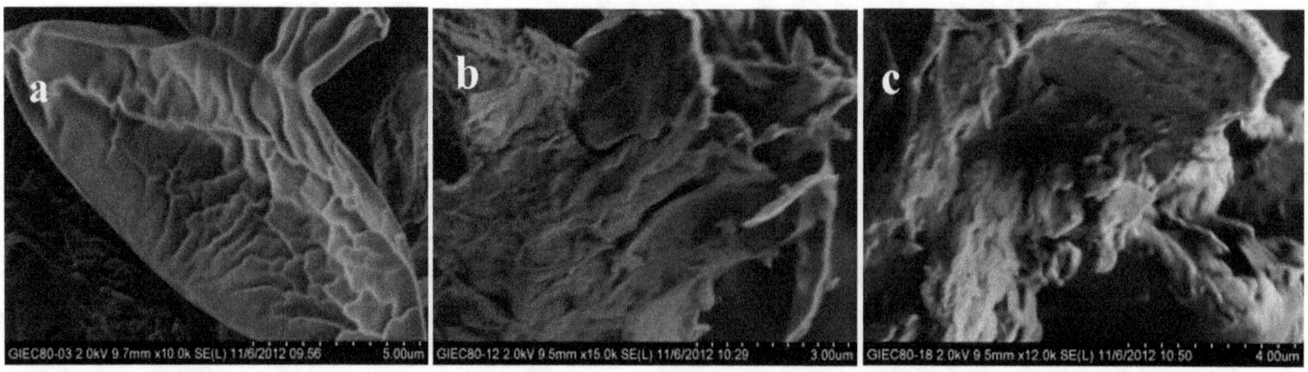

Figure 2. Scanning electron microscope (SEM) micrographs of *Scenedesmus* sp. G4 cultures: (**a**) before enzymatic hydrolysis; (**b**) hydrolyzed by enzyme mixtures 60 h, pH = 3.5; and (**c**) hydrolyzed by enzyme mixtures 60 h, pH = 4.5.

After hydrolysis by enzyme mixtures, the cell surface was rough and significant cell lysis occurred. The enzyme mixtures significantly damaged the intactness of microalgae cells, destroying the cell wall structure which was good for the next step of oil extraction.

3.3. The X-ray Diffraction Analysis of the Enzymatic Hydrolysis Products

As shown in Figure 3, the changes in the microalgae *Scenedesmus* sp. G4 after enzymatic hydrolysis were not significant according to the XRD results. The characteristic peaks at the positions of diffraction angles at 14.7°, 16.8°, 20.5° and 22.7° were attributed to the diffraction planes of 101, $10\overline{1}$, 021 and 002 in cellulose I. Since the composition of green algae basically do not contain lignin, and the content of cellulose Iα was relatively high, while typical lignocellulosic biomass has a relatively higher content of cellulose Iβ [18–20], the hydrogen bond in molecules of cellulose Iα was relatively weak, making the enzymolysis of endoglucanase easy [21] to facilitate the next step of oil extraction process. The crystallization index of *Scenedesmus* sp. G4 was increased after enzymatic hydrolysis. Figure 3 also shows that the crystallization index of *Scenedesmus* sp. G4 changes from 30.7% before enzyme hydrolysis to 36.0% after enzymatic hydrolysis.

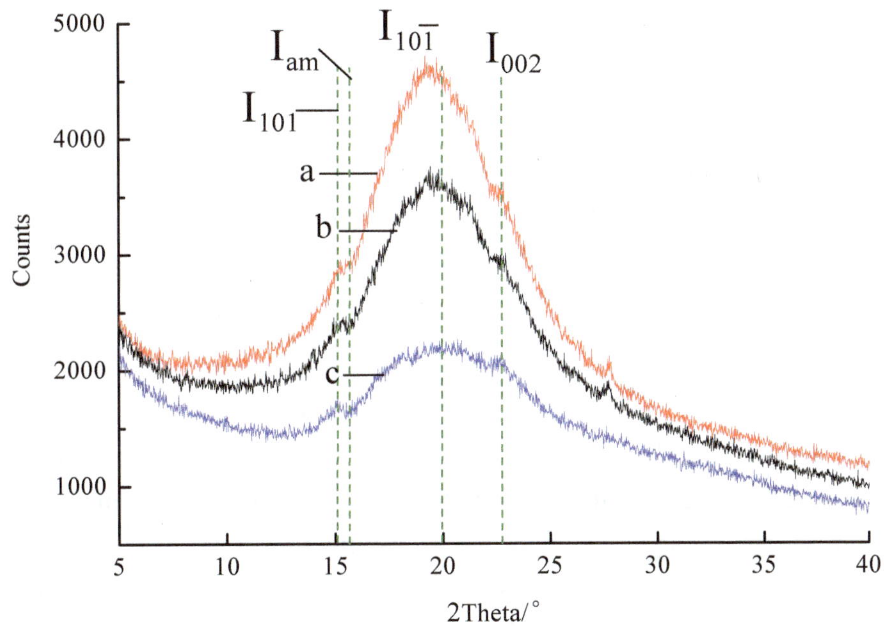

Figure 3. X-ray diffraction (XRD) patterns of *Scenedesmus* sp. G4 cultures: (**a**) before enzyme hydrolysis; (**b**) hydrolyzed by enzyme mixtures 60 h, pH = 3.5; and (**c**) hydrolyzed by enzyme mixtures 60 h, pH = 4.5.

3.4. The Fourier Transform Infrared Spectroscopy Analysis of Enzymatic Hydrolysis

The FT-IR spectra of the algae *Scenedesmus* sp. G4 before and after enzymatic hydrolysis are shown in Figure 4 [22,23]. The main component of the cell walls of green algae is cellulose, which is a glucose polymer connected by β-D-,4-glucosidic bonds. As shown in Figure 4 the broad peak at 3500–3300 cm^{-1} was attributed to –OH under the atomic force of H and O or –NH stretching vibrations. The carboxyl group absorption was located in the range of 2400–3300 cm^{-1}. The small shoulder at 2929 cm^{-1}

might be the vibration peak of associated carboxyl/phenolic groups. Compared with the microalgae *Scenedesmus* sp. G4 before enzymatic hydrolysis, the absorptions at 2929 cm^{-1}, 2855 cm^{-1}, 1743 cm^{-1}, 1158 cm^{-1} were increased after enzymatic hydrolysis. The cellulose and hemicellulose were exposed by the enzymatic hydrolysis. The cell wall bonds were changed by the mixed enzymatic hydrolysis, which improved the oil extraction significantly.

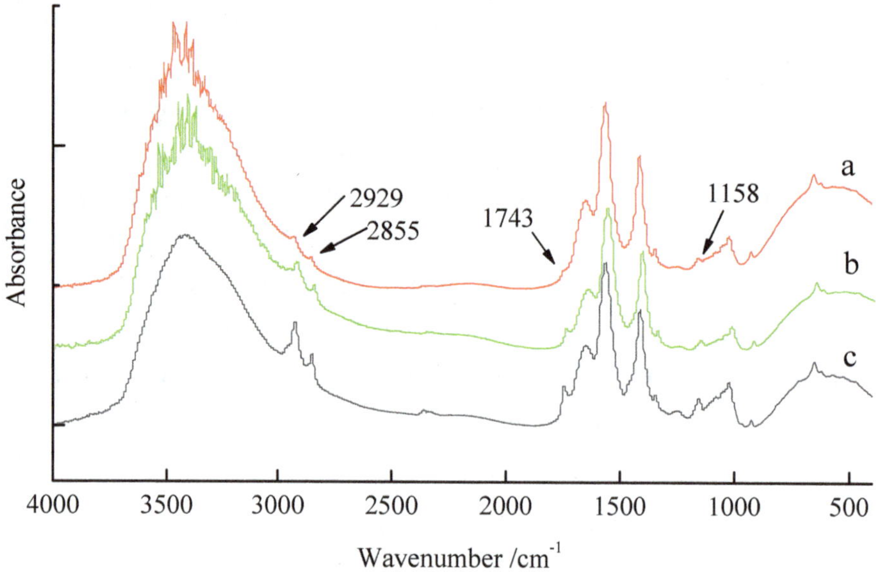

Figure 4. The changes of organic functional groups on *Scenedesmus* sp. G4 cultures: (**a**) before enzyme hydrolysis; (**b**) hydrolyzed by enzyme mixtures 60 h, pH = 4.5; and (**c**) hydrolyzed by enzyme mixtures 60 h, pH = 3.5.

4. Conclusions

Experimental results showed that oil extraction from wet microalgae can be directly carried out by enzyme-assisted extraction. The algae concentration had the greatest influence on oil extraction yield, followed by the temperature and the ratio of enzyme mixtures. The enzyme mixtures had a significant impact on the integrity of microalgae cells and the crystallization index was increased from 30.7% to around 36.0% after enzymatic hydrolysis. The cell wall of *Scenedesmus* sp. G4 with its high content of cellulose Iα and low crystallization index, was beneficial to the oil extraction by the enzyme-assisted hydrolysis method. The enzyme-assisted extraction of oil from wet microalgae could be a valid alternative to the conventional ways.

Acknowledgments

This research was funded by the China Postdoctoral Science Foundation (2014M551519 and 2015T80502), the 12th Five Year Support Plan of the Ministry of Science and Technology, China (2011BAD14B03), the Natural Science Foundation of Jiangsu Province (BK20140540), the Senior Talent Scientific Research Initial Funding Project of Jiangsu University (14JDG024), the State Grid Science and Technology Program (No. SGECS56-2014), the key Laboratory of Renewable Energy, Chinese Academy of Sciences (No. y507k11001) and the Priority Academic Program Development of Jiangsu Higher Education Institutions.

Author Contributions

Shuhao Huo wrote the main part of the paper and performed the experiments. Zhongming Wang conceived the experiments. Fengjie Cui and Bin Zou revised the paper. Pengxiang Zhao and Zhenhong Yuan read and approved the manuscript.

References

1. Demirbas, A.; Demirbas, M.F. *Algae Energy: Algae as a New Source of Biodiesel*; Springer-Verlag: London, UK, 2010.
2. Huo, S.; Dong, R.; Wang, Z.; Pang, C.; Yuan, Z.; Zhu, S.; Chen, L. Available resources for algal biofuel development in China. *Energies* **2011**, *4*, 1321–1335.
3. *National Algal Biofuels Technology Roadmap*; U.S. Department of Energy: Washington, DC, USA, 2010.
4. Rombaut, N.; Tixier, A.S.; Bily, A.; Chemat, F. Green extraction processes of natural products as tools for biorefinery. *Biofuels Bioprod. Biorefining* **2014**, *8*, 530–544.
5. Chemat, F.; Vian, M.A.; Cravotto, G. Green extraction of natural products: Concept and principles. *Int. J. Mol. Sci.* **2012**, *13*, 8615–8627.
6. Mason, T.J. The extraction of natural products using ultrasound or microwaves. *Curr. Org. Chem.* **2011**, *15*, 237–247.
7. Liu, Y.; Zhang, X.; Tan, T. Enzyme-aided extraction of lipid from wet microalgae chlorella vulgaris. *J. Bioprocess Eng. Biorefinery* **2013**, *4*, 193–197.
8. Romero García, J.M.; Acién Fernández, F.G.; Fernández Sevilla, J.M. Development of a process for the production of L-amino-acids concentrates from microalgae by enzymatic hydrolysis. *Bioresour. Technol.* **2012**, *112*, 164–170.
9. Rippka, R.; Deruelles, J.; Waterbury, J.; Herdman, M.; Stanier, R. Generic assignments, strain histories and properties of pure cultures of cyanobacteria. *J. Gen. Microbiol.* **1979**, *111*, 1–61.
10. Bigogno, C.; Khozin-Goldberg, I.; Boussiba, S.; Vonshak, A.; Cohen, Z. Lipid and fatty acid composition of the green oleaginous alga *Parietochloris incisa*, the richest plant source of arachidonic acid. *Phytochemistry* **2002**, *60*, 497–503.
11. Chu, W.L.; Phang, S.; Goh, S.H. Environmental effects on growth and biochemical composition of *Nitzschia inconspicua* Grunow. *J. Appl. Phycol.* **1996**, *8*, 389–396.
12. Yemm, E.W.; Willis, A.J. The estimation of carbohydrates in plant extracts by anthrone. *Biochem. J.* **1954**, *57*, 508–514.
13. Ghose, T.K. Ghose Measurements of cellulase activities. *Pure Appl. Chem.* **1987**, *2*, 257–268.
14. Meneghel, L.; Reis, G.P.; Reginatto, C.; Malvessi, E.; da Silveira, M.M. Assessment of pectinase production by Aspergillus oryzae in growth-limiting liquid medium under limited and non-limited oxygen supply. *Process Biochem.* **2014**, *49*, 1800–1807.
15. Diogo, J.A.; Hoffmam, Z.B.; Zanphorlin, L.M.; Cota, J.; Machado, C.B.; Wolf, L.D.; Squina, F.; Damásio, A.R.; Murakami, M.T.; Ruller, R. Development of a chimeric hemicellulase to enhance the xylose production and thermotolerance. *Enzym. Microb. Technol.* **2015**, *69*, 31–37.

16. Bernhardt, H.; Hoyer, O.; Schell, H.; Lusse, B. Reaction mechanisms involved in the influence of algogenic matter on flocculation. *Wasser Abwasser Forsch.* **1985**, *18*, 67–75.
17. Henderson, R.K.; Parsons, S.A.; Jefferson, B. The impact of differing cell and algogenic organic matter (AOM) characteristics on the coagulation and flotation of algae. *Water Res.* **2010**, *44*, 3617–3624.
18. Hayashi, N.; Sugiyama, J.; Okano, T.; Ishihara, M. Selective degradation of the cellulose Iα component in *Cladophora* cellulose with Trichoderma viride cellulase. *Carbohydr. Res.* **1997**, *305*, 109–116.
19. Atalla, R.H.; Van der Hart, D.L. Native cellulose: A composite of two distinct crystalline forms. *Science* **1984**, *223*, 283–285.
20. Van der Hart, D.L.; Atalla, R.H. Studies of microstructure in native celluloses using solid-state carbon-13 NMR. *Macromolecules* **1984**, *17*, 1465–1472.
21. Daroch, M.; Geng, S.; Wang, G. Recent advances in liquid biofuel production from algal feedstocks. *Appl. Energy* **2013**, *102*, 1371–1381.
22. Zhang, J.; Ma, X.; Yu, J.; Zhang, X.; Tan, T. The effects of four different pretreatments on enzymatic hydrolysis of sweet sorghum bagasse. *Bioresour. Technol.* **2011**, *102*, 4585–4589.
23. Zhao, X.; Wang, X.; Liu, D.H. Peracetic acid pretreatment of sugarcane bagasse for enzymatic hydrolysis: A continued work. *J. Chem. Technol. Biotechnol.* **2008**, *83*, 950–956.

4

Quantum Mechanical Modeling: A Tool for the Understanding of Enzyme Reactions

Gábor Náray-Szabó [1,*], Julianna Oláh [2] and Balázs Krámos [2]

1 Laboratory of Structural Chemistry and Biology and HAS-ELTE Protein Modeling Group, Eötvös Loránd University, Pázmány Péter St. 1A, Budapest H-1117, Hungary

2 Department of Inorganic and Analytical Chemistry, Budapest University of Technology and Economics, Gellért tér 4, Budapest H-1111, Hungary; E-Mails: julianna.olah@mail.bme.hu (J.O.); kramosbalazs@ch.bme.hu (B.K.)

* Author to whom correspondence should be addressed; E-Mail: naraysza@chem.elte.hu

---

Abstract: Most enzyme reactions involve formation and cleavage of covalent bonds, while electrostatic effects, as well as dynamics of the active site and surrounding protein regions, may also be crucial. Accordingly, special computational methods are needed to provide an adequate description, which combine quantum mechanics for the reactive region with molecular mechanics and molecular dynamics describing the environment and dynamic effects, respectively. In this review we intend to give an overview to non-specialists on various enzyme models as well as established computational methods and describe applications to some specific cases. For the treatment of various enzyme mechanisms, special approaches are often needed to obtain results, which adequately refer to experimental data. As a result of the spectacular progress in the last two decades, most enzyme reactions can be quite precisely treated by various computational methods.

Keywords: enzyme; reaction; mechanism; model; quantum mechanics; QM/MM

---

1. Introduction

Enzyme reactions are extremely complicated processes involving directly or indirectly a large number of atoms in the chemical rearrangement of the reacting species. Accordingly, their understanding in atomic details is also very difficult and needs combined application of a variety of experimental and molecular modeling techniques. Rate and binding constants, as well as inhibitory power provide direct information on energetic aspects; however, they are bulk descriptors of the reaction and give information on subtleties only in cases when they can be associated with a single step. A direct method, X-ray diffraction, helps in constructing three-dimensional models at the atomic resolution but does not allow drawing conclusions on energetic aspects of the process. Detailed and precise elucidation of an enzymatic reaction is therefore not possible using present-day experimental methods alone; application of molecular modeling seems to be a must in order to clarify subtle details.

Because of the very large number of atoms involved in the reaction enzymatic processes, in general, cannot be modeled as accurately as interactions between small molecules with a few atoms. Though in some specific cases, by the application of high-level quantum mechanical methods quantitative agreement with experiment could be achieved [1], this is presently not routine. A promising way to obtain a detailed and adequate description of the process is to combine experimental evidence with modeling techniques. The reliability of a model can be checked by comparing calculated and experimental data and if an appropriate agreement is achieved, one may be confident on its validity.

With the spectacular development of computer hardware and software a substantial progress has been made in the development of quantum chemical methods. High-performance models of solvation, as well as combination of *a priori* quantum mechanical and empirical molecular mechanical approaches have become an integral part of commercial molecular modeling program packages. Along with the dramatic increase of the computer performance to price ratio, computational chemistry became suitable even for predictions, and thus it complements experimental findings. It seems that direct mapping between calculated and experimentally observed properties, as well as molecular structures represents a major advantage of quantum chemical modeling since this type of information is very difficult to obtain experimentally. If the observed properties, related to enzyme reactions, can be quantitatively reproduced by calculations we may consider reaction profiles and other calculated parameters to be reliable, too. Agreement between theory and experiment may imply that the reaction mechanism of the studied enzyme can be considered as clarified.

In this review we give an overview first on models, which can be used for the understanding of an enzyme reaction. Then, up-to-date and popular methods of calculation will be surveyed without laying emphasis on their mathematical background and technical details, rather focusing on their performance. The main part of the paper is written on case studies on some enzyme reactions, either involving special effects or dealing with a widely studied class of processes. First we discuss serine proteases, an important enzyme family for which three-dimensional structures have been available very early [2]. While the key reaction, general base assisted catalysis, is well-known in organic chemistry, a variety of factors contributing to rate enhancement has to be considered, and it seems that by now most of them are understood in detail [3]. Next we treat biological phosphate ester hydrolysis, which is a key step e.g., in the transfer of the phosphoryl group from a phosphate ester or anhydride to a nucleophile [4]. Understanding of this process is closely linked to the elucidation of mechanisms for the corresponding non-enzymatic reactions in solution. A wide variety of experimental data is available; however, it is difficult to summarize indirect conclusions in the absence of explicit molecular models. Accordingly, quantum mechanical calculations provide an important option to obtain a reasonable guess for the reaction energy profile. An interesting process is long-range electron transfer in heme peroxidases, its understanding in detail involves special experimental techniques. Therefore, to study enzyme-catalyzed electron transfer reactions we need to perform sophisticated quantum mechanical calculations, classical molecular mechanics alone is not appropriate (cf. Section 4.3). Cytochrome P450 enzymes form a very large superfamily of heme enzymes, their regioselectivity, oxidizing power and reactivity are therefore of outstanding interest, we treat them in Section 4.4. At last we deal with a special case, xylose isomerase catalysis [5], where quantum effects, namely proton tunneling, play a certain role in determining the reaction rate. Tunneling could be relatively precisely reproduced by direct calculations on other enzyme reactions, too.

2. Models

An enzyme has thousands of atoms, therefore the whole molecule and its transformations cannot be treated by direct quantum mechanical methods. Instead of striving for complete models, including all atoms of the enzyme and its surroundings, the system is better partitioned into various regions, which can be described at various levels of sophistication. The active site (**A**) is embedded in the amino-acid residues of the protein core (**P**) with ionizable surface, eventually buried, side chains and the whole protein is dissolved in the bulk (**B**). For most enzymes studied, this latter contains, beside water molecules, counter ions, partly or fully shielding the electrostatic field of the positively or negatively charged side chains. However, in some specific cases, the bulk is not necessarily aqueous, it can be formed by e.g., the atoms of a membrane, where the enzyme is located. The above three regions, schematically depicted in Figure 1, can be calculated at different levels of sophistication.

Figure 1. Hierarchical composition of a full enzyme system. Active site or quantum mechanical region (**A**); protein core or molecular mechanical region (**P**); bulk or dielectric continuum shell (**B**) (figure drawn on the basis of the crystal structure of human aromatase, 3EQM).

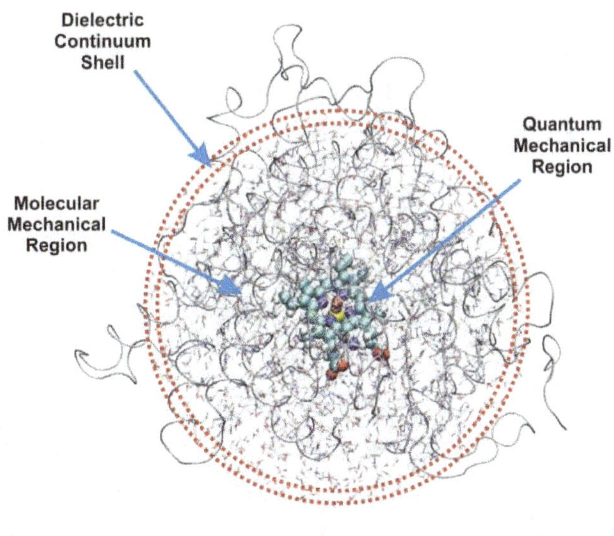

For most enzymes **A** includes the catalytic machinery, *i.e.*, key amino-acid side chains, one or more water molecules and substrate, these fragments are directly involved in catalysis. The minimum number of non-hydrogen atoms in this model may range from 10 to 200, which should be extended by a further number of atoms belonging to essential prosthetic groups, if present. Up-to-date quantum mechanical methods are available even for the high-level treatment of such systems, even if these contain one or more third or higher row atoms, e.g., sulfur, phosphorus or some transition metal. Special electronic effects, like excitation or transfer can be accounted for using sophisticated computational methods, which make use of high-performance software and hardware. The basic reason, why quantum mechanics is indispensable for the adequate description of these models, is that bond fission and formation taking place during the vast majority of enzymatic processes needs such an exact treatment. We may obtain structural parameters for **A** mainly from the Protein Data Bank [6], in some specific cases from other sources. If geometry optimization is necessary, atoms at the boundary of **A** must be fixed in order to avoid artificial distortions from the experimental, chemically relevant structure, which may essentially influence final results and lead to artifacts in the modeling procedure. The atoms of the active site must be kept within the geometric frame of the protein core, which is absent from **A**, therefore sometimes spurious effects may arise as a result of geometry optimization potentially leading to lethal distortion of the active-site structure. In a concrete model covalent bonds linking side chains or other groups of the active site to atoms in **P** must be split, while the resulting dangling bonds can be saturated by hydrogen atoms, eventually methyl groups. This is quite feasible for apolar single C–C bonds, but for polar links, like C–N or C–O replacement of an atom with another, for which the electronegativity is essentially different from that of hydrogen, it causes spurious charge accumulation at the boundary. Similar effects may occur if multiple bonds are cut. Most preferably, the active site should be constructed by cutting C^{α}–C^{β} bonds of the amino-acid residues. A further problem is that hydrogen atoms, used for the saturation of dangling bonds emerging after cutting, may be in steric conflict with some atoms of **P** and this may even lead to convergence problems in quantum mechanical calculations as well as spurious terms in the energy expression within a force field applied in the molecular mechanical calculation. In case of some electron transfer reactions it is very difficult to define the active site atoms appropriately, since localization of an unpaired electron may be uncertain. In these cases, results are very sensitive to the definition of the model, adding or dropping one or two atoms may considerably influence, e.g., the calculated spin density distribution. In such cases, the gradual extension of the size of the model may bring certain saturation in calculated sensitive properties, like charge or spin distribution. In a series of calculations on models with gradually increasing size, the smallest model for which these properties do not change, as compared to previous one, may be appropriate.

Enzymatic mechanisms cannot be understood quantitatively on the basis of active-site models alone. Even in cases, like phosphoryl transfer (cf. Section 4.2), where the basic reaction step seems to be determined by covalent bond fission and formation within the active site, electrostatic, steric and hydration effects influence the formation, structure and protonation state of the active site, which may have some or even basic importance. At least two important phenomena, electrostatics and protein fluctuation cannot be reduced solely to the active site; distant protein residues may and quite often do play a role. In case of electrostatics incorporation of atomic monopoles in the model provides often quite good results; however, sophisticated non-quantum mechanical methods are available for the consideration of such effects. Atomic monopoles can be treated as transferable from one protein to the other; this means, however, that mutual polarization between the active site and protein residues is neglected. This approach is quite feasible; but even semi-quantitative agreement with experiment is rarely achieved. Backbone and side-chain fluctuation can be treated on the basis of molecular dynamics (see Section 3); restriction of the number of protein atoms in the model is allowed only, if local effects (e.g., separated side-chain fluctuation) are investigated.

In most cases the overwhelming majority of the bulk, **B**, is water; however, dissolved inorganic ions and eventually other components are quite important when estimating its effect on the reaction process. The dielectric constant of water is large; therefore the effect of **B** on electrostatic factors, influencing the reaction, may be quite important. In a precise model it is not sufficient to consider only **A** and **P**, the influence of **B** must also be estimated some way. Water molecules may influence the outcome of the enzyme reaction by three different ways. One or more molecules may act as proton donor or proton acceptor during reaction, these have to be explicitly included in **A**. Structural water molecules bind quite strongly to the protein core or the surface, because of their reduced mobility they have to be explicitly included in **P**. Locations can be obtained from X-ray diffraction studies, for which results are deposited in the Protein Data Bank. Like for other atoms, belonging to the protein core, electrostatic effect of structural water molecules can be considered by including appropriate point charges in the Hamiltonian. In case of a molecular dynamics study **P** must include all structural water molecules explicitly. The most complicated type of water in biological systems is bulk, for which a manifold of models with varying adequacy is available. It is possible to model the bulk by a finite set of point charges, however, their electrostatic effect converges very slowly, thus a very large number of molecules has to be included. An implicit way is to use an empirically selected dielectric constant in the force field. Because of the strong effect, *i.e.*, the large dielectric constant of bulk water on protein electrostatics, energy differences for various intermediates and transition states of the reaction are reduced to a quite large extent as compared to the corresponding reaction in the gas phase. Accordingly, great care is needed when comparing quantum mechanically computed values to experimental ones. The bulk may contain small inorganic ions (e.g., Na^+, K^+, Ca^{2+}, Cl^-, HO^-), which partly shield the charge of surface side chains, their concentration is characterized by the ionic strength. Counter ions are not fixed; rather they form a loose distribution of charges. Their effect can be best simulated by high-performance methods like the Poisson-Boltzmann equation, to be discussed in the next section. In some cases when the enzyme reaction is diffusion-controlled the process can be described by Brownian dynamics simulation techniques. These are developed to estimate the rate at which the reactant molecules would collide with the active site in the appropriate orientation [7].

The role of protein dynamics in enzyme catalysis is a controversial question in computational enzymology. Some claim that protein dynamics are essential in understanding enzymatic processes, while others state that dynamics is not an important contribution to catalysis. However computer simulations demonstrate the importance of structural fluctuations and the need to include them in the modeling of certain enzyme reactions [8,9]. In some cases, e.g., in processes with significant changes in solvation it is even impossible to correctly predict activation energies if the dynamics of the protein and its active site is not considered in the calculation. In other reactions fluctuations have only a slight effect on the reaction path (see trypsin in Section 4.1).

It is often supposed that entropy effects play an important role in enzyme catalysis by fixing the reaction partners in the proper orientation and thus reducing translational and rotational entropy in the transition state. This means that an "entropy trap" may contribute very much to rate acceleration. However, these effects may be smaller than anticipated since enzyme molecules are quite flexible. For example in case of serine proteases entropic contribution to rate acceleration is relatively small, rather preorganization of the active site to ensure maximum interaction between reacting partners is the basis of rate enhancement [10]. As we will show in Section 4.1 electrostatic stabilization of the active-site complex plays here a prominent role, while in other cases the liberation of water molecules from the active site into bulk solvent may be also crucial.

3. Methods

In this section we give an overview on some popular methods for the quantum mechanical treatment of enzyme reactions, more details can be found in recent reviews [11,12]. It must be stressed that presently available methods cannot be considered as black-box applications, those who apply them must have some experience in judging the reliability of some of their special features. Otherwise, potential artefacts may appear, which become especially dangerous, if remain hidden. For example, in case of an ill-defined model of the boundary between **A** and **P** iteration, needed to reach the quantum mechanical result, may diverge. This is easily detectable, however, if the iteration converges to a false value, the mistake may remain hidden.

Depending on the size of the active site (**A**) and the required accuracy various quantum mechanical methods can be selected for a calculation on the restricted model, **A**, which contains some dozens of atoms. Several software packages are available for the application of these methods at higher or lower accuracy, like the latest version of the GAUSSIAN program package [13], which incorporates several specific applications. In case of enzyme reactions it is a very appealing feature that it allows to compute the reaction path and the activation energy and to determine the structure of reactants and products, which are connected by a given transition structure. Basically, either the *ab initio* molecular orbital [14] or the density functional [15] method can be applied, as far as the size of **A** allows it. If the size of **A** allows, electron correlation can be accounted for in the calculation, which is highly recommended if a covalent bond is formed or cleaved during the enzymatic process. The quantum chemical treatment of transition metal atoms, which are in a number of enzymes essential parts of **A**, is a notoriously difficult task. The metal-ligand interactions are often highly directional and the selection of the appropriate quantum mechanical method is not always straightforward. The problem is even more difficult if the system may exist in several energetically close-lying spin states which are all characterized by different coordination numbers and local geometries. If **A** is very large with more than 200 non-hydrogen atoms, semiempirical methods can be applied which may provide a rough description of the mechanism and pave the way toward a more precise treatment.

Like in case of partitioning of the whole enzyme system into active site, protein core and surrounding bulk, a layering of the quantum region is also possible. It is the ONIOM method, available within the Gaussian package that allows this [16]. The molecular system under investigation is partitioned into three layers, which are described at successively more accurate levels. The innermost one, where bond formation and breaking takes place, is treated with the most accurate method, referring to electron correlation, if necessary. The outermost layer corresponds to **P** + **B** and is treated with molecular mechanics, a semi-empirical or an *ab initio*, small basis set Hartree-Fock method. The middle layer, which is sometimes dropped from the model, is treated with a method of intermediate accuracy between those applied to the high and low-level layers. It is necessary if electronic effects play a role in **P**, which cannot be treated by any molecular force field.

If the size of **P** is too large for a quantum mechanical method molecular mechanics can be applied for the calculation of various hypothetical or real structures and their energies [17]. In this approximation molecular systems are considered in the frame of the classic Newtonian mechanics approach. The energy is expressed as a sum of bond stretching, bending, torsional, non-bonding and cross terms. These contributions are estimated using mathematically very simple expressions; therefore, calculation of energy changes is very fast, several orders of magnitude faster than in the case of quantum mechanical methods. In earlier versions, the parameters of these simple expressions are fitted either to experimentally determined structural quantities, like bond lengths, bond angles, equilibrium torsional angles, vibration spectra, thermochemical and other data. Up-to-date methods of parameterization refer to high-precision quantum mechanical calculations, which is a better approach since direct correspondence between calculated and estimated parameters can be achieved. It is supposed that parameters are transferable within a relatively large set of molecules, which are constructed of similar fragments. Bonded parameters can be approximated quite precisely; bond lengths and angles, as calculated by molecular mechanics, reproduce experimental values very precisely. Determination of non-bonding parameters often means a problem. Van der Waals parameters can be obtained relatively accurately; however, definition and derivation of atomic point charges is not straightforward, even in modern force fields. The basic problem is that the charge distribution of an atom is not perfectly defined by the simple Coulomb monopole expression, dipole and higher moments. Furthermore, polarization and electron delocalization should be considered, too. In some cases the two-atom term additivity does not hold perfectly, three-body and higher interactions may be essential.

For a wide class of biomolecules, including proteins and nucleic acids the AMBER parameterization and software package became very popular. Since its first publication in 1981 it has been further developed several times until it reached its present version [18]. The original AMBER force field and program package was used for the search for energy minima of separate molecular systems in the gas phase. The current versions are aimed more at the simulation of biomolecules in water solution. Both explicit and implicit models for water as well as for some organic solvents are accounted for. Beside AMBER, several other force fields are available in the literature. For example, GROMOS force fields do not consider aliphatic hydrogen atoms. In contrast to most other ones, this parameterization aims primarily at reproducing the free enthalpies of hydration and apolar solvation for a range of organic compounds [19]. A versatile software GROMACS was developed at the University of Groningen, starting twenty years ago [20]. This is a fast program for molecular dynamics simulation, which does not have a force field of its own, but is compatible with several parameterizations. It was developed and optimized for use on PC-clusters. The CHARMM software is also widely used; it has been developed with focus on biomolecules, like proteins. With its newest parameter set it can be used with various energy functions and models, e.g., combined quantum mechanical-molecular mechanical methods, all-atom classical potentials, explicit and implicit solvent models, as well as various boundary conditions [21]. The OPLS force field is also often applied to elucidate enzyme mechanisms [22].

For the treatment of time-dependent phenomena molecular dynamics methods can be used [23]. The term molecular dynamics (MD) typically refers to the propagation of atoms or groups combining several atoms according to the laws of classical mechanics. The forces acting on the particles are calculated only at discrete points along the trajectory. This method allows computer simulation of local or global molecular movements. The atoms of a molecular system are allowed to interact and the trajectories describing their movement can be calculated by solving Newton's equations of motion. Forces between the particles and the potential energy of the system are obtained by expressions from molecular mechanics. For a large number of atoms within most systems investigated, it is impossible to solve these equations analytically; numerical methods are needed. Long simulations are often mathematically ill-conditioned thus errors may accumulate, which can be minimized using appropriate algorithms. Molecular dynamics simulations may be used to determine macroscopic thermodynamic properties by calculating statistical averages.

If quantum effects play a substantive role in the enzymatic process, the *ab initio* molecular dynamics approach can be applied [24]. In this method the quantum mechanical effect of the electrons is included in the calculation of the energy and forces for the classical motion of the atomic nuclei, explicitly including the electrons as active degrees of freedom in the calculation. This is a first principles molecular dynamics approach, which usually employs periodic boundary conditions, plane-wave basis sets, and density functional theory. Application of this method ranges from the thermodynamics of solids and liquids to the study of chemical reactions in solution and quantum mechanical design of enzymes [25]. Referring to the present-day computer capacity, the size of the system studied is limited to about 250 atoms.

Application of classical force fields to enzyme reactions is problematic if covalent bonds are formed or broken during the process. In such cases quantum mechanical (QM) and molecular mechanical (MM) approaches are combined in a variety of hybrid, quantum classical methods denoted by the acronym QM/MM [11,12]. The active site, **A**, is treated by a quantum mechanical method, while its environment, **P** + **B**, can be described by molecular mechanics. Both Hartee-Fock and density functional methods can be applied in various QM/MM schemes; the latter allows us to include dispersion effects, which is an important feature. The basic assumption in the QM/MM approach is that the system can be partitioned in four regions; accordingly, the Hamiltonian becomes a sum of four terms. Two of them are the QM and MM terms, while the interaction between them (*i.e.*, interaction between **A** and **P** + **B** and their mutual polarization), as well as the boundary between QM and MM regions are treated separately. In most versions polarization of **P** by **A** (the response) is neglected, thus the mathematical procedure becomes much simpler. The boundary, introduced to mimic the rest of the system, is formed of saturated dangling bonds, which remain after cutting the quantum region out of its environment. Several boundary atom schemes have been developed to avoid the artifacts due to them. Thus, the saturation may be done by hydrogen atoms, dummy centers or special, strictly localized molecular orbitals [26]. A boundary center appears as a normal atom in the molecular mechanical calculation.

In order to provide an accurate thermodynamic description of reaction processes in solution and in enzymes free energy perturbation schemes can be combined with QM/MM calculations [27]. The smooth connection between the QM and MM regions in the boundary is crucial to avoid instabilities during the iterative optimization as well as in the free energy calculations. The basic idea is to sample only the molecular mechanics degrees of freedom while the quantum mechanically related atoms are kept fixed. This allows a significant reduction of the computational effort. Warshel and co-workers proposed that the sampling is based on the empirical valence bond (EVB) scheme [28,29], where parameters are fitted to *ab initio* data [30].

A variety of enzyme processes is determined, or at least influenced, by the electrostatic field of the protein core and the bulk. Quite often, the polarity of transition-state structures is different from those of the initial or final ones, which means that their electrostatic interaction with the environment is also different. Protonation and deprotonation, almost always determined by electrostatics, are also important part of the enzymatic reaction process. Thus, for a precise description of the energetic aspects of the reaction, considering the electrostatic effect of the environment on the various steps of the reaction is crucial. One of the most effective and popular methods for the adequate estimation of protein electrostatics is the solution of the Poisson-Boltzmann equation for the whole system containing **A**, **P** and **B**, as well [31]. The method is based on the Poisson equation, which relates the spatial variation of the electrostatic potential as a function of the charge density and the dielectric polarizability. In case of a uniform polarizability, represented by a single dielectric constant, and a point charge distribution the Poisson equation is reduced to the simple Coulomb's law. If the polarizability is not uniform, the dielectric constant is a function of spatial co-ordinates. It is possible to consider the effect of ions dissolved in the bulk via representing the mobile ions by a mean-field approximation. The ionic concentration is determined by the Boltzmann equation and replacing the charge density in the Poisson equation by the sum of densities due to dissolved ions and the protein a combination, the Poisson-Boltzmann equation is obtained. This equation can be applied to systems including **A**, **P** and **B**, which is described by a non-uniform dielectric constant. If the protein does not accumulate a high amount of charges, the equation can be linearized allowing a numeric solution. Advanced finite difference methods can be applied with success and a commercial software is available, which can be applied even by the non-specialist [32].

The combined effect of the **P** and **B** regions is recently more and more frequently treated by the Polarizable Continuum Model (PCM) developed by Tomasi and coworkers [33]. This calculates the molecular free energy in solution as the sum of electrostatic and dispersion-repulsion contributions extended by the cavitation energy, which is needed to form the solvent cavity embedding the solute. The electrostatic energy is calculated from point charges located on the molecular surface, which is constructed from spheres representing non-hydrogen atoms with appropriate van der Waals-radii. The dispersion-repulsion contribution is calculated using the solvent accessible surface. The PCM is clearly less precise than the sophisticated Poisson-Boltzmann method; however, it allows much faster calculations. Its use may be amended in cases where the protein core surrounding the active site can be considered as uniform to at least a certain extent. In cases if **P** consists of several ionizable side chains and ionic strength effects in **B** are thought to be important, the Poisson-Boltzmann method is recommended.

4. Case studies

4.1. Electrostatic Catalysis: Serine Proteases

Over one third of known proteolytic enzymes are serine proteases [34]. They are found both in prokaryotes as well as in eukaryotes and cleave covalent peptide bonds of proteins. At the molecular level this is a very important event in case of several life processes. For example, serine proteases digest food for bacteria and help viruses to infect cells. These enzymes co-ordinate various physiological functions in humans, e.g., digestion, immune response, blood coagulation as well as reproduction. They were given this name because their active site includes a highly reactive serine side chain, which is crucial for catalytic activity and is part of a Ser-His-Asp triad, which is buried in the protein core. The mechanism of action is identical in the various members of the serine protease family; they differ only in substrate specificity. The spatial arrangement of amino acids in the triad is optimized for catalysis by convergent evolution at the molecular level [35].

Serine protease-catalyzed hydrolysis of the peptide bond is a reaction involving a tetrahedral intermediate (see Figure 2). During catalysis a nucleophilic attack by the serine hydroxyl group on the carbonyl carbon atom of the substrate is helped by the histidine imidazole group as a general base. This leads to formation of a tetrahedral intermediate and an imidazolium ion. In a subsequent step the tetrahedral intermediate is broken down by general acid catalysis to an acyl-enzyme, an imidazole base and an amine product. During the acylation step, the imidazole group transfers the proton of the serine hydroxyl to the amine leaving group. The acyl-enzyme is then deacylated through the reverse reaction pathway of acylation, but in the second addition-elimination reaction a water molecule instead of the serine residue is the attacking nucleophile.

Figure 2. Reaction steps during serine protease catalyzed cleavage of the peptide bond (**left**). The acyl-enzyme intermediate hydrolyses via the reverse route (**right**).

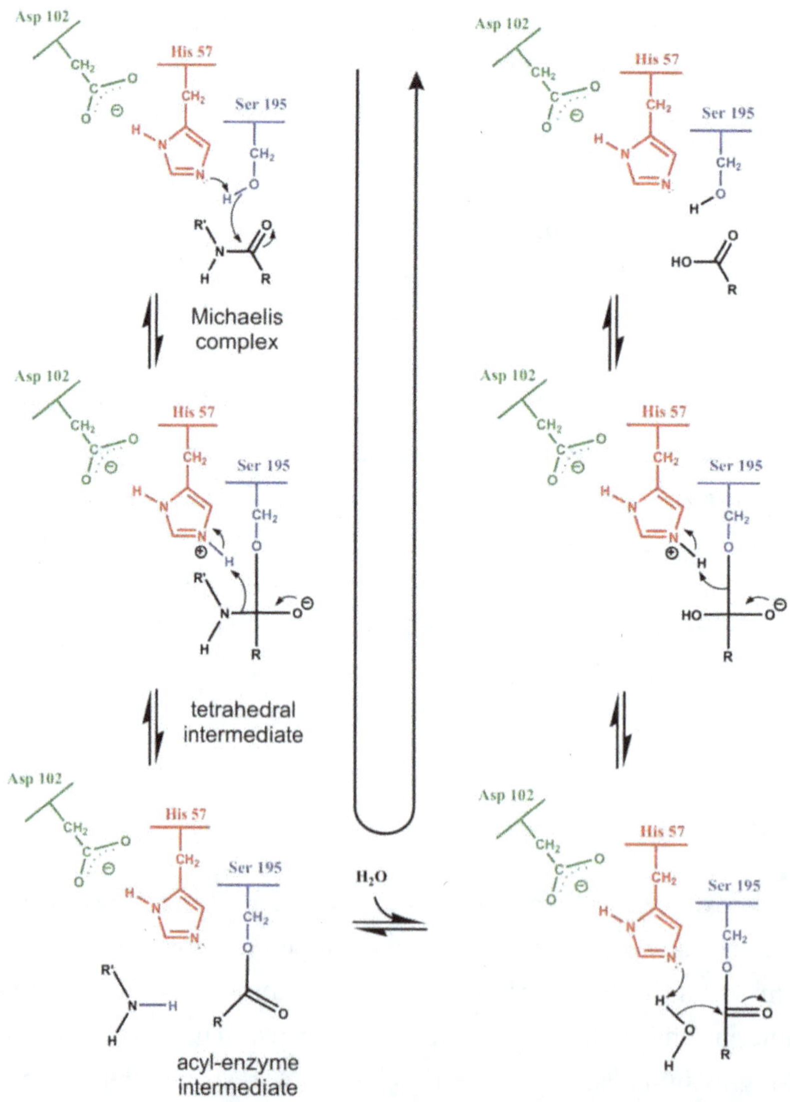

Crucial step of the reaction is the formation of the tetrahedral intermediate, facilitated by the surrounding oxyanion hole, stabilizing the intermediate by hydrogen bonds (see Figure 3). Backbone amide groups form hydrogen bonds with the strongly polarized oxygen atom, keeping the carbonyl bond in the right position for a nucleophilic attack and stabilizing the intermediate structure. Formation of the oxyanion hole is a general feature of serine proteases and it is precisely tailored for the oxygen atom. Thionoester substrates, containing sulfur with a somewhat larger atomic radius than the oxygen atom, are not hydrolyzed by chymotrypsin and subtilisin, although they do not differ significantly from esters in reactivity [36]. The contribution of the oxyanion binding site to the catalytic rate acceleration was estimated by replacing the hydrogen-bond donor Asn155 of subtilisin with a neutral glycine residue. This lowered k_{cat} for a specific substrate by a factor of 150 to 300, however, did not alter the Michaelis constant, K_M at all. Thus, it can be concluded that the binding site contributes only to transition-state stabilization and leaves the binding power unaffected [37].

Figure 3. The oxyanion hole stabilizing the tetrahedral intermediate in α-chymotrypsin. Backbone amide NH groups of Gly193 and Ser195 form hydrogen bonds with the amide oxygen atom of the substrate.

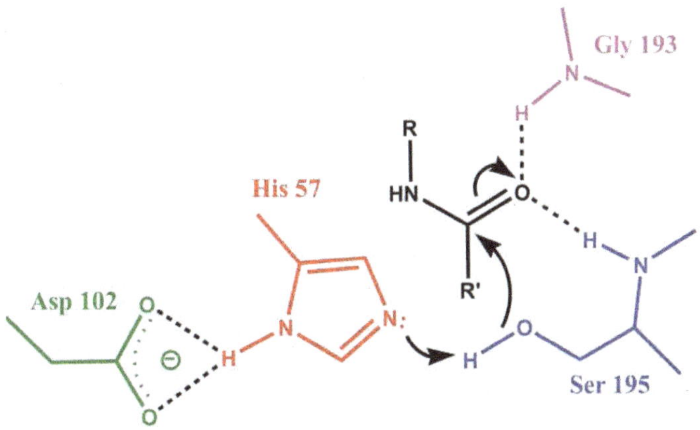

X-ray diffraction studies have shown that catalysis is assisted by a hydrogen-bonded triad of amino-acid side chains. A buried aspartate is linked to the imidazole moiety of histidine, which binds to the catalytic serine hydroxyl group (see Figure 2.) [38]. The geometric relation of the Asp, His and Ser side chains allows assuming that histidine serves for transferring the proton from Ser to Asp in a mechanism called charge relay. However, since proton transfer from the highly basic serine hydroxyl group to the acidic carboxylate side chain of aspartate is unlikely, it was supposed that the role of the buried aspartate is the stabilization of the ion-pair formed by the positive imidazoliun ion and the negatively charged-tetrahedral intermediate [39]. Nuclear magnetic resonance [40] and neutron diffraction [41] studies have confirmed that it is the imidazole and not the aspartate that is protonated.

The stabilizing role of the buried aspartate is also supported by site-directed mutagenesis. Replacement of Asp102 of trypsin with a neutral Ala residue results in a reduction of four orders of magnitude in the rate constant. It is evident from the three-dimensional structure of the mutant enzyme that the catalytic His57 is unable to accept proton from Ser195 [42]. Studies on the D641A mutant of prolyl oligopeptidase, belonging to another family of serine proteases, also indicate the importance of the negative charge of the buried aspartate [43]. It is thus quite clear that it electrostatically stabilizes the ion pair formed in the transition state. This has been shown by empirical valence bond free energy calculations, too [44]. It has been found that in trypsin and subtilisin the buried aspartate contributes to the stabilization of the transition state by 26 (25) and 17 (25) kJ/mol, respectively (experimental values are in parentheses). These calculations clarified the role of the buried aspartate side chain, which is electrostatic stabilization of the ion pair formed between the positive imidazolium side chain and the negative tetrahedral intermediate rather than proton transfer to its carboxylate.

It is interesting to note that mutation of all three members of the catalytic triad to alanine does not reduce activity to zero. This implies that the catalytic triad is not the sole source of the enzyme activity since even in its absence the mutant enzyme hydrolyzes substrates three orders of magnitude faster relative to the reaction in water. This remaining activity may arise from the contribution of stabilization by the preorganized binding pocket and the oxyanion hole. Calculations by Warshel and coworkers provide ample evidence for the crucial role of electrostatic effects due to the preorganization of active sites, which strongly stabilize the transition states of enzymatic reactions [45,46]. Such a preorganization could happen via convergent evolution organizing the hydrogen-bond network around the catalytic triad in such a way that it stabilizes the (- + -) charge distribution. Poisson-Boltzmann electrostatic calculations provided relative stabilization energies, calculated as the electrostatic interaction energy of atoms of the catalytic triad with the protein environment. These numbers reflect changes in relative experimental enzyme activities, proportional to log k_{cat}/k_M for the same substrate, correctly (cf. Table 1) [47].

Table 1. Calculated relative electrostatic interaction energies, *vs.* relative experimental enzyme activities (kJ/mol) of the catalytic triad in some serine proteases.

Enzyme	Calculated	Experimental
subtilisin Carlsberg	−102.6	−36.5
α-chymotrypsin	−71.9	−34.2
subtilisin NOVO	−62.1	−31.6
β-trypsin	−40.7	−26.7
α-lytic protease	9.3	0.3

On the basis of the above and other calculations we may say that enzymes, acting as a "supersolvent", can strongly stabilize polar structures, like ion pairs or the charge distribution located in their active sites. Stabilization in enzymes is stronger than in water, since preorganized and fixed protein dipoles are almost optimally oriented at the active site. The enzyme, therefore, provides a preoriented environment that stabilizes the transition state of the reaction. In contrary, water dipoles are distributed randomly and have to reorient in order to stabilize transition states. Reduction of the preorganization free energy in the enzyme is due to its folding into its final configuration, which precedes the catalytic process and takes place independently, during protein synthesis.

Even if we have diverse experimental evidence for the elucidation of the mechanism of action, direct information, especially on energetic aspects, can be obtained only from molecular modeling. Results should be in line with experimental observation, furthermore, providing subtle details, which cannot be obtained from experiments. For serine proteases, the crucial role of protein electrostatics in enzymatic rate acceleration can be discussed in appropriate detail only by calculation [48]. Early quantum mechanical approaches were based on gas-phase models containing the catalytic triad and the minimum-size substrate [49]. Such models could be extended by consideration of protein electrostatic effects on the reaction process [50]. It was shown by *ab initio* calculations that the electrostatic interaction originating from the hydrogen bond network involving residues around the buried aspartate plays a significant role in lowering the barrier height of the proton transfer. Later on, the basic statements of the above calculations were supported by further studies. Ishida and Kato [51] performed calculations on the entire reaction path of the acylation step catalyzed by trypsin using *ab initio* QM/MM methods combined with molecular dynamics. It was found that the rate-determining step is the formation of the tetrahedral intermediate, while breakdown of the intermediate has a small energy barrier. Their calculated activation free energy is 75 kJ/mol. It was shown that the proton transfer from Ser to His and the nucleophilic attack of Ser to the carbonyl carbon of the scissile bond are concerted. The most important catalytic factor of stabilizing the tetrahedral intermediate is the electrostatic interaction between the active site and the protein core. Topf and Richards [52] studied the deacylation step of serine protease catalysis using extended quantum mechanical calculations on a model containing all protein atoms and the bulk as a continuum. The calculations show that in the enzyme the tetrahedral intermediate is a relatively stable species lying 30 kJ/mol lower in energy than the transition state. The stabilization is mainly due to the electrostatic effect provided by the oxyanion hole and the buried aspartate. The rate-determining step of the reaction is proton transfer from the attacking water molecule to the histidine side chain; the activation energy is about 92 kJ/mol, 21 kJ/mol, less than for the reference reaction in water, and close to the value obtained by Ishida and Kato.

Hudáky and Perczel [53] built a model of the catalytic triad of chymotrypsin containing 18 amino acid residues of the enzyme and its substrate. They found that this model forms a molecule ensemble, which is stable in itself. They obtained an activation barrier of 85 kJ/mol, just between those obtained by the above two groups. Calculations allowed locating elements of the catalytic machinery, which are responsible for the stabilization of the transition state. Most important of these are the catalytic aspartate (by 27 kJ/mol) and the oxyanion hole (by 57 kJ/mol). The full reaction path is displayed in Figure 4. The acylation mechanism of trypsin and its complete free energy reaction profile have been determined also by Born-Oppenheimer *ab initio* QM/MM molecular dynamics simulations with umbrella sampling [54].

Figure 4. Reaction energy profile of the chymotrypsin-catalyzed cleavage of a model substrate (after Hudáky and Perczel [53]). Note that the tetrahedral intermediates as well as the acyl enzyme represent local minima.

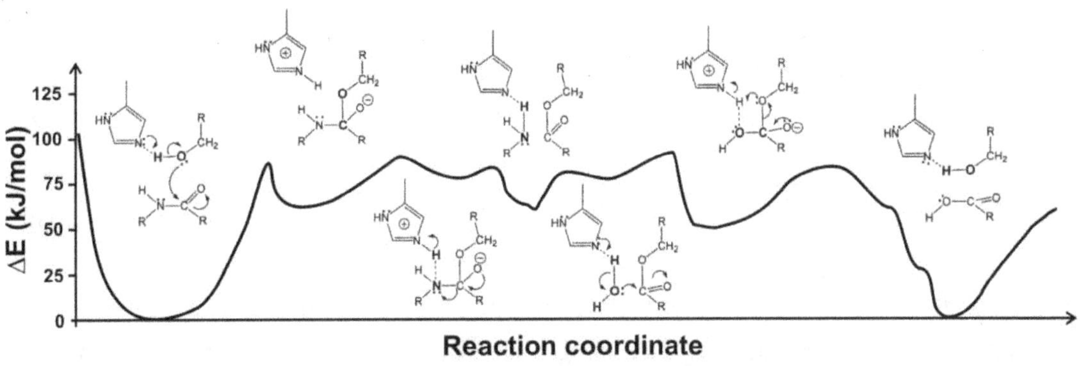

4.2. Phosphoryl Transfer

Phosphoryl transfer belongs to the most important molecular reactions in living systems. These are involved in DNA replication, signal transduction, metabolism and transcription, furthermore they ensure the production of chemical energy, which is required as a driving force for important processes within living cells [4]. Phosphoryl transfer may be catalyzed by kinases, mutases, phosphatases or endonucleases. Along such reactions enormous, even as large as 10^{27}-fold, rate acceleration is observed as compared to the corresponding non-enzymatic process. Despite their central importance, because of the complicated structure of the catalyst, no final information is available about the details of the process although intensive research has been conducted since decades.

Three different types of reaction paths can be considered for enzymatic and non-enzymatic phosphoryl transfer reactions, these are outlined in Figure 5. In the dissociative mechanism (Figure 5, top) a stepwise displacement takes place via a solvated planar metaphosphate intermediate, while in the associative mechanism (Figure 5, bottom) the reaction proceeds via a trigonal bipyramidal phosphorane intermediate. In both cases the intermediates are locally stable; they represent a local minimum on the reaction energy path. In contrast, in case of the S_N2-type concerted displacement the pentacoordinated phosphorane structure represents a transition state represented by a maximum on the reaction energy profile.

Figure 5. Reaction paths for phosphoryl transfer reactions. (**a**) Dissociative (top), (**b**) associative (middle), (**c**) S_N2-type (bottom) mechanism. Square brackets represent the transition state.

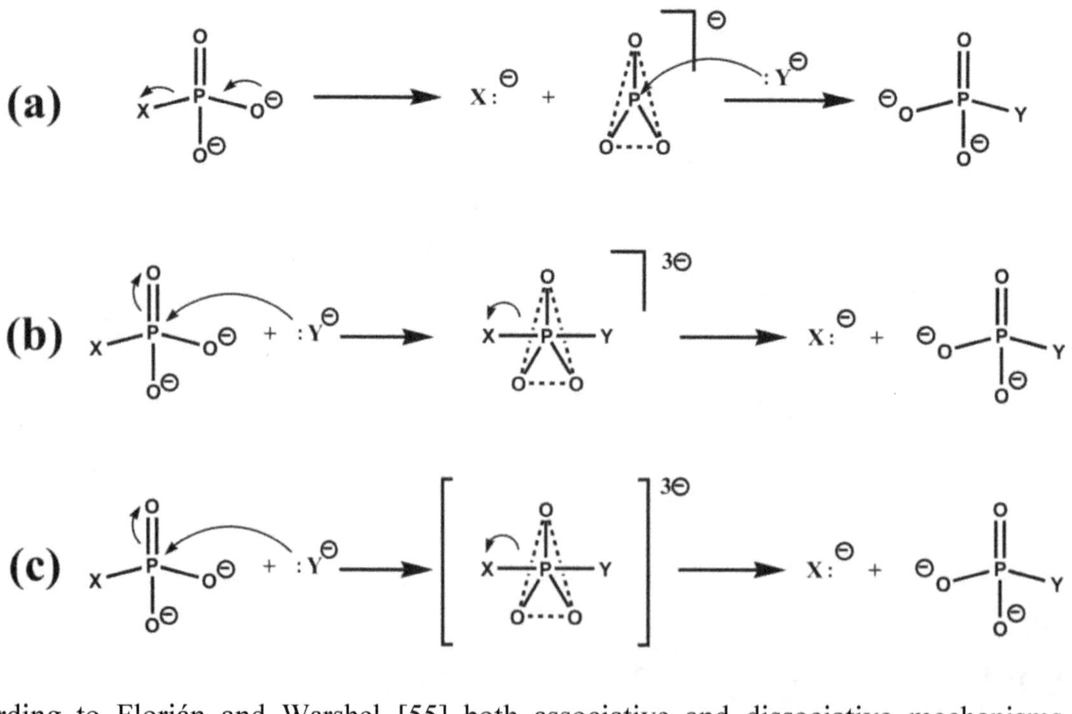

According to Florián and Warshel [55] both associative and dissociative mechanisms may be operative in the aqueous phase. On the other hand, as stated by Herschlag *et al.* [4] for phosphate monoesters considerable experimental evidence supports a concerted reaction mechanism, where no stable intermediate, rather a transition state can be located on the reaction path. Furthermore, they state that enzymatic catalysis generally does not influence the structure of transition states as compared to those in solution. While this might be true in some cases, we found on the basis of density functional calculations on a model of HIV integrase that metal ions, located at the active site, as well as the electrostatic effect of the protein core considerably reduces the activation energy [56]. The high (295 kJ/mol) reaction barrier in the gas-phase reduces to less than the half, 141 kJ/mol, within the enzyme. This means that the protein environment has an important impact on the reaction, which is partly due to the fact that electrostatic repulsion is reduced in the active site due to a diffuse charge distribution. The nucleophile is only partially charged in the transition state. A further effect is that of the doubly positively charged magnesium ion at the active site, which reduces the barrier height by 42 kJ/mol. A similar result using the density functional theory was obtained for the hydrolysis of guanosine triphosphate [57]. Accordingly, combined geometric and electrostatic effects may play an important role in rate acceleration, like in case of hydrolytic reactions catalyzed by serine proteases (see Section 4.1).

While several experimental arguments can be mentioned strengthening the hypothesis that the concerted mechanism may be operative in enzymes, two different X-ray studies seemed to support the associative mechanism, *i.e.*, the formation of a relatively stable phosphorane intermediate during the catalytic process. Lahiri and coworkers thought to locate such a pentacovalent structure (an XPO3Y moiety) in the phosphoryl transfer reaction catalyzed by *β*-phosphoglucomutase [58]. However, their results have been questioned by Blackburn *et al.* [59] who suggested that this structure represents an MgF_3^- transition-state analogue, which is hydrogen-bonded to ligands at the binding site. Subsequent evidence from ^{19}F and ^{31}P nuclear magnetic resonance, further X-ray experiments, and kinetic data [60] as well as quantum mechanical calculations [61,62] provided strong support for this latter assignment. A second example of a potentially trapped high-energy intermediate has been published by Barabás *et al.* [63]. Crystal structure of the complex with a special substrate, *α,β*-imino-dUTP, was determined. In this case the catalytic process slowed down considerably; therefore the reaction path could be followed by localizing snapshot structures. A relatively stable intermediate was thought to be trapped, for which the crystal structure is available. However, the resolution of the active site structure was not enough to present a convincing argument, so the final proof that the associative mechanism is operative in these enzymes is not yet available.

Quantum mechanical studies on various enzymes catalyzing phosphoryl transfer support different mechanisms. Our early calculations for HIV integrase [56] indicate that an S_N2-type mechanism may be effective. We calculated the reaction path carefully and found a very shallow minimum, rather a shoulder on the reactant side (cf. Figure 6). This means that if environmental effects, mainly electrostatics, influences the stability of the transition-state complex, under certain circumstances a locally stable phosphorane intermediate can be formed. Thus, the associative reaction route may be realized. Studies for the DNA repair enzyme endonuclease IV [64] as well as for cyclin-dependent kinase [65,66], DNA polymerase [67] and phosphodiesterase [68] support S_N2-type concerted mechanisms, too. A recent study stresses that a water mediated and substrate assisted mechanism is followed in DNA polymerase, and provides an asymmetric intermediate, which is stabilized by nearby amino-acid residues [69]. This intermediate transforms to the product via a pentacovalent phosphorane transition state, which remains stable for over a nanosecond of MD simulation. This study provides further support for the concerted mechanism because the lifetime of a true phosphorane intermediate should be much longer.

Figure 6. Quantum mechanically calculated reaction profile for the phosphoryl transfer reaction catalyzed by HIV integrase (figure drawn on the basis of Ref. 56).

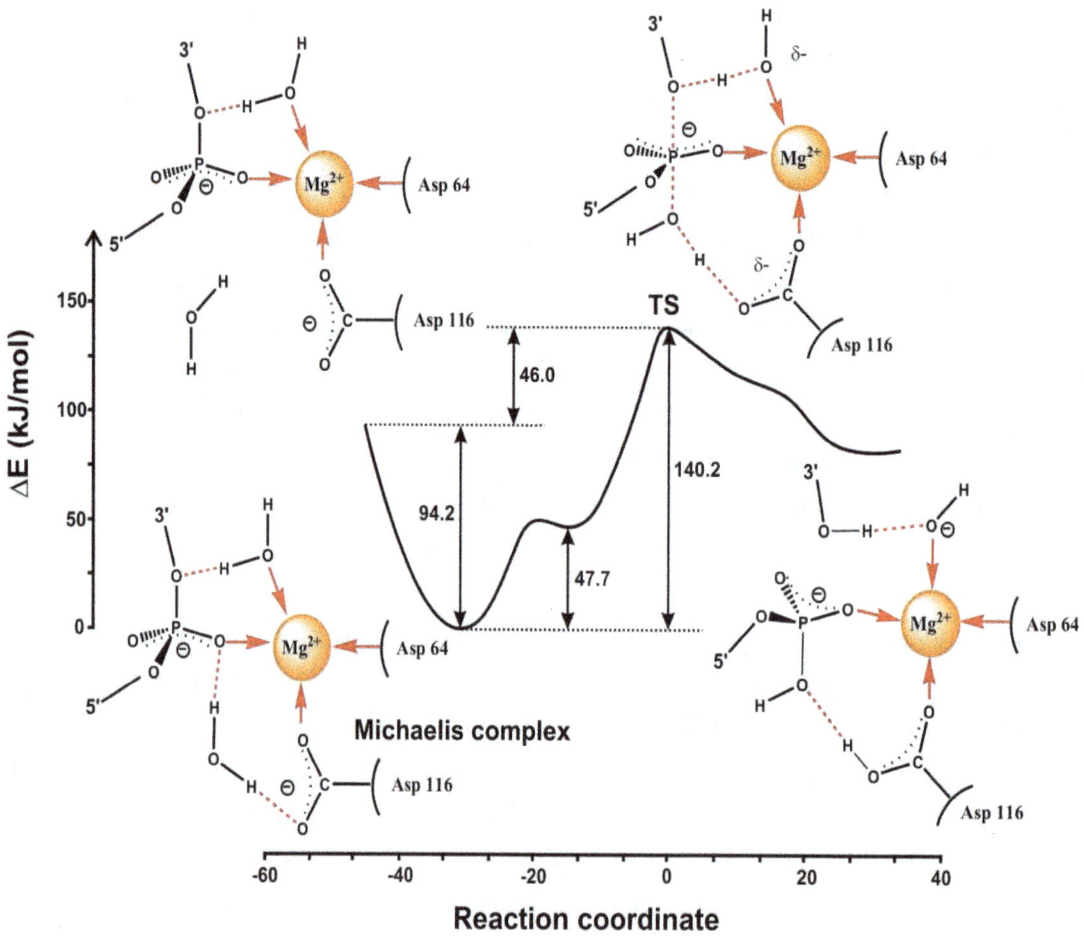

Ab initio QM/MM studies on phoshoenol-pyruvate (PEP) mutase suggest that the catalytic reaction follows a concerted mechanism with a planar trigonal transition state representing metaphosphate (see Figure 7) [70]. This mechanism is consistent with the failure in detecting the putative intermediate in rapid quench experiments and is in line with the statement of Herschlag and coworkers, who call for a concerted reaction mechanism in case of phosphate monoesters [4]. It is interesting to note that the transition state is planar trigonal, in contrast to other cases, where it is thought to be trigonal bipyramidal. Making a distinction between potentially existing structures of transition states complicates appropriate selection of a given mechanism, since it often leaves out of consideration, whether the given structure represents a local minimum or a maximum on the reaction energy profile. At present it is not possible to collect direct experimental evidence for the existence of a given structure in the transition state, since its lifetime is practically zero. Quantum mechanical calculations can provide concrete models, like in Figure 6; however, these may not reproduce experimental data sufficiently accurately, and therefore conclusions on such basis are often questioned. York and coworkers studied the uncatalyzed conversion of PEP to phosphonopyruvate and they have found that due to the large negative charge located at the active site, solvation has a crucial effect on the reaction pathway [71].

Figure 7. Molecular graphics model of the transition state in the reaction catalyzed by phoshoenol-pyruvate (PEP) mutase (on the basis of Figure 3 by Xu and Guo [70]). Note the planar metaphosphate intermediate stabilized by hydrogen bonds to amino-acid residues of the active site.

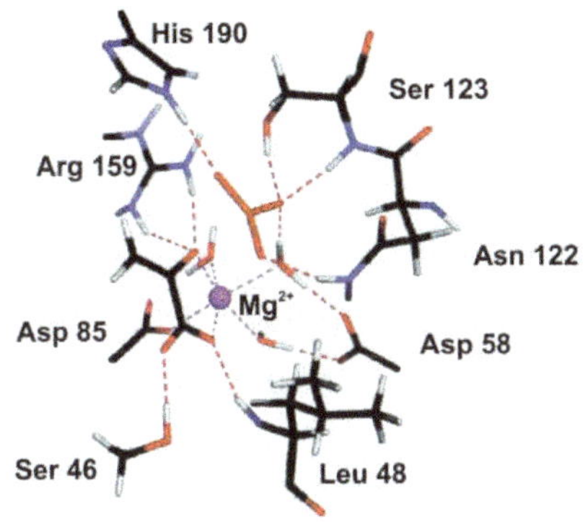

Figure 8. Two possible reaction routes for the hydrolysis of ribonuclease H. Top: schematic structure of the active site, bottom: blue line: attack by a water molecule, red line: attack by a hydroxyl group (figure drawn on the basis of Figure 3 of Ref. 72).

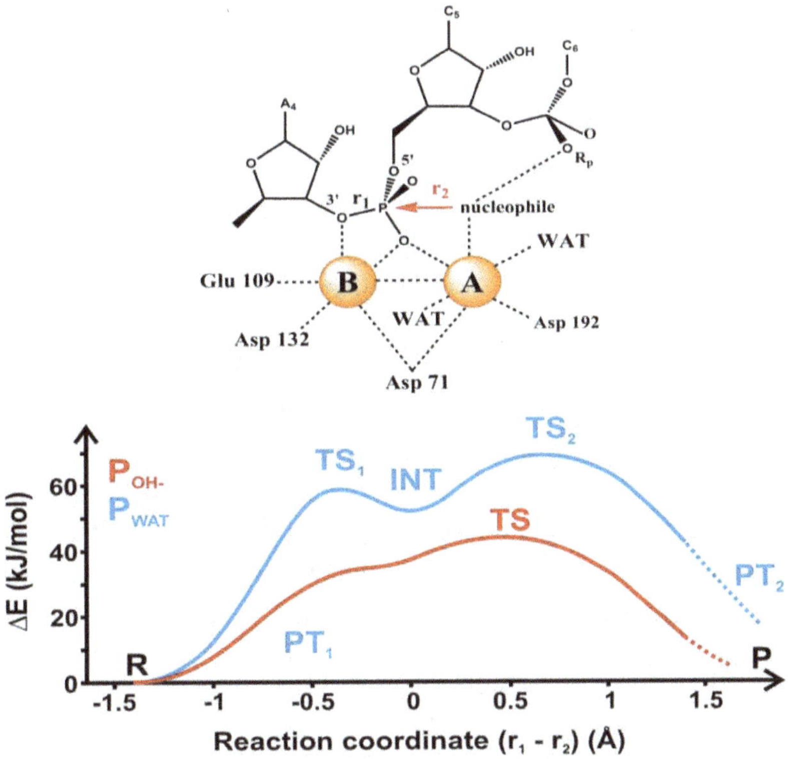

Two recent sophisticated quantum mechanical studies indicate that in ribonuclease H the phosphodiester cleavage occurs via an associative mechanism [72,73]. These calculations indicate that as the reaction approaches the barrier, one of the protons of the attacking water molecule transfers to one of the oxygen atoms of the phosphate group and a penta-coordinated phosphorane intermediate is formed. The calculated energy barrier is consistent with the experimental rate found for the human enzyme. De Vivo and coworkers made a distinction between the attack of a neutral water molecule and a negatively charged hydroxyl group [72]. They have found that in both cases an in-line S_N2-like nucleophilic attack takes place on the central phosphorus atom. This results in an associative mechanism with phosphorane-like transition states, in agreement with crystal structures of transition- state analogues [74]. It is interesting that the attack of a water molecule leads to the formation of a meta-stable pentavalent phosphorane intermediate, as depicted in Figure 8. This is a unique characteristic of the energy profile of the reaction, which has not been observed in previous computational studies. Like in e.g., xylose isomerase (see below), one or more structural water molecules near the active site may facilitate the reaction [75,76].

Warshel and coworkers [77] compared the associative and dissociative mechanisms of phosphate monoester hydrolysis on the example of the methyl phosphate dianion and the methyl pyrophosphate trianion in aqueous solution. They have found that, in good agreement with experimental findings, both associative and dissociative transition states have near-zero entropies of activation. This means that near-zero activation entropy is not indicative of a dissociative pathway, as supposed earlier.

4.3. Long-Range Electron Transfer in Heme Peroxidases

Peroxidases oxidize a variety of substrates by reacting with hydrogen peroxide. In most cases the typical substrates are small aromatic molecules, however, in case of cytochrome *c* peroxidase (CCP) a protein, cytochrome *c*, is the redox partner. The overall reaction mechanism is given in Scheme 1. Several peroxidase structures were determined by X-ray diffraction (e.g., yeast cytochrome *c* peroxidase [78], lignin peroxidase (LIP) [79], pea cytosolic ascorbate peroxidase (APX) [80], and horseradish peroxidase (HRP) [81]. On this basis it is possible to construct a model of the active site (see Figure 9). A histidine and an aspartic acid side chain are present in the proximal heme pocket in all these structures, whereas the proximal tryptophan is only present in CCP and APX.

Scheme 1. Reaction mechanism of heme peroxidases. P is the porphyrin group of the enzyme whose heme iron is indicated, S is the substrate.

$$\mathbf{Fe^{3+}P + H_2O_2 \rightarrow Fe^{4+}{=}OP\bullet\ (Compound\ I) + H_2O}$$
$$\mathbf{Fe^{4+}{=}OP\bullet + S \rightarrow Fe^{4+}{=}OP\ (Compound\ II) + S\bullet}$$
$$\mathbf{Fe^{4+}{=}OP + S \rightarrow Fe^{3+}P + H_2O + S\bullet}$$

Figure 9. Schematic active-site models of cytochrome *c* peroxidase (CCP) (**left**) and ascorbate peroxidase (APX) (**right**). Top: distal position, bottom: proximal position, a separated red dot represents a water molecule (figure drawn on the basis of the crystal structures 1ZBZ and 2XI6).

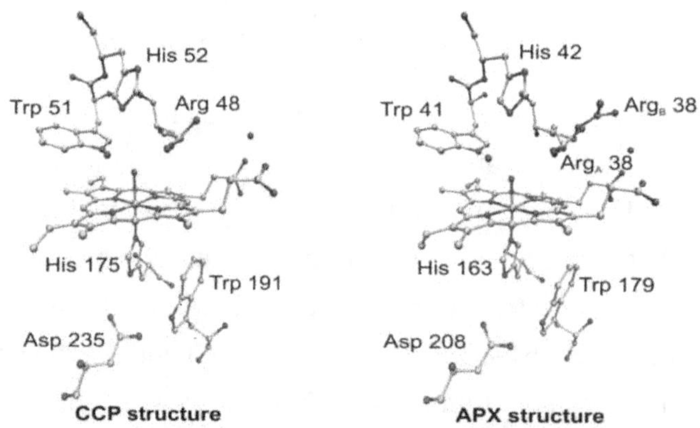

In the first step of the catalytic process (cf. Scheme 1 and Figure 10) the peroxide removes two electrons from the enzyme and the so-called Compound I is formed. During this process the peroxide O–O bond is broken, water is produced and the second oxygen atom of the peroxide remains coordinated to the metal, which loses one electron to give an oxy-ferryl intermediate ($Fe^{4+} = O$). A second electron is removed from the heme providing a cation radical. In the next step, P• is reduced and a substrate radical, S• is formed. Finally, in the last step Compound II is reduced by a second substrate molecule.

Figure 10. Catalytic mechanism of the formation of Compound I. The distal His assists in removing a proton from the incoming peroxide and delivering it to the peroxide O_2 atom.

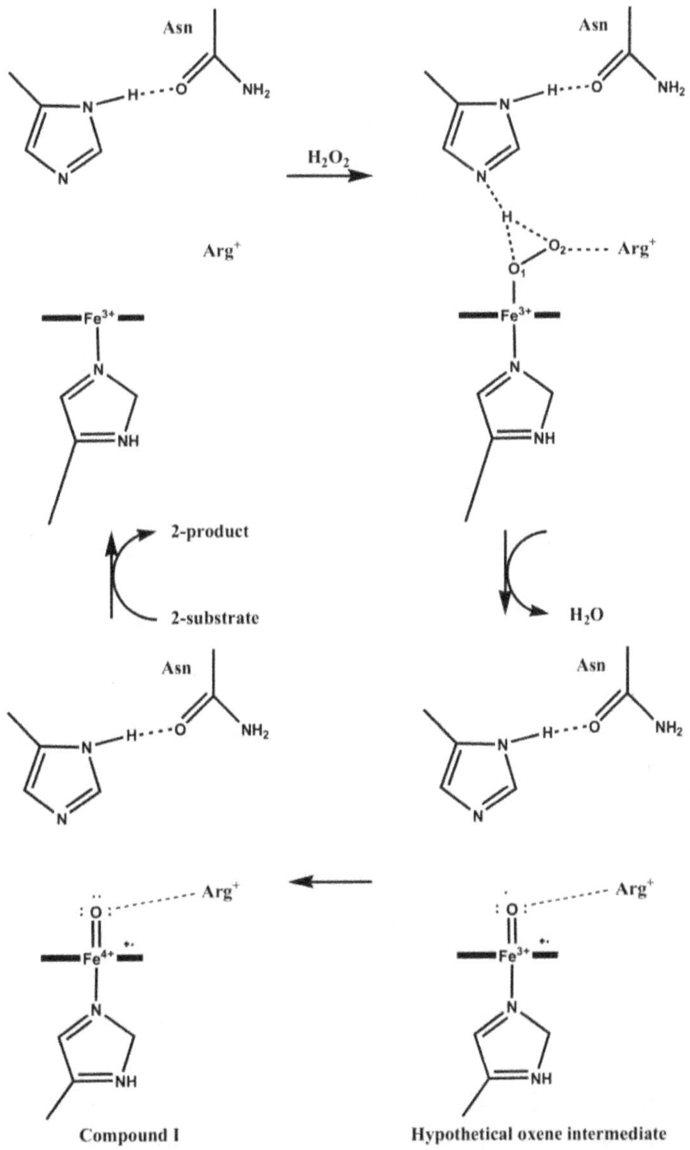

Discussing the mechanism at the computational level an interesting problem arises. APX shares 33% global sequence identity with CCP and has a very similar active site structure [82]. Despite very similar protein structures, APX and CCP stabilize different radical species during enzyme turnover. At variance with CCP, in APX the free electron is shifted to the porphyrin ring of the active site during reaction. This difference is thought to contribute to the different substrate specificities of the two proteins. Earlier, we suggested, on the basis of *ab inito,* minimal basis set molecular orbital and Poisson-Boltzmann electrostatic potential calculations that both in APX and CCP proton transfer involving the proximal histidine (His163 or His175), the radical-forming tryptophan (Trp179 and Trp191), and a nearby aspartic acid (Asp208 and Asp235) side chains may control the location of the radical state [83]. Both molecular orbital and electrostatic potential calculations suggest that the spin distribution depends on the protonation state of the proximal His...Asp...Trp triad. If the transferable proton is shifted from the Trp side chain to Asp, the free electron is localized on the indole group, while if it remains on Trp the unpaired electron transfers to the heme group. Therefore, in Compound I of CCP Trp is supposed to be deprotonated, while in case of APX it is supposed to be protonated and neutral. Protonation state of the side chains is influenced by the electrostatic effect of the protein environment, which differs in these enzymes, especially in the immediate vicinity of the Asp side chain.

The above hypothesis is not supported by some other calculations. Warshel and coworkers applied the empirical Protein Dipoles Langevin Dipoles method and found that the electrostatic effect of the protein environment of CCP stabilizes the tryptophan cation radical by 330 mV relative to that in APX. They proposed that mainly the cation binding site contributes to radical stabilization, but is not the sole cause [84]. The study by Siegbahn and coworkers [85] supports the mechanism, which has been proposed on the basis of X-ray diffraction studies. They have found that if the proximal His-Asp-Trp triad of both enzymes is included in the computational model the free electron of the cationic radical is shared between the porphyrin ring and the protonated histidine side chain. The difference in ionization potential between tryptophan and porphyrin ring in the oxyferryl form of APX and CCP seems to be small. This strengthens the statement that location of the free electron depends on subtle differences of the protein environment in both enzymes. It has been shown that location of the protein radical in APX is most probably Trp179 [86]. Quantum mechanical optimization of the crystal structure of the active site in CCP show that the geometric differences of the two states are quite small and it cannot be decided where the proton is in the crystal structure [87]. QM/MM calculations indicate that the proton is located on the His ligand in all states in the reaction mechanism. According to de Visser [88] the quartet-doublet energy splitting is strongly dependent on local perturbations. Even a point charge far away from the tryptophan radical can transfer the system from a mixed porphyrin-tryptophan radical into a pure porphyrin or tryptophan radical. These perturbations cause varying quartet-doublet energy splitting, which eventually may influence reactivity. Calculations by Mulholland and coworkers [89] reproduced the observed difference in electronic structure, and called the attention to the subtle electrostatic effects which may affect the ionization state of both the tryptophan and porphyrin groups. Their calculations did not support the deprotonation of the tryptophan group, or protonation of the oxoferryl oxygen atom. The remarkable difference in electronic structure between the compound I intermediates in CCP and APX seems to be due to differences in the electrostatic potential around the key groups in the two enzymes.

On the basis of their QM/MM calculations Thiel and coworkers concluded that in another heme peroxidase, horseradish peroxidase the proximal ligand is imidazole and not imidazolate [90].Similar conclusions were drawn by Jensen and Ryde for reduced models of heme proteins [91]. A recent review on QM/MM calculations for heme proteins is also available [92], while the catalase activity of heme peroxidases is treated by Vlasitsa *et al.* [93]. Harris and Loew [94] performed density functional calculations and showed that the imidazolate/Asp-H and imidazole/Asp^- tautomers are very close in energy (the difference is 4 kJ/mol), but only the first representation reproduces the observed shift in the Soret band of Compound I in horseradish peroxidase as compared to chloroperoxidase.

4.4. Cytochrome P450 Enzymes

Cytochrome P450 enzymes (P450s) form a huge superfamily of thiolate-ligated heme enzymes with more than 10,000 members sequenced up-to date. They can be found in almost all living organisms, in bacteria and archaea, in fungi, plants and animals indicating the ancient nature of these enzymes. The great variety of P450s can be attributed to their vital role: they defend the organisms from xenobiotics by oxidizing them to products that can be excreted more easily. As the structure and properties of xenobiotics may differ almost indefinitely, during evolution a very large number of P450s have evolved [95], in some cases acquiring new functions as well. P450s also play a vital role in hormone biosynthesis in mammals, e.g., in man they are responsible for the aromatization of androstenedione and testosterone to estrone and estradiol [96]. In the following we present three case studies, which may shed light on some interesting issues related to P450s.

4.4.1. Is the Oxidizing Power of Active Species of Various P450 Isoforms the Same?

The catalytic cycle of P450s consists of many electron or proton transfer steps [97], which leads to the formation of Compound I, the ultimate oxidizing species carrying out the oxidation reaction. The species is very reactive and elusive that has only been recently observed [98]. Compound I has low-lying quartet and doublet spin states, which differ in energy by less than 4 kJ/mol. In both states there are two electrons in the Fe–O π^* antibonding orbitals which can be coupled via an anti- or ferromagnetic scheme to the porphyrin a_{2u} orbital with significant contribution from the sulfur *p* orbital giving rise to the doublet and quartet states. (Figure 11.) As several P450 isoforms are responsible for the metabolism of drugs taken by man several crucial questions have been raised: (1) does the reactivity of the different isoforms depend on the oxidizing power of the isoform; (2) is the oxidizing power of Compound I of the different isoforms very similar? As the active site of the different isoforms shows great variability it may tune the reactivity of the oxidizing species and influence the regioselectivity of the enzyme [99,100], which has been called "chameleon" behavior [101]. Therefore, a systematic study involving three human P450 isoforms (P450 2D6, P450 2C9, P450 3A4) and a bacterial enzyme ($P450_{cam}$) has been conducted in the presence and absence of substrate in the active site in order to reveal the similarities and differences between the active species of these isoforms [102]. Dextromethorphan was used as a substrate of P450 2D6 and P450 3A4, warfarin as the substrate of P450 2C9, and propene as the substrate of $P450_{cam}$. 5ns long molecular dynamics simulations were run on the ligand-free and ligand bound enzymes and every 200 ps a snapshot has been taken (altogether 26 for each system) that was subjected to QM/MM optimization, in the quartet spin state. After the QM/MM optimization of each snapshot, the oxo group bonded to the iron was deleted and a single point calculation was carried out to estimate the energy of the resting state of the enzyme.

Figure 11. Singly occupied molecular orbitals of Compound I.

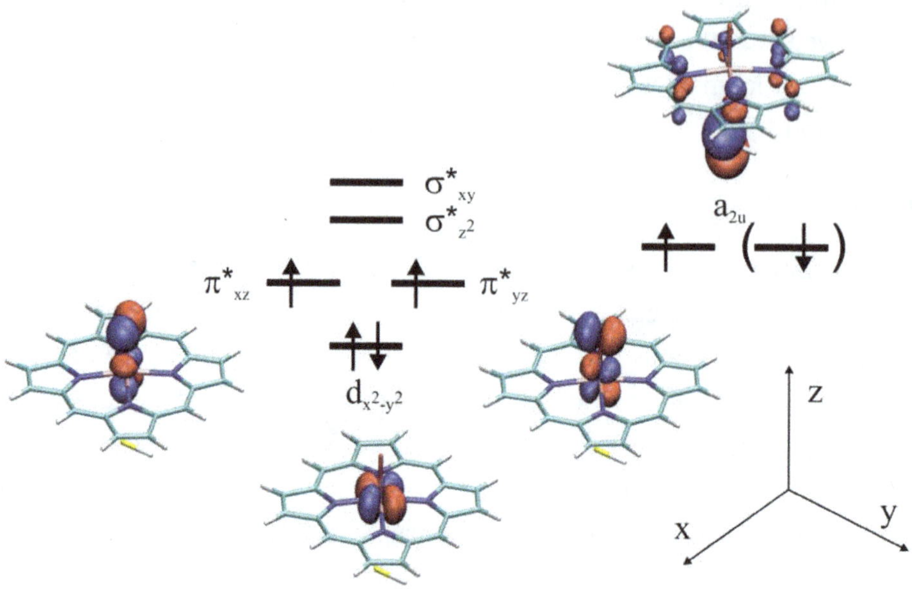

The thermodynamic cycle, shown in Figure 12, depicts the importance of the Fe–O bond enthalpy (ΔE_2) calculated in the study. ΔE_1 includes the energy of oxidation of the substrate by Compound I to form the product and the resting state of the heme. Its value is dependent on the substrate, and could be expressed as $\Delta E_1 = \Delta E_2 + \Delta E_3$, where ΔE_2 is the energy required to break the Fe–O bond in Compound I, and ΔE_3 is the energy released upon addition of the oxygen atom to the substrate to yield the product. ΔE_2 is independent of the substrate, and smaller values correspond to a more reactive and oxidizing state of Compound I. In our study ΔE_2 was estimated as $\Delta E_2 = E_{QM}(O) + E_{QM}(Fe) - E_{QM}(Fe^*) + E_{QM/MM}(Fe^*) - E_{QM/MM}(FeO)$, where $E_{QM}(O)$ is the energy of oxygen *in vacuo* in the triplet state, $E_{QM}(Fe) - E_{QM}(Fe^*)$ accounts for the relaxation energy of the resting state of the heme complex compared to the structure in Compound I and for the energy difference between the sextet and doublet states of the resting state (103 kJ/mol). $E_{QM/MM}(Fe^*)$ was obtained by a single-point calculation on the QM/MM optimized geometry of Compound I by deleting the oxygen atom from it $E_{QM/MM}(FeO)$ is the QM/MM energy of Compound I. The last two quantities were calculated for each snapshot and averaged over each system.

Figure 12. Thermodynamic cycle showing the relationship between ΔE_1, ΔE_2, (Fe-O bond enthalpy) and ΔE_3.

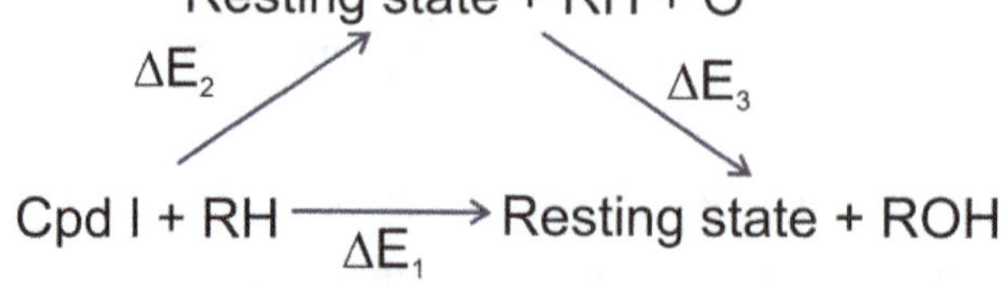

In order to assess the electronic configuration of the oxidizing species, structural data (especially the Fe–S bond length), and charge and spin density data (on the Fe, O, S atoms and the porphyrin ring) have been collected and compared. All these data indicated that Compound I of the P450s is very similar in the studied human isoforms, as variation of the data for the different snapshots taken for a given isoform was larger than the variation between isoforms. However, the bacterial $P450_{cam}$ isoform showed an increased spin density on sulfur compared to the human isoforms. It has been suggested that the different hydrogen bonding environment of the sulfur atom could be the main reason behind it. In P450s the axial cysteinate ligand of the heme is hydrogen bonded to the amide groups of the three consecutive amino acids of the protein chain as depicted for P450 2D6 in Figure 13, and one of the hydrogen bonds to sulfur was found to be much stronger in $P450_{cam}$ than in the human isoforms. Upon substrate binding a slight decrease in the spin density of sulfur was observed in all cases, which was attributed to the displacement of water molecules from the active site, thereby changing its polarity. These results imply that the properties of Compound I are sensitive to changes in the hydrogen bonding environment and also to polarization effects exerted by the active site.

Figure 13. Quantum mechanical/molecular mechanical (QM/MM) optimized snapshot of Compound I with residues hydrogen-bonded to the axial cysteinate in P450 2D6.

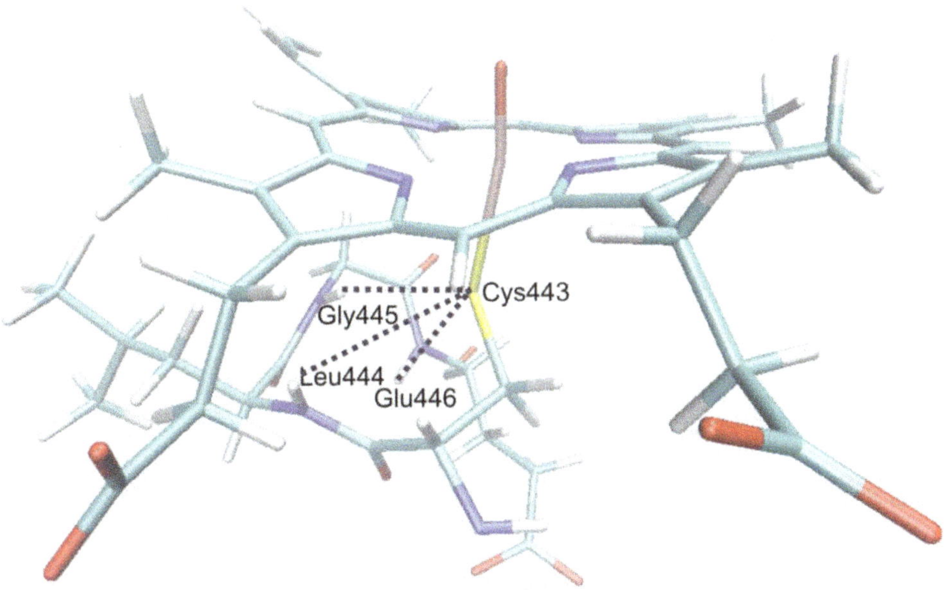

The Fe–O bond enthalpy does not show great variability over the various isoforms either, as its variation is on the order of the same magnitude for the different snapshots of a given isoform as those between the various isoforms. This result suggests a similar oxidizing power for the different isoforms, which is slightly lowered in the presence of the substrate in the active site. This latter effect is most likely due to the fact that the substrate expels the water out of the active site and decreases the number of hydrogen bonds to the ferryl oxygen. As a consequence the spin density on oxygen increases and the charge associated to it decreases. It was also shown that structures with largest spin density on oxygen have the lowest Fe–O bond enthalpy.

The overall conclusion of this study is that active species of the studied P450 isoforms exhibited very similar properties; therefore, it is reasonable to assume that the reactivity of Compounds I of different isoforms will be very similar. However, it was also shown that the electronic structure of Compound I varies with thermal fluctuations, therefore conclusions drawn on a single QM/MM optimized snapshot may not be necessarily reliable and it could be more appropriate to average the data over a number of structures.

4.4.2. Metabolism of Dextromethorphan by P450 2D6: What Drives the Regioselectivity of the Enzymes?

Most P450s are promiscuous enzymes: a substrate molecule may be metabolized by various P450 isoforms leading to different products. A good example of this is the case of dextromethorphan, a common antitussive compound, which is even frequently used in various experimental studies of the P450 2D6 enzyme. In the human body, dextromethorphan is metabolized by P450 3A4 yielding an N-demethylated product (see Figure 14), while O-demethylation is catalyzed by P450 2D6 [103]. In contrast, the aromatic ring of this compound is not oxidized, despite the fact that aromatic oxidation of other aromatic ethers have been observed in rat [104] and rabbit [105]. The interesting metabolic properties of dextromethorphan make it a good candidate for the computational study of the factors influencing the regioselectivity of P450 isoforms. Earlier computation studies showed that in general the barrier of N-demethylation is lower than that of O-demethylation [106], therefore it is not astonishing that P450 3A4, which has a large active site capable of accommodating two substrate molecules and in which the substrate may be oriented in any position catalyzes the thermodynamically favored N-demethylation. However, it is more surprising that P450 2D6 only catalyzes the O-demethylation despite that fact that the aromatic ring is located close to the heme iron as well.

Figure 14. Metabolic routes of dextromethorphan in man as indicated by arrows.

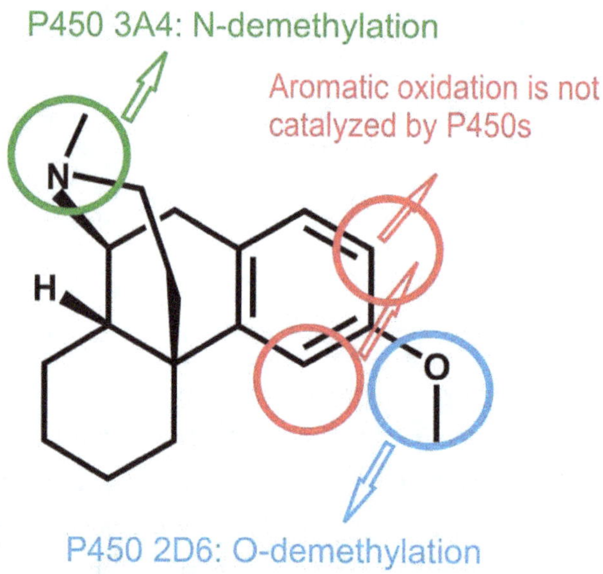

In order to clarify this problem, we conducted a study to investigate the factors influencing the regioselectivity of P450 2D6 [107]. First we carried out gas-phase quantum mechanical calculations on a model system (shown in Figure 15) to get an idea about the relative energy barriers of hydrogen abstraction from the methoxy group. This is the rate-determining step of O-demethylation, and of aromatic oxidation. Both reactions have a lower barrier in the doublet spin state than in the quartet spin state, 50.2 kJ/mol for hydrogen abstraction and 64.0 kJ/mol for aromatic oxidation. Although the barrier for O-demethylation is lower, the difference between the barriers of the two reaction channels does not explain why only O-demethylation is observed.

Figure 15. (**a**) QM region used in the calculations (**b**) Barriers of O-demethylation and aromatic carbon oxidation obtained from quantum mechanical and QM/MM calculations.

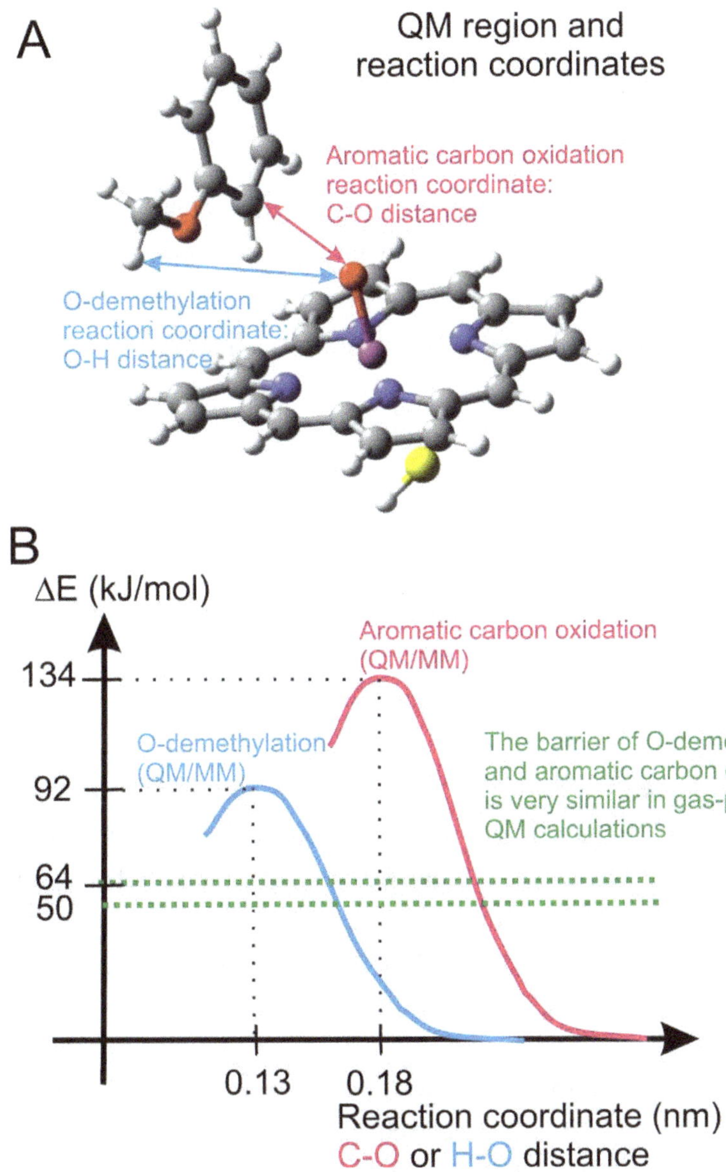

The above result indicates that the regioselectivity of P450 2D6 might be also modified by its active site architecture. For this reason we turned our attention towards docking, molecular dynamics and QM/MM methods which are capable of taking into account the enzymatic structure. There are two important pharmacophores of the ligands of P450 2D6: all of them contain basic nitrogen and an aromatic ring [108]. Therefore the N-protonated form of dextromethorphan was used throughout together with the crystallographic structure (PDB code 2F9Q) [109] of P450 2D6.

First, the ligand was docked into the active site of P450 2D6 in five different orientations, two of which seemed to be allowing for the O-demethylation reaction. In one of these two positions the protonated nitrogen of dextromethorphan interacts with the acidic Glu216 residue, and in the other one with Asp301, both of which have been implicated as important for catalysis [110]. In both structures the aromatic ring of the ligand participates in a π–π interaction with Phe120 whose mutation also leads to altered regioselectivity [111]. The structure in which interaction with Glu-216 was observed was chosen for further MD simulations and QM/MM calculations. During the course of a 2ns MD simulation, the docked structure maintained its major interactions with the active site residues, and it was observed that both the methoxy group and the aromatic ring of dextromethorphan remained in the close proximity of the heme iron during the whole time span of the simulation, indicating the possibility to be oxidized, in contrast to the fact that experimentally only the former one is known to occur.

As neither gas-phase quantum mechanical calculations nor docking alone could explain the experimental regioselectivity, it seemed to be essential to combine in the calculations the accuracy of quantum mechanical calculations with the steric restraints exerted by the protein structure in the frame of QM/MM calculations. The trajectory of the MD run was thoroughly analyzed and 3–3 suitable starting structures were chosen to model the two reaction channels. The QM region in the QM/MM calculations consisted of the same atoms as in the quantum mechanical calculations (Figure 15), but the axial thiolate ligand was modeled by an SCH_3^- moiety. Dangling bonds were saturated by hydrogen-type link atoms. Using adiabatic mapping, which gives an estimate of the enthalpy component of the Gibbs-free energy of the reaction, we obtained 92 kJ/mol for the barrier of hydrogen abstraction (O-demethylation) and 134 kJ/mol for the barrier of aromatic carbon oxidation. The difference between the activation energies of the two reaction channels increased to about 40 kJ/mol in contrast to the 14 kJ/mol calculated for the gas-phase (see the green dotted lines in Figure 15). The reason for the great increase of the barrier of the aromatic carbon oxidation is that the active site of P450 2D6 is relatively tight and the movement of dextromethorphan is hindered by several non-polar active site residues, e.g., Phe120, Ala305, Val308 (see Figure 16). Therefore the favorable transition state structure for aromatic oxidation cannot be formed in the active site, and its barrier considerably increases. The most important conclusion of the study was that gas-phase calculations or docking studies alone or even when they are combined are not enough to predict the regioselectivity of the P450 2D6 enzyme and more sophisticated models are needed. It was suggested that the metabolite predicting algorithms might be improved by docking the approximate transition state structures into the active site of the enzyme and assessing the feasibility of its formation [101].

Figure 16. Active site of P450 2D6 with dextromethorphan. The movement of dextromethorphan in the active site is hindered by its salt bridge to Glu216 and by the steric hindrance of the bulky amino acid side-chains.

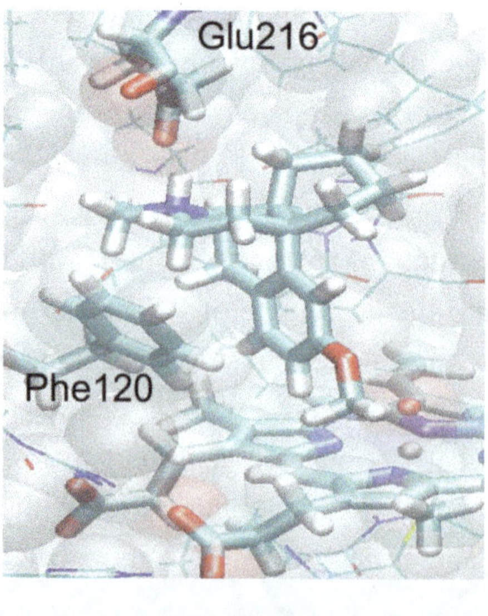

For a comprehensive study of the oxidation of further three pharmacologically important molecules see Lonsdale *et al.* [112].

4.4.3. Reactivity of the $P450_{nor}$ Enzyme

Nitric oxide reductase ($P450_{nor}$) is a very special member of the superfamily of P450s. Instead of the most generally catalyzed oxidation reactions of P450s, it catalyzes the reduction of nitric oxide to dinitrogen oxide and for its proper functioning it does not require a redox mediator, but contains its own NADH binding site [113]. $P450_{nor}$ is a soluble protein found in the cytosol of the fungus *Fusarium oxysporum,* a denitrifying organism, which can generate N_2O from NO_3^- ions in three main steps using three different enzymes. The last step of the conversion is nitric oxide reduction to N_2O by nitric oxide reductase. Since N_2O has a 300 times larger greenhouse effect than carbon dioxide [114], it should be taken into account in chemically fertilized nitrate abundant area [115]. Nitric oxide plays an important role in mammals for instance as a neurotransmitter or as part of the immune response against pathogens [116]. *Histoplasma capsulatum* is a human pathogen, which can avoid damage using the Nor1p enzyme, which shows great similarity to $P450_{nor}$ (61% sequence identity, 79% similarity) [117], and a similar mode of action can be assumed for converting NO.

The conversion catalyzed by $P450_{nor}$ takes place in three main steps (see Scheme 2). The first step is the binding of nitric oxide in the active site of $P450_{nor}$, which is followed by hydride transfer to $(Fe^{III}P450_{nor})NO$ from NADH to form the intermediate (**I** in Scheme 2.). After the binding of a second molecule of nitric oxide, the final product, N_2O, is generated and the resting state of the enzyme is restored. Several suggestions have been put forward for the structure for the intermediate based on experimental and theoretical investigations [118–123]. According to the most accepted and experimentally proven mechanism direct hydride transfer occurs from the NADH co-factor to the oxygen or nitrogen atom of nitric oxide. DFT calculations performed by Lehnert *et al.* suggested that the hydride is most likely transferred to the nitrogen atom of nitric oxide in the key step, and intermediate (**I**) is the doubly protonated $[Fe(Porph)(CH_3S)(NHOH)]$ species [123]. Based on the calculated relative energies at the B3LYP/LanL2DZ* and BP86/TZVP levels the $[Fe(P)(CH_3S)(NHO)]^-$ species is 109.7 kJ/mol more stable than the $[Fe(P)(CH_3S)(NOH)]^-$ species, and the protonation of $[Fe(P)(CH_3S)(NHO)]^-$ is a slightly exothermic process ($\Delta G = -36.0$ kJ/mol).

Scheme 2. Conversion reaction catalyzed by $P450_{nor}$.

$$\mathbf{Fe^{III}P450_{nor} + NO = (Fe^{III}P450_{nor})NO}$$

$$\mathbf{(Fe^{III}P450_{nor})NO + NADH = I + NAD^+}$$

$$\mathbf{I + NO + H^+ = Fe^{III}P450_{nor} + N_2O + H_2O}$$

In the previous sections it was shown that steric and electronic factors significantly influence the reactivity of enzymes, and that QM/MM calculations may provide a better description of enzymatic processes than gas phase calculations alone. Two QM/MM studies addressing the reactivity of $P450_{nor}$ were published recently, however dealing with partially different aspects of its reactivity. Our study focused on two key questions, (1) to which atom of nitric oxide will the hydride be transferred, and (2) how does the origin (docked or co-crystallized) influence the outcome of QM/MM calculations [124]. To answer the questions we used quantum mechanical and QM/MM methods, and based on experimental findings we assumed that the spin multiplicity of the system is singlet [125,126] and applied a closed-shell formalism in the calculations.

The results suggest that in the active site pocket of $P450_{nor}$ hydride transfer from NADH occurs to the nitrogen end of nitric oxide. In this respect all of our calculations (quantum mechanical or QM/MM) provided the same result, but the barriers were significantly higher in the QM/MM calculations. For instance the calculated barrier for hydride transfer from NADH to the nitrogen atom of nitric oxide is 37.3 kJ/mol in quantum mechanical and 113.5 kJ/mol in QM/MM calculations including zero-point energy correction in contrast to the 139.0 kJ/mol (quantum mechanical) and 161.2 kJ/mol (QM/MM) values obtained for the hydride transfer to the oxygen end of nitric oxide. The obtained high QM/MM barriers were most likely due to the incomplete sampling of initial complexes, and the chosen complexes did not provide an ideal starting structure for the reaction to occur. Nevertheless, the relative energy of the barriers convincingly showed that the hydride transfer will occur to the nitrogen of nitric oxide.

Figure 17. The position of the NADH ligand in the docked structure (structure **A**) and in the crystal structure (structure **B**).

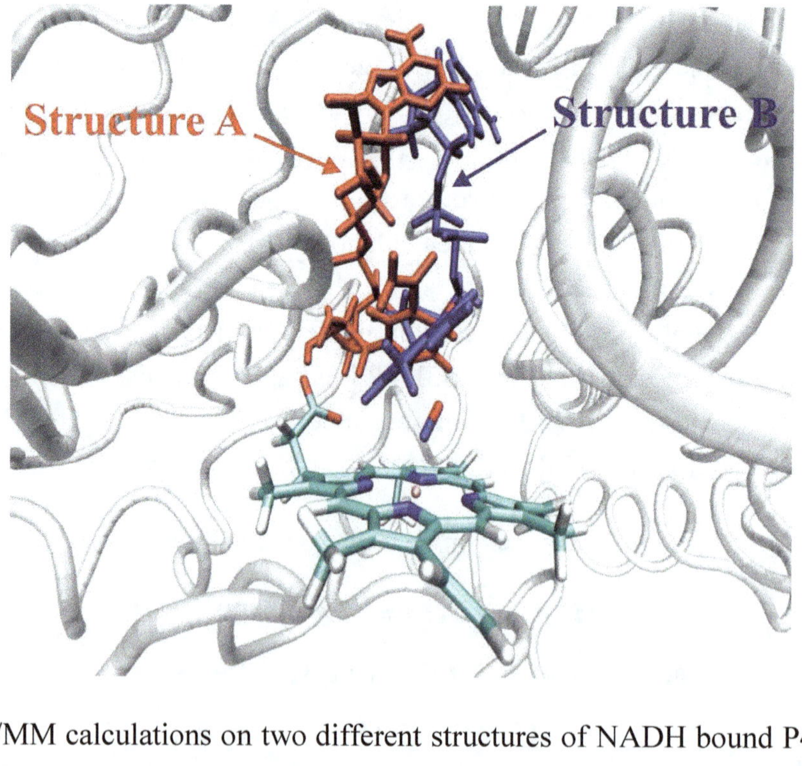

Comparing QM/MM calculations on two different structures of NADH bound $P450_{nor}$ brought also interesting results. Two structures were used. In structure **A** NADH has been docked to the active site of the NO-bound $P450_{nor}$ enzyme (Protein Data Bank code: 1CL6) [127], while structure **B** was obtained from the crystal structure of $P450_{nor}$ in complex with NAAD (nicotinic acid adenine dinucleotide), an analogue of NADH (Protein Data Bank code: 1XQD) [128]. In the latter structure nitric oxide was manually inserted. The position of the ligand and its interaction with active site residues was significantly different in the two models (see Figure 17). In structure **A** the pro-S hydrogen, while in structure **B** the pro-R hydrogen atom of NADH could be transferred as hydride to nitric oxide. We obtained significantly lower barriers using structure **B**, in accordance with the results of experimental kinetic isotope effect measurements, which also suggested the transfer of the pro-R hydrogen [121]. Therefore our results imply that the selection of docked structures as models should be avoided, since upon ligand binding a major rearrangement of the active site may take place. Unfortunately, present docking methods cannot take these rearrangements into account, and therefore the obtained results on the regioselectivity of the enzyme may contradict to experiments, furthermore, barrier heights may also be considerably influenced.

In another QM/MM study hydride shift directed to the nitrogen atom of nitric oxide was investigated [129]. This study resulted in considerably lower energy barriers: 38.9 kJ/mol for a closed-shell system at the DFT(RIJCOSX-B3LYP)/MM(OPLS-AA) level including zero-point energy correction. The lower barrier could be explained by the presence of a stabilizing hydrogen bond interaction between the amide group of the NADH molecule and the Ser286, Thr243 and Ala239 residues (see Figure 18), which are responsible for stabilization of the transition state. These were absent in models used in the work discussed above. When the authors applied an unrestricted formalism to obtain the wave function of the broken-symmetry singlet state, the calculated barrier further dropped to 27.2 kJ/mol, and showed that that the open-shell singlet state of the $[Fe(Porph)(Met)(NHO)]^-$ species is by about 16.7 kJ/mol more stable than the closed-shell one. The study also suggested that the N–N bond of N_2O is formed in a spin-recoupling process, followed by the release and spontaneous decomposition of the $N_2O_2H_2$ ligand to N_2O and H_2O.

Figure 18. Energy profiles for closed-shell and open-shell singlet pathways for hydride transfer in the $P450_{nor}$ and the schematic structure of the transition state.

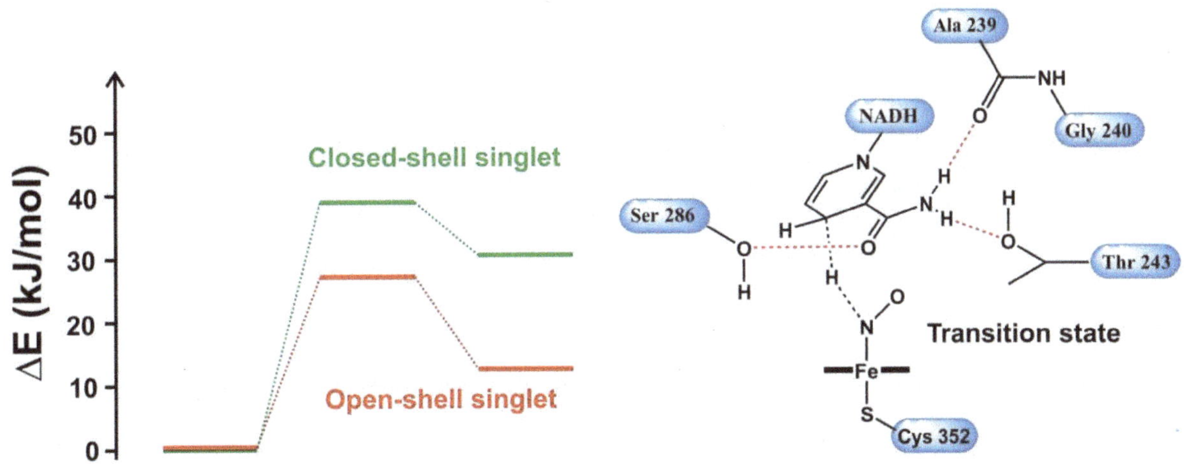

4.5. Xylose Isomerase

D-Xylose isomerase catalyzes the conversion of D-xylose to D-xylulose and the transformation of other sugars from aldose to ketose. It is widely used in industry for the conversion of D-glucose to D-fructose. The enzyme is activated by divalent metal cations (Mg^{2+}, Co^{2+} or Mn^{2+}), whereas other cations (e.g., Ca^{2+}, Ba^{2+} and Al^{3+}) inhibit the reaction [130]. The 3D structures of xylose isomerases isolated from different species have been extensively investigated by X-ray diffraction techniques [131–133]. On this basis the following mechanism of action has been proposed (see Figure 19 and Ref. 5). Binding the ring form of the sugar substrate to the structural metal ion is followed by a ring-opening step in which the hydrogen atom of the hydroxyl group on C1 is transferred to O5. This step is accompanied by the formation of a (– + –) charge distribution, involving the substrate hydroxide; His54 and Asp57, similar to those found in serine proteases (see Section 4.1). In the extended conformation, 02 and 04 of the sugar substrate are bound to the structural metal ion (M_{str}), while the catalytic metal ion (M_{cat}) is not in direct contact with the substrate. A structural water molecule coordinated to M_{cat} removes the OH2 proton of the substrate. Isomerization takes place via hydride-shift, in which the C2 hydrogen is transferred to C1 via an anionic transition state, assisted by M_{cat}. This is the rate-limiting step of the reaction.

Figure 19. Catalytic mechanism of xylose isomerase. Top: ring opening, middle: substrate deprotonation, bottom: hydride shift.

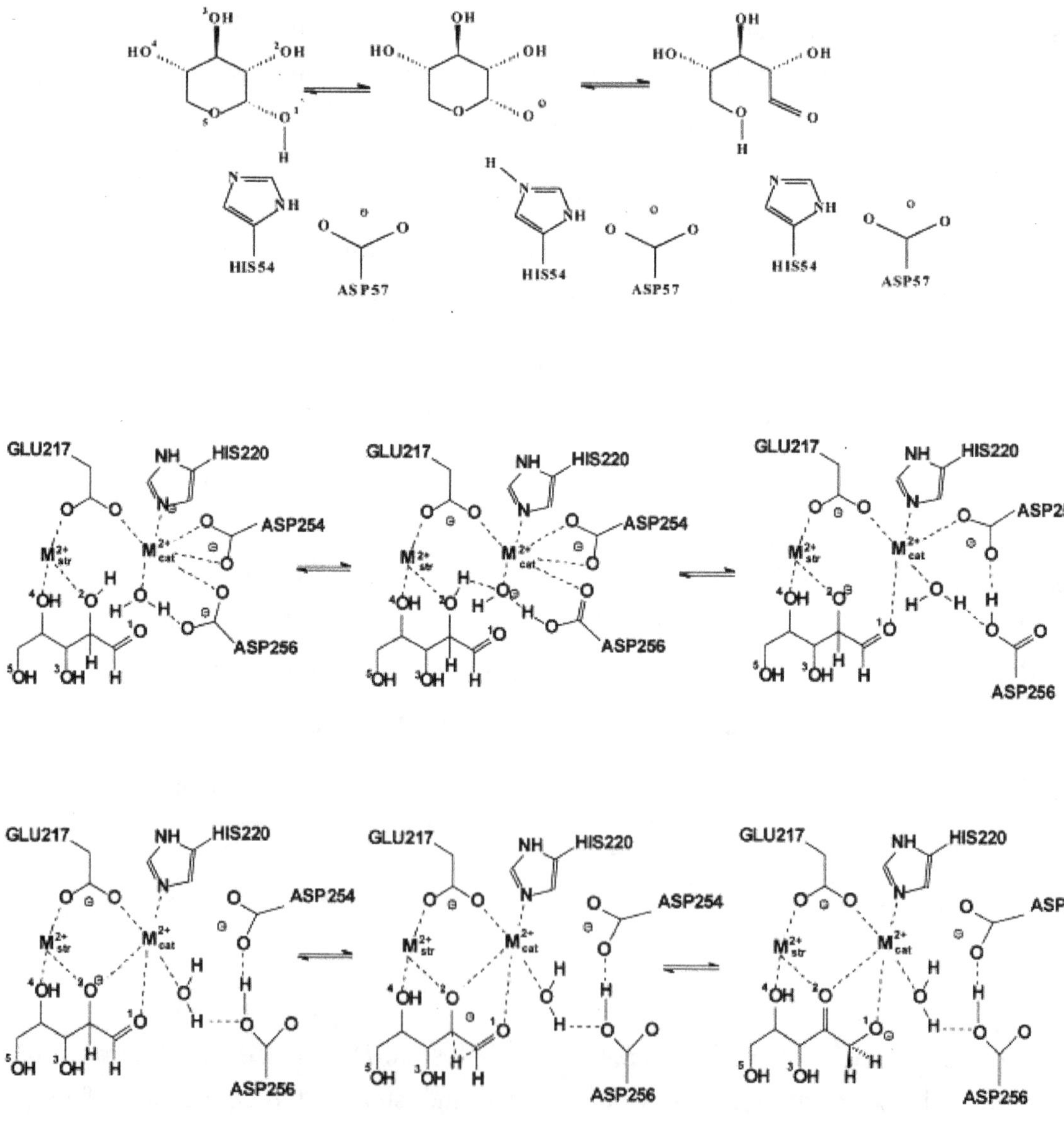

Extended calculations have been done on the validity of the above mechanism by Hu, Shiu and Li [134]. Their results indicate that in the enzyme M_{cat} moves from one site to another, which is close to the substrate and stays there. M_{cat} acts as a Lewis acid, polarizes the substrate and catalyzes the hydride shift step. Calculations showed that Lys183 plays an important role in the isomerization reaction, because its protonated terminal ammonium group provides a proton to the carboxide ion of the substrate to form a hydroxyl group after the hydride shift step. If this side chain is dropped from the active-site, model proton transfer instead of hydride shift is found to be rate limiting [135]. The calculated activation energies for ring opening and hydride shift steps were found to be 42 and 71 kJ/mol. respectively. If the activating metal cation, Mg^{2+}, was replaced by an inhibiting one, Zn^{2+}, calculated activation energies increased to 63 and 138 kJ/mol, respectively.

Our early calculations [136] indicated that the sugar ring opens only after the proton becomes shared between NE2 of His54 and O5 of the sugar. Ring opening is initiated by the approach of Asp57 towards His54 enhancing the basicity of the imidazole side chain of the latter. After the transition state has been reached, the movement in the opposite direction has been predicted by the calculations. The hypothesis that the charge-relay mechanism may be effective in catalysis, like in case of serine proteases, could be ruled out, because the electrostatic potential pattern provided by the protein environment and the bound metal cation (M_{str}) stabilizes the ion-pair form of the $Asp57^{-}...NE2^{+}...O1^{-}$ triad. The negatively charged side chains of Asp57, as well as two bound structural water molecules assist in stabilizing the polar transition states. As it is seen in Figure 20, the activation energies of the proton transfer from O1 to NE2, as well as from NE2 to O5 become quite high if we do calculations on a small model, excluding Asp57 and the structural water molecules.

Figure 20. Initial, transition and final structures in the proton transfer steps of the ring opening reaction (relative energies are given in kJ/mol).

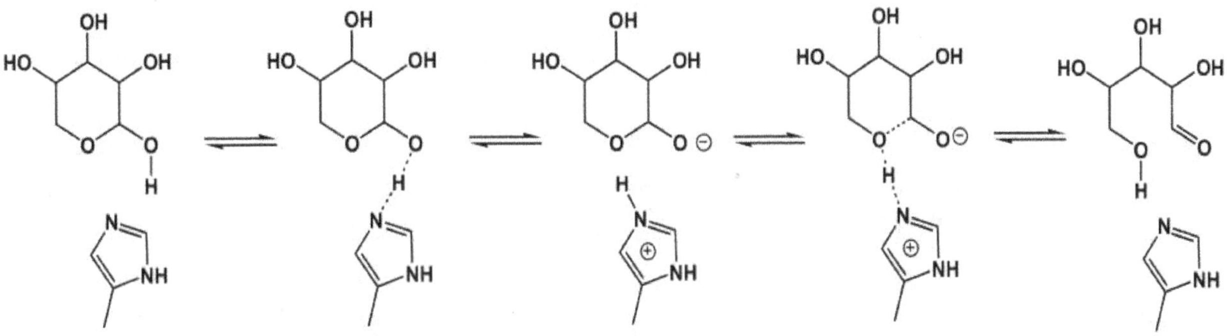

0 94 80 178 18

The anomeric specificity of the enzyme was also accounted for, since only the α-anomer is properly positioned for a proton transfer. Similarly, Lambeir *et al.* found that it is the His54 residue, which is responsible for the conformational specificity of xylose isomerases [137]. After abstracting a proton from the structural water by Asp256 the hydroxyl ion formed binds to the O2H moiety of the substrate, the structural water breaks the M_{cat}...Asp256 link and a hydrogen-bond forms between them.

A specific effect, hydrogen tunneling takes place in xylose isomerases. This is especially interesting, since it cannot be described by other methods; specific quantum mechanics is needed. Gao and coworkers combined molecular dynamics with quantum mechanics including consideration of hydrogen tunneling to calculate the reaction rate of the transformation of xylose catalyzed by xylose isomerase in the presence of two divalent magnesium cations. Their model includes more than 25,000 explicit atoms. The simulation confirms the essential features of a mechanism involving a rate-determining 1,2-hydride shift with prior and post proton transfers. They have found that inclusion of quantum mechanical vibrational energy is important for computing the free energy of activation, and quantum mechanical tunneling effects are essential for computing kinetic isotope effects. It is predicted that 85% of the hydride shift reaction proceeds by tunneling. The computed kinetic isotope effect was found to be in good agreement with experimental results. The molecular dynamics simulations reveal that proton and hydride transfer reactions are assisted by breathing motions of the mobile Mg^{2+} ion in the active site during the hydride transfer step [138,139].

Further studies have been reported on tunneling in various other enzymes, too. The reaction pathway for tryptamine oxidation by aromatic amine dehydrogenase was studied by Masgrau *et al.* [140]. It has been shown by combining experiment and computer simulation, that proton transfer occurs in a reaction, which is dominated by tunneling. In this enzyme system instead of long-range coupled motions tunneling is promoted by a short-range motion modulating the proton-acceptor distance. Hydrogen tunneling was studied in a model system that represents the active site of soybean lipoxygenase-1. Calculations showed that the tunneling process has a three-dimensional nature [141].

5. Conclusions

Enzyme reactions involve a large number of atoms, of which some, belonging to the active site, are crucial in understanding the process, while others are of less, however, not negligible importance. Owing to this special situation, special models and computational methods are needed for an adequate description. The active site can be modeled by small organic structures and some basic features of the reaction mechanisms can be elucidated by these. Calculations on this region must be at the quantum mechanical (Hartree-Fock, density functional or Empirical Valence Bond) level since bonds are broken and formed during the reaction, this needs a special treatment. More distant regions may be also of some importance, especially if the electrostatic field induced by them at the active site is strong. This is the case in point mutants where an ionizable side chain, which is close to the active site becomes charged or neutral by mutation. Those reactions, which involve a change in the polarity of the active site during transformation to the transition state, are strongly influenced, in most cases accelerated by the protein electrostatic field. In some cases subtle changes in the structure of the protein core may trigger essential changes in the mechanism. X-ray diffraction studies provide adequate models for the calculations.

Acknowledgments

This work was supported by grants from the Hungarian Scientific Research Fund (OTKA NK101072). J.O. and B. K. thank the financial support of the New Széchenyi Plan (TÁMOP-4.2.2/ B-10/1-2010-0009).

References

1. Claeyssens, F.; Harvey, J.N.; Manby, F.R.; Mata, R.A.; Mulholland, A.J.; Ranaghan, K.E.; Schütz, M.; Thiel, S.; Thiel, W.; Werner, H.J. High-accuracy computation of reaction barriers in enzymes. *Angew. Chem.* **2006**, *118*, 7010–7013.
2. Matthews, B.W.; Sigler, P.B.; Henderson, R.; Blow, D.M. Three-Dimensional structure of tosyl α-chymotrypsin. *Nature* **1967**, *214*, 652–656.
3. Hedström, L. Serine protease mechanism and specificity. *Chem. Rev.* **2002**, *102*, 4501−4523.
4. Lassila, J.K.; Zalatan, J.G.; Herschlag, D. Biological phosphoryl-transfer reactions: Understanding mechanism and catalysis. *Annu. Rev. Biochem.* **2011**, *80*, 669–702.
5. Asbóth, B.; Náray-Szabó, G. Mechanism of action of d-xylose isomerase. *Curr. Protein Pept. Sci.* **2000**, *1*, 237–254.
6. Berman, H.M.; Westbrook, J.; Feng, Z.; Gilliland, G.; Bhat, T.N.; Weissig, H.; Shindyalov, I.N.; Bourne, P.E. The protein data bank. *Nucleic Acids Res.* **2000**, *28*, 235–242.
7. Wade, R.C.; Luty, B.A.; Demchuk, E.; Madura, J.D.; Davis, M.E.; Briggs, J.M.; McCammon, J.A. Simulation of enzyme-substrate encounter with gated active sites. *Nat. Struct. Biol.* **1994**, *1*, 65–69.
8. Hammes-Schiffer, S.; Benkovic, S.J. Relating protein motion to catalysis. *Annu. Rev. Biochem.* **2006**, *75*, 519–541.
9. McGeagh, J.D.; Ranaghan, K.E.; Mulholland, A.J. Protein dynamics and enzyme catalysis: Insights from simulations. *Biochim. Biophys. Acta* **2011**, *1814,* 1077–1092.
10. Warshel, A. *Computer Modeling of Chemical Reactions in Enzymes and Solutions*; Wiley: New York, NY, USA, 1991.
11. Ranaghan, K.E.; Mulholland, A.J. Investigations of enzyme-catalysed reactions with combined quantum mechanics/molecular mechanics (QM/MM) methods. *Int. Rev. Phys. Chem.* **2010**, *29*, 65–133.
12. Senn, H.M.; Thiel, W. QM/MM methods for biomolecular systems. *Angew. Chem. Int. Ed.* **2009**, *48*, 1198–1229.
13. Frisch, M.J.; Trucks, G.W.; Schlegel, H.B.; Scuseria, G.E.; Robb, M.A.; Cheeseman, J.R.; Scalmani, G.; Barone, V.; Mennucci, B.; Petersson, G.A. *et al. Gaussian 09*; Revision B.01; Gaussian, Inc.: Wallingford, CT, USA, 2010.
14. Szabo, A.; Ostlund, N.S. *Modern Quantum Chemistry: Introduction to Advanced Electronic Structure Theory*; Dover Publications: Mineola, NY, USA, 1996.
15. Jacobsen, H.; Cavallo, L. Directions for Use of Density Functional Theory: A Short Introduction Manual for Chemists. In *Handbook of Computational Chemistry*; Ed. Leszczynski, J. Springer: Berlin, Germany, 2011; pp. 97–133.
16. Vreven, T.; Morokuma, K. Hybrid methods: ONIOM(QM:MM) and QM/MM. *Annu. Repts Comput. Chem.* **2006**, *2*, 35–51.
17. Poltev, V. Molecular Mechanics: Method and Applications. In *Handbook of Computational Chemisry*; Ed. Leszczynski, J. Springer: Berlin, Germany, 2011; pp. 260–291.

18. Case, D.A.; Darden, T.A.; Cheatham, T.E., III; Simmerling, C.L.; Wang, J.; Duke, R.E.; Luo, R.; Walker, R.C.; Zhang, W.; Merz, K.M.; *et al.* AMBER 11, 2010, University of California, San Francisco.
19. Christen, M.; Hunenberger, P.H.; Bakowies, D.; Baron, R.; Bürgi, R.; Geerke, D.P.; Heinz, T.N.; Kastenholz, M.A.; Krautler, V.; Oostenbrink, C.; *et al.* The gromos software for biomolecular simulation: Gromos05. *J. Comp. Chem.* **2005**, *26*, 1719–1751.
20. Available online: http://www.gromacs.org/ (accessed on 6 August 2013).
21. Brooks, B.R.; Brooks, C.L., III; Mackerell, A.D.; Nilsson, L.; Petrella, R.J.; Roux, B.; Won, Y.; Archontis, G.; Bartels, C.; Boresch, S.; *et al.* CHARMM: The biomolecular simulation program. *J. Comp. Chem.* **2009**, *30*, 1545–1615.
22. Jorgensen, W.L.; Tirado-Rives, J. Potential energy functions for atomic-level simulations of water and organic and biomolecular systems. *Proc. Natl. Acad. Sci. USA* **2005**, *102*, 6665–6670.
23. Nowak, W. Applications of Computational Methods to Simulations of Proteins Dynamics. In *Handbook of Computational Chemistry*; Ed. Leszczynski, J. Springer: Berlin, Germany, 2011; pp. 1129–1155.
24. Marx, D.; Hutter, J. *Ab Initio Molecular Dynamics: Basic Theory and Advanced Methods*; Cambridge University Press: Cambridge, UK, 2009.
25. Röthlisberger, D.; Khersonsky, O.; Wollacott, A.M.; Jiang, J.; DeChancie, J.; Betker, J.; Gallaher, J.L.; Althoff, E.A.; Zanghellini, A.; Dym, O.; *et al.* Kemp elimination catalysts by computational enzyme design. *Nature* **2008**, *453*, 190–195.
26. Náray-Szabó, G.; Surján, P.R. Bond orbital framework for rapid calculation of environmental effects on molecular potential energy surfaces. *Chem. Phys. Lett.* **1983**, *96*, 499–501.
27. Hu, H.; Yang, W. Free energies of chemical reactions in solution and in enzymes with ab initio quantum mechanics/molecular mechanics methods. *Annu. Rev. Phys. Chem.* **2008**, *59*, 573–560.
28. Warshel, A.; Florian, J. *Empirical Valence Bond and Related Approaches*; Wiley: New York, NY, USA, 2004.
29. Kamerlin, S.C.L.; Cao, J.; Rosta, E.; Warshel, A. On unjustifiably misrepresenting the EVB approach while simultaneously adopting it. *J. Phys. Chem. B* **2009**, *113*, 10905–10915.
30. Bentzien, J.; Muller, R.P.; Florián, J.; Warshel, A. Hybrid ab initio quantum mechanics/molecular mechanics calculations of free energy surfaces for enzymatic reactions: The nucleophilic attack in subtilisin. *J. Phys. Chem B* **1998**, *102*, 2293–2301.
31. Rocchia, W.; Alexov, E.; Honig, B. Extending the applicability of the nonlinear poisson-boltzmann equation: Multiple dielectric constants and multivalent ions. *J. Phys. Chem. B* **2001**, *105*, 6507–6514.
32. Sarkar, S.; Witham, S.; Zhang, J.; Zhenirovskyy, M.; Rocchia, W.; Alexov, E. DelPhi Web Server: A comprehensive online suite for electrostatic calculations of biological macromolecules and their complexes. *Comm. Comp. Phys.* **2013**, *13*, 269–284.
33. Tomasi, J.; Mennucci, B.; Cammi, R. Quantum mechanical continuum solvation models. *Chem. Rev.* **2005**, *105*, 2999–3094.
34. Polgár, L. *Mechanisms of Protease Action*; CRC Press: Boca Raton, FL, USA, 1989.
35. Brenner, S. The molecular evolution of genes and proteins: A tale of two serines. *Nature* **1988**, *334*, 528–530.

36. Asbóth, B.; Polgár, L. Transition-state stabilization at the oxyanion binding sites of serine and thiol proteinases: Hydrolyses of thiono and oxygen esters. *Biochemistry* **1983**, *22*, 117–122.
37. Carter, P.; Wells, J.A. Functional interaction among catalytic residues in subtilisin BPN'. *Proteins Struct. Funct. Genet.* **1990**, *7*, 335–342.
38. Blow, D.M.; Birktoft, J.J.; Hartley, B.S. Role of a buried acid group in the mechanism of action of chymotrypsin. *Nature* **1969**, *221*, 337–340.
39. Polgár, L.; Bender, M.L. The nature of general base-general acid catalysis in serine proteases. *Proc. Natl. Acad. Sci. USA* **1969**, *64*, 1335–1342.
40. Bachovchin, W.W. Confirmation of the assignment of the low-field proton resonance of serine proteases by using specifically nitrogen-15 labeled enzyme. *Proc. Natl. Acad. Sci. USA* **1985**, *82*, 7948–7951.
41. Kossiakoff, A.A.; Spencer, S.A. Direct determination of the protonation states of aspartic acid-102 and histidine-57 in the tetrahedral intermediate of the serine proteases: Neutron structure of trypsin. *Biochemistry* **1981**, *20*, 6462–6474.
42. Sprang, S.; Standing, T.; Fletterick, R.J.; Stroud, R.M.; Finer-Moore, J.; Xuong, N.H.; Hamlin, R.; Rutter, W.J.; Craik, C.S. The Three-dimensional structure of Asn102 mutant of trypsin: Role of Asp102 in serine protease catalysis. *Science* **1987**, *237*, 905–909.
43. Szeltner, Z.; Rea, D.; Renner, V.; Fülöp, V.; Polgár, L. Electrostatic effects and binding determinants in the catalysis of prolyl oligopeptidase. Site specific mutagenesis at the oxyanion binding site. *J. Biol. Chem.* **2002**, *277*, 42613–42622.
44. Warshel, A.; Náray-Szabó, G.; Sussman, F.; Hwang, J.K. How do serine proteases really work? *Biochemistry* **1989**, *28*, 3629–3637.
45. Warshel, A. Electrostatic origin of the catalytic power of enzymes and the role of preorganized active sites. *J. Biol. Chem.* **1998**, *273*, 27035–27038.
46. Kamerlin, S.C.L.; Sharma, P.K.; Chu, Z.T.; Warshel, A. Ketosteroid isomerase provides further support for the idea that enzymes work by electrostatic preorganization. *Proc. Natl. Acad. Sci. USA* **2010**, *107*, 4075–4080.
47. Gérczei, T.; Asbóth, B.; Náray-Szabó, G. Conservative electrostatic potential patterns at enzyme active sites. The anion-cation-anion triad. *J. Chem. Inf. Comput. Sci.* **1999**, *39*, 310–315.
48. Náray-Szabó, G.; Fuxreiter, M.; Warshel, A. Electrostatic Basis of Enzyme Catalysis. In *Computational Approaches to Biochemical Reactivity*; Warshel, A., Náray-Szabó, G., Eds.; Kluwer: Dordrecht, The Netherlands, 1997; pp. 237–293.
49. Umeyama, H.; Nakagawa, S.; Kudo, T. Role of Asp102 in the enzymatic reaction of bovine beta-trypsin. A molecular orbital study. *J. Mol. Biol.* **1981**, *150*, 409–421.
50. Náray-Szabó, G. Electrostatic effect on catalytic rate enhancement in serine proteinases. *Int. J. Quant. Chem.* **1982**, *22*, 575–582.
51. Ishida, T.; Kato, S. Theoretical perspectives on the reaction mechanism of serine proteases: The reaction free energy profiles of the acylation process. *J. Am. Chem. Soc.* **2003**, *125*, 12035–12048.
52. Topf, M.; Richards, W.G. Theoretical studies on the deacylation step of serine protease catalysis in the gas phase, in solution, and in elastase. *J. Am. Chem. Soc.* **2004**, *126*, 14631–14641.
53. Hudáky, P.; Perczel, A. A self-stabilized model of the chymotrypsin catalytic pocket. The energy profile of the overall catalytic cycle. *Proteins Struct. Funct. Bioinf.* **2006**, *62*, 749–759.

54. Zhou, Y.; Zhang, Y. Serine protease acylation proceeds with a subtle re-orientation of the histidine ring at the tetrahedral intermediate. *Chem. Commun.* **2011**, *47*, 1577–1579.
55. Florián, J.; Warshel, A. Phosphate ester hydrolysis in aqueous solution: Associative versus dissociative mechanisms. *J. Phys. Chem. B* **1998**, *102*, 719–734.
56. Bernardi, F.; Bottoni, A.; de Vivo, M.; Garavelli, M.; Keserű, G.; Náray-Szabó, G. A hypothetical mechanism for HIV-1 integrase catalytic action: DFT modelling of a bio-mimetic environment. *Chem. Phys. Lett.* **2002**, *362*, 1–7.
57. Franzini, E.; Fantucci, P.; de Gioia, L. Density functional theory investigation of guanosine triphosphate models. Catalytic role of Mg2+ ions in phosphate ester hydrolysis. *J. Mol. Catal. A* **2003**, *204*, 409–417.
58. Lahiri, S.D.; Zhang, G.; Dunaway-Mariano, D.; Allen, K.N. The pentacovalent phosphorus intermediate of a phosphoryl transfer reaction. *Science* **2003**, *299*, 2067–2071.
59. Blackburn, G.M.; Williams, N.H.; Gamblin, S.J.; Smerdon, S.J. The pentacovalent phosphorus intermediate of a phosphoryl transfer reaction. *Science* **2003**, *301*, 1184c.
60. Baxter, N.J.; Bowler, M.W.; Alizadeh, T.; Cliff, M.J.; Hounslow, A.M.; Wu, B.; Berkowitz, D.B.; Williams, N.H.; Blackburn, G.M.; Waltho, J.P. Atomic details of near-transition state conformers for enzyme phosphoryl transfer revealed by MgF3- rather than by phosphoranes. *Proc. Natl. Acad. Sci. USA* **2010**, *107*, 4555–4560.
61. Webster, C.E. High-energy intermediate or stable transition state analogue: Theoretical perspective of the active site and mechanism of β-phosphoglucomutase. *J. Am. Chem. Soc.* **2004**, *126*, 6840–6841.
62. Berente, I.; Beke, T.; Náray-Szabó, G. Quantum mechanical studies on the existence of a trigonal bipyramidal phosphorane intermediate in enzymatic phosphate ester hydrolysis. *Theor. Chem. Acc.* **2007**, *118*, 129–134.
63. Barabás, O.; Pongrácz, V.; Kovári, J.; Wilmanns, M.; Vértessy, B.G. Structural insights into the catalytic mechanism of phosphate ester hydrolysis by dUTPase. *J. Biol. Chem.* **2004**, *279*, 42907–42915.
64. Ivanov, I.; Tainer, J.A.; McCammon, J.A. Unraveling the three-metal-ion catalytic mechanism of the DNA repair enzyme endonuclease IV. *Proc. Natl. Acad. Sci. USA* **2007**, *104*, 1465–1470.
65. Cavalli, A.; de Vivo, M.; Recanatini, M. Computational study of the phosphoryl transfer catalyzed by a cyclin-dependent kinase. *Chem. Eur. J.* **2007**, *13*, 8437–8444.
66. Smith, G.K.; Ke, Z.; Guo, H. Insights into the phosphoryl transfer mechanism of cyclin-dependent protein kinases from ab initio QM/MM free-energy studies. *J. Phys. Chem. B* **2011**, *115*, 13713–13722.
67. Kamerlin, S.C.L.; McKenna, C.E.; Goodman, M.F.; Warshel, A. A computational study of the hydrolysis of dGTP analogues with halomethylene-modified leaving groups in solution: Implications for the mechanism of DNA polymerases. *Biochemistry* **2009**, *48*, 5963–5971.
68. Wong, K.-Y. Gao, J. Insight into the phosphodiesterase mechanism from combined QM/MM free energy simulations. *FEBS J.* **2011**, *278*, 2579–2595.
69. Lior-Hoffmann, L.; Wang, L.; Wang, S.; Geacintov, N.E.; Broyde, S.; Zhang, Y. Preferred WMSA catalytic mechanism of the nucleotidyl transfer reaction in human DNA polymerase κ elucidates error-free bypass of a bulky DNA lesion. *Nucleic Acids Res.* **2012**, *40*, 9193–9205.

70. Xu, D.; Guo, H. Ab initio QM/MM studies of the phosphoryl transfer reaction catalyzed by PEP Mutase suggest a dissociative metaphosphate transition state. *J. Phys. Chem. B* **2008**, *112*, 4102–4108.
71. Xu, D.; Guo, H.; Liu, Y.; York, D.M. Theoretical studies of dissociative phosphoryl transfer in interconversion of phosphoenolpyruvate to phosphonopyruvate: Solvent effects, thio effects, and implications for enzymatic reactions. *J. Phys. Chem. B* **2005**, *109*, 13827–13834.
72. De Vivo, M.; Dal Peraro, M.; Klein, M.L. Phosphodiester cleavage in ribonuclease H occurs via an associative two-metal-aided catalytic mechanism. *J. Am. Chem. Soc.* **2008**, *130*, 10955–10962.
73. Rosta, E.; Woodcock, H.L.; Brooks, B.R.; Hummer, G. Artificial reaction coordinate "Tunneling" in free-energy calculations: The catalytic reaction of RNase H. *J. Comput. Chem.* **2009**, *30*, 1634–1641.
74. Schlichting, I.; Reinstein, J. pH influences fluoride coordination number of the AlFx phosphoryl transfer transition state analog. *Nat. Struct. Biol.* **1999**, *6*, 721–723.
75. De Vivo, M.; Ensing, B.; Dal Peraro, M.; Gomez, G.A.; Christianson, D.W.; Klein, M.L. Proton shuttles and phosphatase activity in soluble epoxide hydrolase. *J. Am. Chem. Soc.* **2007**, *129*, 387–394.
76. Ram Prasad, B.; Plotnikov, N.V.; Warshel, A. Addressing open questions about phosphate hydrolysis pathways by careful free energy mapping. *J. Phys. Chem. B* **2013**, *117*, 153–163.
77. Kamerlin, S.C.L.; Florián, J.; Warshel A. Associative versus dissociative mechanisms of phosphate monoester hydrolysis: On the interpretation of activation entropies. *ChemPhysChem* **2008**, *9*, 1767–1773.
78. Finzel, B.C.; Poulos, T.L.; Kraut, J. Crystal structure of yeast cytochrome c peroxidase refined at 1.7 Å resolution. *J. Biol. Chem.* **1984**, *269*, 32759–32767.
79. Poulos, T.L.; Edwards, S.L.; Wariishi, H.; Gold, M.H. Crystallographic Refinement of lignin peroxidase at 2 Å. *J. Biol. Chem.* **1993**, *268*, 4429–4440.
80. Patterson, W.R.; Poulos, T.L.; Goodin, D.B. Identification of a porphyrin Pi cation radical in ascorbate peroxidase compound I. *Biochemistry* **1995**, *34*, 4342–4345.
81. Gajhede, M.; Schuller, D.J.; Henriksen, A.; Smith, A.T.; Poulos, T.L. Crystal structure of horseradish peroxidase C at 2.15 Å resolution. *Nat. Struct. Biol.* **1997**, *4*, 1032–1038.
82. Patterson, W.R.; Poulos, T.L. Crystal structure of recombinant pea cytosolic ascorbate peroxidase. *Biochemistry* **1995**, *34*, 4331–4341.
83. Menyhárd, D.K.; Náray-Szabó, G. Electrostatic effect on electron transfer at the active site of heme peroxidases: A comparative molecular orbital study on Cytochrome C peroxidase and ascorbate peroxidase. *J. Phys. Chem. B* **1999**, *103*, 227–233.
84. Jensen, G.M.; Bunte, S.W.; Warshel, A.; Goodin, D.B. Energetics of cation radical formation at the proximal active site tryptophan of cytochrome *c* peroxidase and ascorbate peroxidase. *J. Phys. Chem.* **1998**, *102*, 8221–8228.
85. Wirstam, M.; Blomberg, M.R.A.; Siegbahn, P.E.M. Reaction mechanism of compound i formation in heme peroxidases: A density functional theory study. *J. Am. Chem. Soc.* **1999**, *121*, 10178–10185.
86. Hiner, A.N.; Martínez, J.I.; Arnao, M.B.; Acosta, M.; Turner, D.D.; Lloyd Raven, E.; Rodríguez-López, J.N. Detection of a tryptophan radical in the reaction of ascorbate peroxidase with hydrogen peroxide. *Eur. J. Biochem.* **2001**, *268*, 3091–3098.

87. Heimdal, J.; Rydberg, P.; Ryde, U. Protonation of the proximal histidine ligand in heme peroxidases. *J. Phys. Chem. B* **2008**, *112*, 2501–2510.
88. De Visser, S.P. What affects the quartet-doublet energy splitting in peroxidase enzymes? *J. Phys. Chem. A* **2005**, *109*, 11050–11057.
89. Bathelt, C.M.; Mulholland, A.J.; Harvey, J.N. QM/MM studies of the electronic structure of the compound I intermediate in cytochrome c peroxidase and ascorbate peroxidase *J. Chem. Soc. Dalton Trans.* **2005**, 3470–3476.
90. Derat, E.; Cohen, S.; Shaik, S.; Altun, A.; Thiel, W. Principal active species of horseradish peroxidase compound I: A hybrid quantum mechanical/molecular mechanical study. *J. Am. Chem. Soc.* **2005**, *127*, 13611–13621.
91. Jensen, K.P.; Ryde, U. Importance of proximal hydrogen bonds in haem proteins. *Mol. Phys.* **2003**, *101*, 2003–2018.
92. Guallar, V.; Wallrapp, F.H. QM/MM methods: Looking inside heme proteins biochemisty. *Biophys. Chem.* **2010**, *149*, 1–11.
93. Vlasitsa, J.; Jakopitscha, C.; Bernroitnera, M.; Zamockya, M.; Furtmüllera, P.G.; Obingera, C. Mechanisms of catalase activity of heme peroxidases. *Arch. Biochem. Biophys.* **2010**, *500*, 74–81.
94. Harris, D.L.; Loew, G.H. Proximal ligand effects on electronic structure and spectra of compound I of peroxidases. *J. Porphyr. Phthalocyanins* **2001**, *5*, 334–344.
95. Nelson, D.R.; Goldstone, J.V.; Stegeman, J.J. The cytochrome P450 genesis locus: The origin and evolution of animal cytochrome P450s. *Phil. Trans. R. Soc. B* **2013**, *368*, 20120474.
96. Guengerich, F.P. Chapter 10 (6.45). In *Cytochrome P450: Structure, Mechanism, and Biochemistry*, 2nd ed.; Ortiz de Montellano, P.R., Ed.; Plenum Press: New York, NY, USA, 1995; pp. 450–452.
97. Shaik, S.; Kumar, D.; de Visser, S.P.; Altun, A.; Thiel, W. Theoretical perspective on the structure and mechanism of cytochrome P450 enzymes. *Chem. Rev.* **2005**, *105*, 2279–2328.
98. Rittle, J.; Green, M.T. Cytochrome P450 Compound I: Capture, characterization, and C-H Bond activation kinetics. *Science* **2010**, *330*, 933–937.
99. De Visser, S.P.; Ogliaro, F.; Sharma, P.K.; Shaik, S. Hydrogen bonding modulates the selectivity of enzymatic oxidation by P450: A chameleon oxidant behavior by compound I. *Angew. Chem. Int. Ed.* **2002**, *41*, 1947–1951.
100. De Visser, S.P.; Ogliaro, F.; Sharma, P.K.; Shaik, S. What factors affect the regioselectivity of oxidation by cytochrome p450? A DFT study of allylic hydroxylation and double bond epoxidation in a model reaction. *J. Am. Chem. Soc.* **2002**, *124*, 11809–11826.
101. Ogliaro, F.; Cohen, S.; de Visser, S.P.; Shaik, S. Medium polarization and hydrogen bonding effects on compound I of cytochrome P450: What kind of a radical it really is? *J. Am. Chem. Soc.* **2000**, *122*, 12892–12893.
102. Lonsdale, R.; Oláh, J.; Mulholland, A.J.; Harvey, J.N. Does compound I vary significantly between isoforms of cytochrome P450? *J. Am. Chem. Soc.* **2011**, *133*, 15464–15474.
103. Schmider, J.; Greenblatt, D.J.; Fogelman, S.M.; von Moltke, L.L.; Shader, R.I. Metabolism of dextromethorphan *in vitro*: Involvement of cytochromes P450 2D6 and 3A3/4, with a possible role of 2E1. *Biopharm. Drug. Dispos.* **1997**, *18*, 227–240.
104. Daly, J. Metabolism of acetanilides and anisoles with rat liver microsomes. *Biochem. Pharmacol.* **1970**, *19*, 2979–2993.

105. Bray, H.G.; James, S.P.; Thorpe, W.V.; Wasdell, M.R. The metabolism of ethers in the rabbit. 1. anisole and diphenyl ether. *Biochem. J.* **1953**, *54*, 547–551.
106. Olsen, L.; Rydberg, P.; Rod, T.H.; Ryde, U. Prediction of activation energies for hydrogen abstraction by cytochrome P450. *J. Med. Chem.* **2006**, *49*, 6489–6499.
107. Oláh, J.; Mulholland, A.J.; Harvey, J.N. Determinants of selectivity in drug metabolism by cytochrome P450: QM/MM modeling of dextromethorphan oxidation by CYP2D6. *Proc. Natl. Acad. Sci. USA* **2011**, *108*, 6050–6055.
108. Ingelman-Sundberg, M. Genetic polymorphisms of cytochrome P450 2D6 (CYP2D6): Clinical consequences, evolutionary aspects and functional diversity. *Pharmacogenomics J.* **2005**, *5*, 6–13.
109. Rowland, P.; Blaney, F.E.; Smyth, M.G.; Jones, J.J.; Leydon, V.R.; Oxbrow, A.K.; Lewis, C.J.; Tennant, M.G.; Modi, S.; Eggleston, D.S.; *et al.* Crystal structure of human cytochrome P450 2D6. *J. Biol. Chem.* **2006**, *281*, 7614–7622.
110. McLaughlin, L.A.; Paine, M.J.I.; Kemp, C.A.; Maréchal, J.D.; Flanagan, J.U.; Ward, R.; Sutcliffe, M.J.; Roberts, G.C.K.; Wolf, C.R. Why is quinidine an inhibitor of cytochrome P450 2D6? The role of key active-site residues in quinidine binding. *J. Biol. Chem.* **2005**, *280*, 38617–38624.
111. Flanagan, J.U.; Maréchal, J.D.; Ward, R.; Kemp, C.A.; McLaughlin, L.A.; Sutcliffe, M.J.; Roberts, G.C.K.; Paine, M.J.I.; Wolf, C.R. Phe120 contributes to the regiospecificity of cytochrome P450 2D6: Mutation leads to the formation of a novel dextromethorphan. *Biochem. J.* **2004**, *380*, 353–360.
112. Lonsdale, R.; Houghton, K.T.; Żurek, J.; Bathelt, C.M.; Foloppe, N.; de Groot, M.J.; Harvey, J.N.; Mulholland, A.J. Quantum mechanics/molecular mechanics modeling of regioselectivity of drug metabolism in cytochrome P450 2C9. *J. Am. Chem. Soc.* **2013**, *135*, 8001–8015.
113. Zhang, L.; Kudo, T.; Takaya, N.; Shoun, H. The B′ helix determines cytochrome P450nor specificity for the electron donors NADH and NADPH. *J. Biol. Chem.* **2002**, *277*, 33842–33847.
114. De Vries, S.; Schröder, I. Comparison between the nitric oxide reductase family and its aerobic relatives, the cytochrome oxidases. *Biochem. Soc. Trans.* **2002**, *30*, 662–667.
115. Richardson, D.; Felgate, H.; Watmough, N.; Thomson, A.; Baggs, E. Mitigating release of the potent greenhouse gas N(2)O from the nitrogen cycle - could enzymic regulation hold the key? *Trends Biotechnol.* **2009**, *27*, 388–397.
116. Wink, D.A.; Hines, H.B.; Cheng, R.Y.S.; Switzer, C.H.; Flores-Santana, W.; Vitek, M.P.; Ridnour, L.A.; Colton, C.A.J. Nitric oxide and redox mechanisms in the immune response. *Leukocyte Biol.* **2011**, *89*, 873–891.
117. Chao, L.Y.; Rine, J.; Marletta, M.A. Spectroscopic and kinetic studies of nor1, a cytochrome p450 nitric oxide reductase from the fungal pathogen *Histoplasma capsulatum. Arch. Biochem. Biophys.* **2008**, *480*, 132–137.
118. Shiro, Y.; Fujii, M.; Iizuka, T.; Adachi, S.; Tsukamoto, K.; Nakahara, K.; Shoun, H. Spectroscopic and kinetic studies on reaction of cytochrome P450nor with nitric oxide. Implication for its nitric oxide reduction mechanism. *J. Biol. Chem.* **1995**, *270*, 1617–1623.

119. Harris, D.L. Cytochrome P450nor: A nitric oxide reductase - structure, spectra, and mechanism. *Int. J. Quantum Chem.* **2002**, *288*, 183–200.

120. Tsukamoto, K.; Nakamura, S.; Shimizu, K. SAM1 semiempirical calculations on the catalytic cycle of nitric oxide reductase from *Fusarium oxysporum*, *J. Mol. Struct. (THEOCHEM)* **2003**, *624*, 309–322.

121. Averill, B.A. Dissimilatory nitrite and nitric oxide reductases. *Chem. Rev.* **1996**, *96*, 2951–2964.

122. Daiber, A.; Nauser, T.; Takaya, N.; Kudo, T.; Weber, P.; Hultschig, C.; Shoun, H.; Ullrich, V. Isotope effects and intermediates in the reduction of NO by P450NOR. *J. Inorg. Biochem.* **2002**, *88*, 343–352.

123. Lehnert, N.; Praneeth, V.K.K.; Paulat, F. Electronic structure of Fe(II)-porphyrin nitroxyl complexes: Molecular mechanism of fungal nitric oxide reductase (P450nor). *J. Comput. Chem.* **2006**, *27*, 1338–1351.

124. Krámos, B.; Menyhárd, D.K.; Oláh, J. Direct hydride shift mechanism and stereoselectivity of P450nor confirmed by QM/MM calculations. *J. Phys. Chem. B* **2012**, *116*, 872–885.

125. Suzuki, N.; Higuchi, T.; Urano, Y.; Kikuchi, K.; Uchida, T.; Mukai, M.; Kitagawa, T.; Nagano, T. First synthetic no-heme-thiolate complex relevant to nitric oxide synthase and cytochrome P450nor. *J. Am. Chem. Soc.* **2000**, *122*, 12059–12060.

126. Shimizu, H.; Obayashi, E.; Gomi, Y.; Arakawa, H.; Park, S.Y.; Nakamura, H.; Adachi, S.; Shoun, H.; Shiro, Y. Proton delivery in no reduction by fungal nitric-oxide reductase. cryogenic crystallography, spectroscopy, and kinetics of ferric-no complexes of wild-type and mutant enzymes. *J. Biol. Chem.* **2000**, *275*, 4816–4826.

127. Menyhárd, D.K.; Keserű, G.M. Binding mode analysis of the NADH cofactor in nitric oxide reductase: A theoretical study. *J. Mol. Graphics Model.* **2006**, *25*, 363–372.

128. Oshima, R.; Fushinobu, S.; Su, F.; Zhang, L.; Takaya, N.; Shoun, H. Structural evidence for direct hydride transfer from NADH to cytochrome P450nor. *J. Mol. Biol.* **2004**, *342*, 207–217.

129. Riplinger, C.; Neese, F. The reaction mechanism of cytochrome P450 NO reductase: A detailed quantum mechanics/molecular mechanics study. *ChemPhysChem* **2011**, *12*, 3192–3203.

130. Callens, M.; Tomme, P.; Kerstens-Hilderson, H.; Cornelis, W.; Vangrysperre, W.; de Bruyne, C.K. Metal ion binding to d-xylose isomerase from streptomyces violaceoruber. *Biochem. J.* **1988**, *250*, 285–291.

131. Farber, G.K.; Machin, P.; Almo, S.C.; Petsko, G.A.; Hajdu, J. X-ray laue diffraction from crystals of xylose isomerase. *Proc. Natl. Acad. Sci. USA* **1988**, *85*, 112–118.

132. Collyer, C.A.; Henrick, K.; Blow, D.M. Mechanism for aldose-ketose interconversion by D-Xylose isomerase involving ring opening followed by a 1,2-hydride shift. *J. Mol. Biol.* **1990**, *212*, 211–235.

133. Jenkins, J.; Janin, J.; Rey, F.; Chidmi, M.; Tilbeurgh, H.; Lasters, I.; Maeyer, M.; Belle, D.; Wodak, S.; Lauwereys, M.; *et al.* Protein engineering of xylose (Glucose) isomerase from actinoplanes missouriensis. 1. Crystallography and site-directed mutagenesis of metal binding sites. *Biochemistry* **1992**, *31*, 5449–5458.

134. Hu, H.; Liu, H.; Shi, Y. The reaction pathway of the isomerization of *d*-xylose catalyzed by the enzyme d-xylose isomerase: A theoretical study. *Proteins Struct. Funct. Genet.* **1997**, *27*, 545–555.

135. Fuxreiter, M.; Farkas, Ö.; Náray-Szabó, G. Molecular modeling of xylose isomerase catalysis: The role of electrostatics and charge transfer to metals. *Protein Eng.* **1995**, *8*, 925–933.
136. Fábián, P.; Asbóth, B.; Náray-Szabó, G. The role of electrostatics in the ring-opening step of xylose isomerase catalysis. *J. Mol. Struct. Theochem.* **1994**, *307*, 171–178.
137. Lambeir, A.M.; Lauwereys, M.; Stanssens, P.; Mrabet, N.T.; Snauwaert, J.; van Tilbeurgh, H.; Matthyssens, G.; Lasters, I.; de Maeyer, M.; Wodak, S.J.; *et al.* Protein engineering of xylose (glucose) isomerase from actinoplanes missouriensis. 3. Changing metal specificity and the PH profile by site-directed mutagenesis. *Biochemistry* **1992**, *31*, 5459–5466.
138. Garcia-Viloca, M.; Alhambra, C.; Truhlar, D.G.; Gao, J. Quantum dynamics of hydride transfer catalyzed by bimetallic electrophilic catalysis: Synchronous motion of Mg2+ and H- in xylose isomerase. *J. Am. Chem. Soc.* **2002**, *124*, 7268–7269.
139. Garcia-Viloca, M.; Alhambra, C.; Truhlar, D.G.; Gao, J. Hydride transfer catalyzed by xylose isomerase: Mechanism and quantum effects. *J. Comput. Chem.* **2003**, *24*, 177–190.
140. Masgrau, L.; Roujeinikova, A.; Johannissen, L.O.; Hothi, P.; Basran, J.; Ranaghan, K.E.; Mulholland, A.J.; Sutcliffe, M.J.; Scrutton, N.S.; Leys, D. Atomic description of an enzyme reaction dominated by proton tunneling. *Science* **2006**, *312*, 237–241.
141. Iyengar, S.S.; Sumner, I.; Jakowski, J. Hydrogen tunneling in an enzyme active site: A quantum wavepacket dynamical perspective. *J. Phys. Chem. B* **2008**, *112*, 7601–7613.

5

A Sensitive DNA Enzyme-Based Fluorescent Assay for Bacterial Detection

Sergio D. Aguirre [1,†], **M. Monsur Ali** [1,†], **Bruno J. Salena** [2] **and Yingfu Li** [1,3,4,*]

1 Department of Biochemistry and Biomedical Sciences, McMaster University, 1280 Main St. W., Hamilton, ON L8S 4K1, Canada; E-Mails: sergio.aguirre454@gmail.com (S.D.A); mmonsurali@gmail.com (M.M.A.)

2 Dvision of Gastroenterology, Department of Medicine, McMaster University, 1280 Main St. W., Hamilton, ON L8S 4K1, Canada; E-Mail: salenab@mcmaster.ca

3 Department of Chemistry and Chemical Biology, McMaster University, 1280 Main St. W., Hamilton, ON L8S 4K1, Canada

4 Michael G. DeGroote Institute for Infectious Disease Research, McMaster University, 1280 Main St. W., Hamilton, ON L8S 4K1, Canada

† These authors contributed equally to this work.

* Author to whom correspondence should be addressed; E-Mail: liying@mcmaster.ca

---

Abstract: Bacterial detection plays an important role in protecting public health and safety, and thus, substantial research efforts have been directed at developing bacterial sensing methods that are sensitive, specific, inexpensive, and easy to use. We have recently reported a novel "mix-and-read" assay where a fluorogenic DNAzyme probe was used to detect model bacterium *E. coli*. In this work, we carried out a series of optimization experiments in order to improve the performance of this assay. The optimized assay can achieve a detection limit of 1000 colony-forming units (CFU) without a culturing step and is able to detect 1 CFU following as short as 4 h of bacterial culturing in a growth medium. Overall, our effort has led to the development of a highly sensitive and easy-to-use fluorescent bacterial detection assay that employs a catalytic DNA.

Keywords: bacterial detection; DNAzyme; fluorescence; biosensor; *Escherichia coli*

---

1. Introduction

Infectious agents, such as foodborne pathogens, have caused numerous large-scale and costly outbreaks in the human history and will continue to be a major public health threat and financial burden for our society [1–4]. Early detection of pathogens, as the first step to prevent such outbreaks, has become increasingly more important today because the globalization of commerce and speedy travel have significantly increased the rate and breadth of the spread of infectious agents. Thus, the demand for faster, simpler, less expensive and more reliable pathogen testing methods has become ever greater.

Although the traditional culture method continues to be the 'gold standard' for bacterial detection, it is time-consuming and requires days or even weeks to complete (depending on the specific pathogen in question) [5]. Modern methods take advantage of well-established biomolecular techniques, such as polymerase chain reaction (PCR) and immunoassay (where an antibody is used as molecular recognition element), to achieve faster and more sensitive pathogen detection [5–11]. Despite the popularity of these techniques, they also come with certain drawbacks, such as the need for costly instrumentation and highly trained personnel to isolate or purify relevant targets (DNA for PCR and proteins for immunoassays). Thus, the entire test using such methods often still needs one or more days to complete. Detection sensitivity (for immunoassay) and tendency to generate false-positive results (for PCR) are also issues of concerns. For these considerations, we recently began to examine the utility of RNA-cleaving fluorogenic DNAzyme (RFD) probes for bacterial detection [12–14]. RFDs can be isolated from random-sequence DNA pools to perform three linked functions: ligand binding, catalysis and fluorescence generation. Each RFD cleaves a synthetic nucleic acid substrate containing a single ribonucleotide as the cleavage site embedded in a DNA sequence, and the cleavage site is located between two nucleotides modified with a matching pair of fluorophore and quencher [12–21]. Because of these two features, these reporter molecules emit an increasing level of fluorescence when they carry out the catalytic cleavage of the RNA linkage. In other words, the cleavage event results in separation of the fluorophore from the quencher, accompanied by the increase of fluorescence intensity in real time.

More recently, we developed a method of isolating novel DNAzyme probes against the crude extracellular mixture (CEM) left behind by a specific type of bacteria in their environment or in the media they are cultured [12]. The CEM is rich in diverse targets, including small molecules and proteins. Thus the use of the crude mixture as the complex target to conduct *in vitro* selection [22–24] experiment circumvents the tedious process of purifying and identifying a suitable target from the microbe of interest for biosensor development, and provides a subsequent assaying procedure that is simple because it does not require steps to purify a target of interest. Using this approach, we have isolated an RFD that cleaves its substrate only in the presence of the CEM produced by *E. coli* (CEM-EC) [12]. This *E. coli*-sensing RFD, named RFD-EC1, was found to be highly selective to CEM-EC but nonresponsive to CEMs from many other Gram-negative and Gram-positive bacteria. We have also shown that the DNAzyme-based assay is capable of reporting the presence of a single *E. coli* cell after 12 h of culturing. These experiments have illustrated the utility of RFDs as fluorogenic bacterial indicators. In this work we carried out a thorough investigation to characterize this bacterial detection system with a goal to further improve the detection sensitivity.

2. Results and Discussion

2.1. Establishing a Trans-*Acting DNAzyme*

Our previously reported RFD-EC1 is a *cis*-acting DNAzyme that cleaves a covalently attached substrate. However, a *trans*-acting DNAzyme where the DNAzyme cleaves a detached substrate has an additional advantage such as ease-of-synthesis, thus lowering the cost and labor. Synthesis of long DNA chain modified with fluorophore, quencher and ribonucleotide is associated with lower yields and higher costs. Therefore, in this study, we first examined the possibility of converting it into a *trans*-acting catalyst by detaching the substrate portion of the sequence, FS1, from the DNAzyme portion, EC1 (Figure 1A). We found that EC1 was indeed able to cleave FS1 in *trans*, even at 1:1 ratio (50 nM each of EC1 and FS1), in a CEM-EC dependent manner (Figure 1B). Note that the reaction mixtures were analyzed by denaturing polyacrylamide gel electrophoresis (dPAGE).

Figure 1. Design of *trans*-acting DNAzymes. (**A**) The sequences of EC1, EC1T, EC1TM and FS1. EC1 is the full length DNAzyme including two primer binding sites (nucleotides in italic) for polymerase chain reaction used in the original *in vitro* selection experiment. EC1T is the shortened version of EC1 with deleted primer binding sites. EC1TM is a mutant of EC1T wherein the nucleotides shown as lower-case letters are altered. The substrate FS1 contains an adenosine ribonucleotide (R) flanked by a fluorescein-dT (F) and a DABCYL-dT (Q). (**B**) dPAGE analysis of the cleavage reaction mixtures of FS1 with EC1, EC1T, or EC1TM in the absence (−) and presence (+) of CEM-EC. P1 represents the 5'-cleavage product, which can be observed by fluorescence scan as it contains the F unit. MK (marker) is a sample of FS1 fully cleaved by NaOH. Clv% for each sample was calculated following our previously reported method [20].

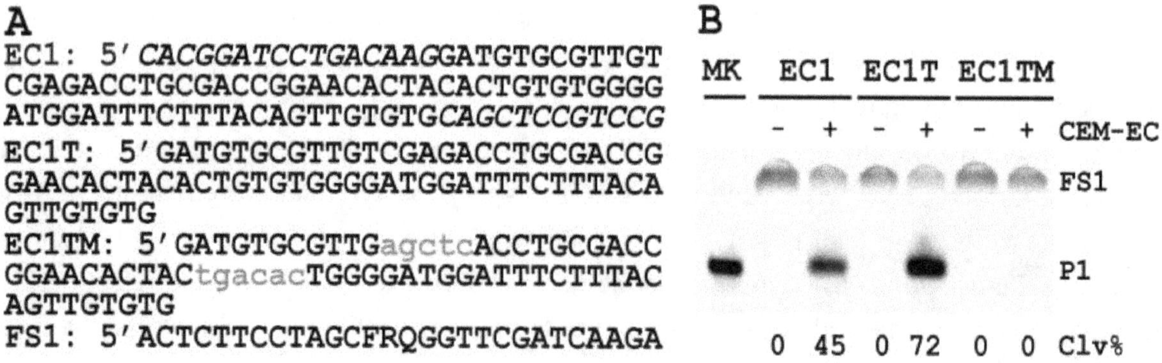

We next tested a second *trans* construct, named EC1T (Figure 1A), by truncating 28 nucleotides from the two ends of EC1 (italic letters, Figure 1A) that were used as the primer-binding sites for polymerase chain reaction during the original *in vitro* selection experiment. Interestingly, EC1T was found to be considerably more active than EC1 (comparing Lanes 3 and 5, Figure 1B; *i.e.*, 45% *vs.* 72%). As a control, we also tested a mutant sequence, EC1TM, with 10 nucleotides (lower-case letters, Figure 1A) mutated from EC1T. These mutations rendered EC1TM completely inactive in the presence of CEM-EC (Figure 1B).

2.2. *Comparing DNAzyme Activity Using Crude Extracellular Mixture (CEM) and Crude Intracellular Mixture (CIM) of* E. coli

The original DNAzyme RFD-EC1 was isolated to cleave in the presence of CEM of *E. coli*. We hypothesized that the target that activates the DNAzyme might be more abundant inside the cellular environment. To test this idea, we made an *E. coli* culture and used it to prepare the CEM-EC and CIM-EC as follows: the cells were precipitated by centrifugation and the supernatant was taken as the CEM-EC. The cell pellet was re-suspended in the reaction buffer, heat-treated, and then centrifuged; the remaining supernatant was taken as the CIM-EC (see experimental section for details). The CEM-EC and CIM-EC were then used to induce the cleavage activity of EC1T towards FS1, and the results are illustrated in Figure 2A. It is clear that the CIM-EC indeed contained a much higher amount of the target than the CEM-EC as it induced much stronger cleavage of FS1 by EC1T (45% *vs.* 1%). Note that much lower cleavage in this experiment with CEM-EC is due to the shorter culture time (7 h) with low number of *E. coli* cells (50,000 colony forming units). For the remaining experiments, the CIM-EC was used as the target of interest.

Figure 2. Cleavage reactions of EC1T/FS1 with (**A**) crude extracellular mixture (CEM)-EC and CIM-EC and (**B**) crude intracellular mixture (CIM)-EC collected from *E. coli* cells grown in various culture broths. NC is a negative control where the reaction was conducted in the absence of CEM-EC and CIM-EC. Each reaction mixture was analyzed by 10% dPAGE, followed by fluorimaging. NC: negative control where the reaction was conducted in RB without CEM-EC or CIM-EC.

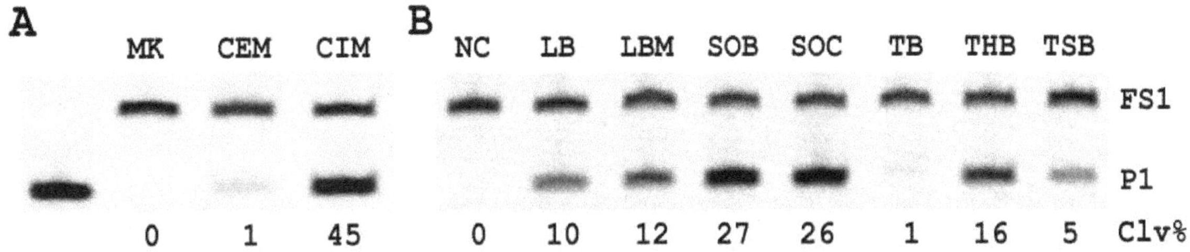

2.3. *Searching for an Optimal Culture Broth*

We next investigated the effect of bacterial growth media on the quality of CIM (as measured by the cleavage activity of EC1T/FS1) in order to establish an optimal culture broth. Seven common growth media were chosen for this analysis and they were: Luria Bertani (LB), Terrific Broth (TB), Todd-Hewitt (TH), Lysogeny Broth Miller (LBM), Tryptic Soy Broth (TSB), Super Optimal Broth (SOB) and Super Optimal Broth with Catabolic repressor (SOC). 250 *E. coli* cells were allowed to grow in 1 mL of each broth for 7 h at 37 °C, from which CIM was prepared and used to induce the cleavage of EC1T/FS1; the results are illustrated as Figure 2B. The CIMs from SOB and SOC produced the highest activity (~26% cleavage), followed by those from LB, LBM, and THB (10–16%). The CEMs from TSB and TB were least effective (≤5%). Based on these results, SOB was chosen as the broth for the remaining experiments.

2.4. Effects of Divalent Metal Ions

Divalent metal ions play crucial roles in catalytic functions of DNAzymes and it has been shown that different metal ions can significantly affect the catalytic activity of a DNAzyme [25–28]. For example, 8-17, a well-studied RNA-cleaving DNAzyme, exhibits the highest activity in presence of lead ions even though it was originally isolated using Mg^{2+} [29] or Zn^{2+} [30]. A recent study has revealed that Pb^{2+} promotes the most favorable folding of 8-17 [31]. Therefore, we sought to compare the effects of various divalent metal ions on the activity of our *E. coli*-sensing DNAzyme although the original DNAzyme RFD-EC1 was obtained by *in vitro* selection in the presence of 15 mM $MgCl_2$ [12]. Nine different divalent metal ions were tested and they were: Ba^{2+}, Cd^{2+}, Co^{2+}, Mg^{2+}, Mn^{2+}, Ni^{2+}, Cu^{2+}, Zn^{2+}, and Ca^{2+}; the results are given in Figure 3A. We found that Ba^{2+}, Ca^{2+}, Mg^{2+} and Mn^{2+} all induced a robust cleavage activity of the DNAzyme (causing 56–68% of cleavage). In contrast, Cd^{2+}, Co^{2+}, Ni^{2+}, Cu^{2+}, and Zn^{2+} resulted in weak cleavage (1–2%). It is possible that Ba^{2+} Mg^{2+}, Mn^{2+} and Ca^{2+} fit into the catalytic core better than the other divalent metal ions. However, this should be experimentally verified.

Figure 3. (**A**) Cleavage activity of EC1T/FS1 in the presence of CEM-EC and various divalent metal ions. (**B**) Effect of the Ba^{2+} concentration.

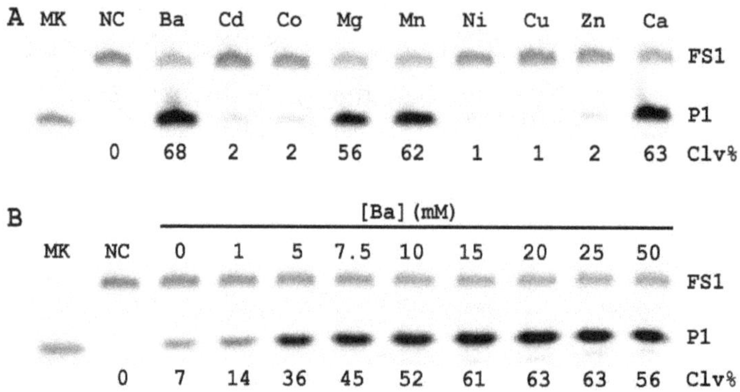

It is noteworthy that we have previously shown that Mn^{2+} exhibits potent fluorescence quenching effect, resulting in significantly reduced signal magnitude when the fluorescence intensity is measured in a fluorimeter [32]. We also found a similar effect of Mn^{2+} in our assay (data not shown). In contrast, Ba^{2+} produced no quenching effect. This observation indicates that Ba^{2+} is a more suitable divalent metal ion for our assay. Thus, Ba^{2+} was chosen for further experiments. In order to establish the optimal Ba^{2+} concentration we investigated the effect of Ba^{2+} concentration on EC1T's activity. The data presented in Figure 3B indicates that the catalytic activity of EC1T reaches a plateau at 15 mM Ba^{2+}.

2.5. Varying Reaction Temperature

We examined the cleavage activity of EC1T/FS1 at different temperatures and the results are provided in Figure 4A. A robust cleavage activity was observed at both 15 and 23 °C. In contrast, reduced activity was observed when the reaction temperature was decreased to 4 °C or increased to 37 °C and 50 °C. Interestingly, although CIM was absolutely required to induce the cleavage at 4,

15 and 23 °C, EC1T can cleave FS1 in the absence of CIM at both 37 °C and 50 °C (grey bars in Figure 4A). Since room temperature is the most ideal condition to conduct assays avoiding the requirements of heating and cooling system, we chose 23 °C as the reaction temperature for the remaining experiments.

Figure 4. Cleavage activity of EC1T/FS1 with varying temperature (**A**), pH (**B**), and EC1T/FS1 ratio (**C**). The data are the average of two independent experiments.

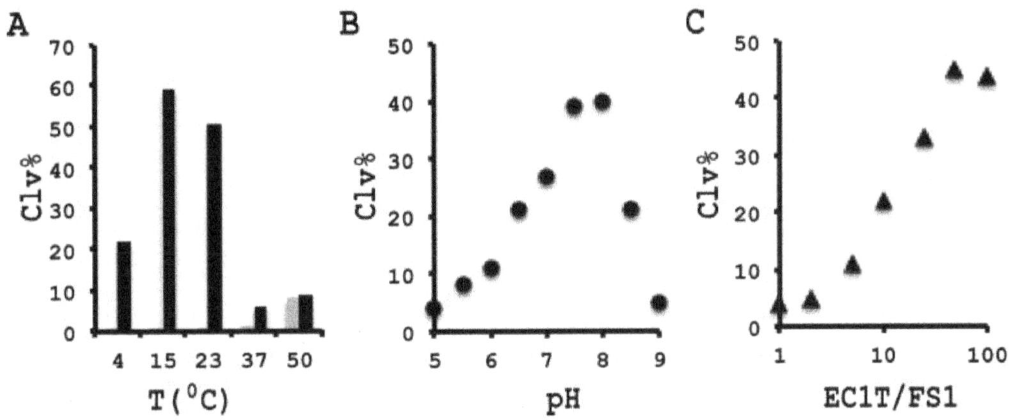

2.6. pH Effect

We next examined the activity of EC1T/FS1 when the reaction pH was varied between 5.0 and 9.0; the results are shown in Figure 4B. Although EC1T was able to cleave FS1 in the entire pH range tested, the highest activity was observed at pH 7.5–8.0. Since the original DNAzyme was isolated at pH 7.5, it is not surprising that EC1T exhibits such a narrow pH preference.

2.7. DNAzyme/Substrate Ratio

We also examined the cleavage activity at different ratios of EC1T/FS1. For this experiment, the concentration of FS1 was kept at 50 nM while the DNAzyme concentration was changed from 0 to 5 μM; the results are shown in Figure 4C. The cleavage activity reached the plateau at a ratio of 50:1. Thus, this ratio was used for the remaining experiments.

2.8. Specificity

With the significant changes of the reaction conditions, we wondered if EC1T was still able to maintain its specificity for *E. coli*. Four other gram-negative bacteria and four gram-positive bacteria were arbitrarily chosen for comparison. Each bacterium was cultured in SOB for a different period of time until the OD_{600} (optical density at 600 nm) of each culture reached ~1. The CIM was then prepared and tested with EC1T/FS1 under the optimal reaction buffer (50 mM HEPES, pH 7.5, 150 mM NaCl and 15 mM $BaCl_2$, room temperature, EC1T/FS1 = 50/1). None of the CIMs from other bacteria was able to induce cleavage (Figure 5), indicating that EC1T/FS1 retained the specificity for *E. coli*.

Figure 5. Specificity of EC1T/FS1 for various gram-negative and gram-positive bacteria. PP: *Pseudomonas peli*, YR: *Yersinia rukeri*, HA: *Hafnea alvei*, AX: *Achromobacter xylosoxidans*, EC: *Escherichia coli*, BS: *Bacillus subtilis*, LM: *Leuconostoc mesenteroides*, LP: *Lactobacillus planturum*, PA: *Pediococcus acidilactici.*

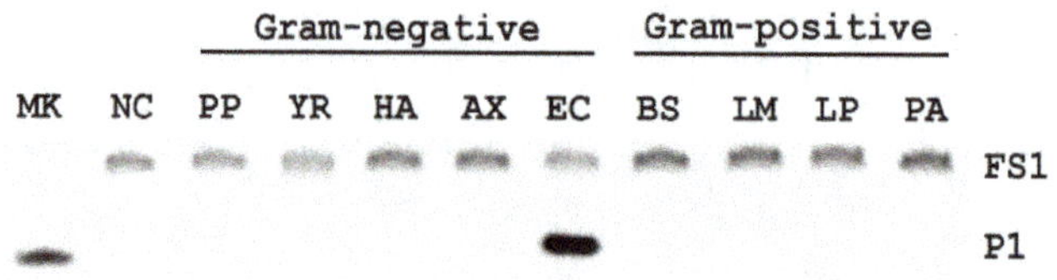

2.9. Detection Sensitivity

To test the detection sensitivity of EC1T/FS1, we prepared a series of *E. coli* stock solutions from which CIM samples were prepared as described in experimental section. These samples were then assessed for inducing the cleavage of EC1T/FS1 under the optimal reaction condition established above. These reactions were monitored in a fluorimeter in real time for 60 min (Figure 6A). The reaction mixtures were also analyzed by dPAGE (Figure 6B). We found that the fluorimeter method was able to detect 10^5 cells while the dPAGE method can detect 10^4 cells.

Figure 6. Sensitivity test. (**A**) Real-time fluorescence monitoring and (**B**) dPAGE analysis of EC1T/FS1 in the presence of CIMs prepared from 10^3–10^7 *E. coli* cells. (**C**) and (**D**) Similar experiments using RNA-cleaving fluorogenic DNAzyme (RFD-EC1) with CIMs prepared from 10^2–10^7 *E. coli* cells. The data in (A) and (C) are the average of two independent experiments.

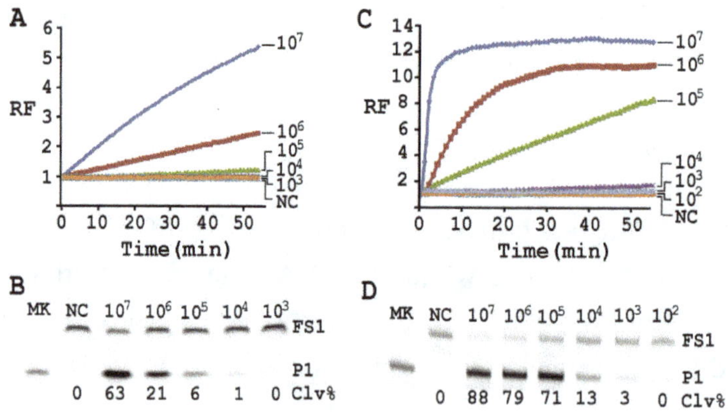

We also tested the detection sensitivity of the original *cis*-acting DNAzyme RFD-EC1 using the optimal reaction condition. Interestingly, RFD-EC1 showed better sensitivities: the fluorimeter method can detect 10^4 cells (Figure 6C) while the dPAGE method was able to detect 10^3 cells (Figure 6D).

2.10. Detection of a Single Cell via Culturing

Finally we determined the time required to enrich a single live bacterium (*i.e.*, one colony forming unit or 1 CFU) via culturing in SOB. Following a previous protocol [12], we inoculated a single *E. coli* cell in SOB and cultured for 2, 4, 6, 8 and 10 h at 37 °C. CIMs were prepared for the samples collected at each time point and tested with both *trans* and *cis* constructs. These samples were then assessed for inducing the cleavage of EC1T/FS1 under the optimal reaction condition. Each reaction was examined both in a fluorimeter (Figure 7A) and by dPAGE (Figure 7B). Using EC1T/FS1, 8 h of culturing was sufficient for detection by the fluorimeter method (Figure 7A) and 6 h by dPAGE method (Figure 7B). Using RFD-EC1, however, only 6 h and 4 h of culturing were required to achieve the detection by the fluorimeter (Figure 7C) and dPAGE (Figure 7D) method, respectively. The lower activity of EC1T/FS1 in comparison to the *cis*-acting RFD-EC1 might be due to the weakened interaction between enzyme and substrate strands when they were separated from each other.

Figure 7. Culturing time required to detect a single *E. coli* cell (1 CFU). (**A**) Monitoring fluorescence of EC1T/FS1 with CIMs prepared from samples taken after a culturing time of 2, 4, 6, 8 and 10 h. (**B**) dPAGE analysis of the reaction mixtures in (A). (**C**) and (**D**) are equivalent experiments in which RFD-EC1 was used to replace EC1T/FS1. The data in (A) and (C) are the average of two independent experiments.

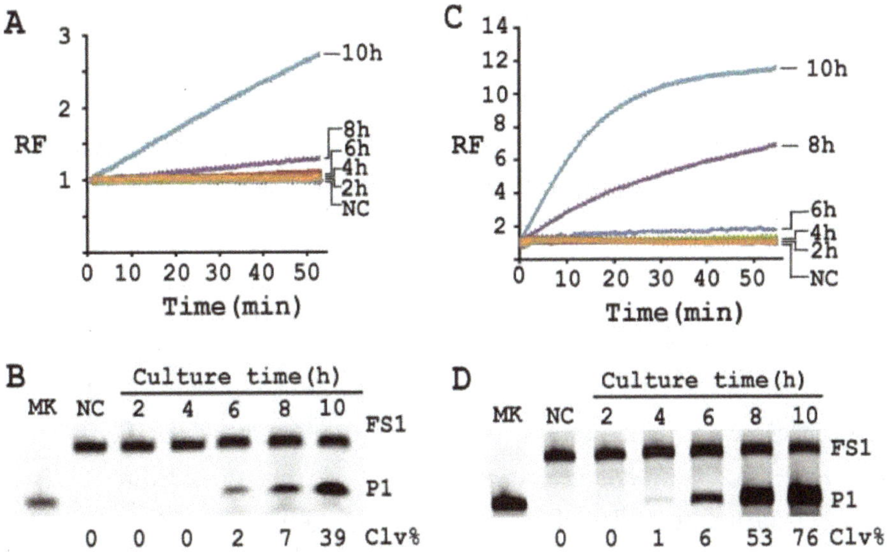

3. Experimental Section

3.1. Synthesis and Purification of Oligonucleotides

The standard DNA oligonucleotides (EC1, EC1T, EC1TM and EC1LT) were purchased from Integrated DNA Technologies (IDT, Coralville, IA, USA) and purified by 10% denaturing polyacrylamide gel electrophoresis (dPAGE). The modified oligonucleotide FS1 was acquired from W. M. Keck Oligonucleotide Synthesis Facilities (Yale University, New Haven, CT, USA), deprotected and purified by 10% dPAGE following a previously reported protocol [15].

3.2. Enzymes and Chemical Reagents

T4 DNA ligase and T4 polynucleotide kinase (PNK) were purchased from MBI Fermentas (Burlington, ON, Canada). Tryptone and yeast extract was acquired from BD Biosciences (Mississauga, ON, Canada). All other chemical reagents were purchased from Sigma-Aldrich (Oakville, ON, Canada) and were used without further purification.

3.3. Growth Media

Luria Bertani (LB), Terrific Broth (TB), and Todd-Hewitt (TH) were purchased from Sigma-Aldrich. Lysogeny Broth Miller (LBM) was obtained from EMD Canada (Mississauga, ON, Canada). Tryptic Soy Broth (TSB) was acquired from BD Biosciences. Super Optimal Broth (SOB) and Super Optimal Broth with Catabolic repressor (SOC) were made in house. SOB contains 2% (w/v) tryptone, 0.5% (w/v) yeast extract, 10 mM NaCl and 2.5 mM KCl. SOC has the same ingredients as SOB but also contains 20 mM glucose and 10 mM $MgCl_2$.

3.4. Preparation of Cis-*Acting RFD-EC1*

RFD-EC1 was generated by template-mediated ligation of FS1 to EC1. In brief, 200 pmol of FS1 were treated with 1× PNK buffer A (MBI Fermentas), 1 mM ATP and 20 U (units) of PNK for 30 min at 37 °C (reaction volume = 50 µL). The reaction was quenched by heating at 90 °C for 5 min. Equimolar EC1 and EC1LT (5'-CTAGG AAGAG TCGGA CGGAG CTG; the ligation template) were then added to this solution and was heated at 90 °C for 30 s and cooled to room temperature for 10 min. Afterwards, 10 µL of 10× T4 DNA ligase buffer (MBI Fermentas), 39 µL of deionized distilled water (ddH_2O) and 1 µL of T4 DNA ligase (10 U/µL) were added. After incubation at room temperature (RT) for 2 h, the ligated EC1-FS1 was purified by 10% dPAGE.

3.5. Bacterial Cells

Gram-negative bacteria *Pseudomonas peli*, *Yersinia rukeri*, *Hafnea alvei*, and *Achromobacter xylosoxidans* were donated by Dr. Gerard Wright (Micheal G. DeGroote Institute for Infectious Disease Research, McMaster University). Gram-positive bacteria *Leuconostoc mesenteroides*, *Lactobacillus planturum* and *Pediococcus acidilactici* (PA) were gifts from Dr. Brian Coombes and Dr. Russel Bishop (Department of Biochemistry and Biomedical Sciences, McMaster University). *E. coli* K12 (MG1655) and *Bacillus subtilis* 168 are regularly maintained in our laboratory.

3.6. Comparison of the Cleavage Activity of EC1, EC1T and EC1TM in the Presence of CEM-EC

E. coli was plated onto a TSB agar (1.5%) plate and grown for 14 h at 37 °C. A single colony was taken and inoculated into 2 mL of TSB and grown for 14 h at 37 °C with shaking at 250 rpm. A 1% fresh culture was made by re-inoculating 20 µL of the above culture into 2 mL of TSB. The re-inoculation was allowed to grow at 37 °C with shaking at 250 rpm until the culture reached an OD_{600} of ~1. 1 mL of this culture was centrifuged at 11,000 g for 5 min at room temperature; the supernatant was taken as the crude extracellular mixture (CEM-EC) and stored at −20 °C.

For each candidate DNAzyme construct, two reactions were set up, a control and a test. For the test, 25 µL of 2× reaction buffer (2× RB; 100 mM HEPES, 300 mM NaCl, 30 mM $MgCl_2$, pH 7.5) was mixed with 23 µL of the CEM-EC prepared above, 1 µL of 2.5 µM FS1 and 1 µL of 2.5 µM EC1, EC1T or EC1TM. For the control, TSB was used to substitute the CEM-EC. Each reaction mixture was incubated at RT for 60 min, followed by quenching with 5 µL of 3 M NaOAc (pH 5.5) and 135 µL of cold ethanol. DNA was recovered by centrifugation and analyzed by 10% dPAGE. DNA bands in the gel were visualized by Typhoon 9200 (GE Healthcare) and quantified by ImageQuant software (Molecular Dynamics).

3.7. Comparison of the Cleavage Activity of EC1T in the Presence of CEM-EC and CIM-EC

100 µL of 50,000 CFU/mL glycerol stock of *E. coli* was inoculated into 2 mL of TSB and grown at 37 °C for 7 h with shaking at 250 rpm. 1 mL of this culture was centrifuged at 11,000 g for 5 min at room temperature; the supernatant was taken as the CEM-EC for this experiment. The cell pellet was suspended in 200 µL of 1× RB and heated at 50 °C for 15 min. The heat-treated cell suspension was then centrifuged at 11,000 g for 5 min at RT. The clear supernatant was taken as the CIM-EC for the experiment.

The cleavage reaction with the CEM-EC was carried out by mixing 25 µL of 2× RB, 23 µL of the CEM-EC prepared above, 1 µL of 2.5 µM FS1 and 1 µL of 2.5 µM EC1T. The reaction concerning the CIM-EC was conducted by mixing 41 µL of 1× RB, 5 µL of CIM-EC, 1 µL of 2.5 µM FS1, 1 µL of 2.5 µM EC1T and 2 µL of 2× RB (note that the CIM-EC was made by suspending the cell pellet from originally 1 mL of *E. coli* culture in 200 µL of 1× RB, which translates into a concentrating factor of 5). A control experiment without the CEM-EC and CIM-EC was also conducted. Each reaction mixture was incubated at RT for 60 min, followed by 10% dPAGE analysis as described above.

3.8. Comparison of the Cleavage Activity of EC1T in the Presence of CIM-EC Obtained from E. coli *Grown in Various Growth Media*

100 µL of 2500 CFU/mL glycerol stock of *E. coli* was inoculated into 2 mL of LB, LBM, SOB, SOC, TB, TH, or TSB. Following 7 h incubation at 37 °C, 1 mL of each culture was taken and centrifuged at 11,000 g for 5 min at RT. The cell pellet was re-suspended in 200 µL of 1× RB. Cleavage reactions were then conducted by mixing 41 µL of 1× RB, 5 µL of each CIM-EC, 1 µL of 2.5 µM FS1, 1 µL of 2.5 µM EC1T and 2 µL of 2× RB. Each reaction mixture was incubated at RT for 60 min, followed by 10% dPAGE analysis as described above.

3.9. Comparison of the Cleavage Activity of EC1T in the Presence of Different Divalent Metals

First, stocks of 2× RB′ (100 mM HEPES, 300 mM NaCl, pH 7.5) and 150 mM MCl_2 (M = Cd, Co, Mg, Mn, Ni, Cu, Zn and Ca) were prepared. The CIM-EC was also prepared from *E. coli* grown in SOB in the same way as described immediately above except for the use of 1× RB′ instead of 1× RB. The cleavage reactions as shown in Figure 3A were set up by mixing 15.5 µL of water, 22.5 µL of 2× RB′, 5 µL of a relevant MCl_2 stock, 1 µL of 2.5 µM FS1, 1 µL of 2.5 µM EC1T, and 5 µL of the CIM-EC. The cleavage reactions as shown in Figure 3B were set up similarly except that the volume of water and 150 mM $BaCl_2$ were co-varied to achieve a final [$BaCl_2$] of 0, 1, 5, 7.5, 10, 15, 20, 25 and 50 mM. Each reaction mixture was incubated at RT for 60 min, followed by 10% dPAGE analysis as described above.

3.10. Comparison of the Cleavage Activity of EC1T at Different Reaction Temperature

A 2× RB_{Ba} stock (100 mM HEPES, 300 mM NaCl, 30 mM $BaCl_2$, pH 7.5) was first prepared. Five cleavage reaction mixtures were then set up by mixing 19.5 µL of water, 22.5 µL of 2× RB_{Ba}, 1 µL of 2.5 µM FS1, 1 µL of 2.5 µM EC1T, and 5 µL of the CIM-EC prepared with 1× RB_{Ba}. These mixtures were incubated, respectively, at 4, 15, 23, 37 and 50 °C for 60 min, followed by 10% dPAGE analysis as described above.

3.11. Comparison of the Cleavage Activity of EC1T at Different pH

A series of 2× RB_{Ba}' stock (300 mM NaCl, 30 mM $BaCl_2$, along with a chosen buffering agent at 100 mM) were first prepared with pH being varied from 5.0 to 9.0 at an increasing interval of 0.5 units. MES was used for pH 5.0, 5.5 and 6.0; HEPES was used for pH 6.5, 7.0, 7.5 and 8.0; Tris was used for pH 8.5 and 9.0. The cleavage reactions were then conducted in a similar fashion as described in the section immediately above. Note that the CIM-EC for a given pH was prepared with a relevant 1× RB_{Ba}'.

3.12. Comparison of the Cleavage Activity of EC1T at Varying FS1/EC1T Ratios

Stocks of EC1T at 2.5, 5, 12.5, 25, 62.5, 125, and 250 µM were first prepared. Cleavage reactions were then conducted by mixing 19.5 µL of water, 22.5 µL of 2× RB_{Ba}, 1 µL of 2.5 µM FS1, 1 µL of a given EC1T stock, and 5 µL of the CIM-EC prepared with 1× RB_{Ba}. Each reaction mixture was incubated at RT for 60 min, followed by 10% dPAGE analysis as described above.

3.13. Specificity Test

Five Gram-negative bacteria (*P. peli*, *Y. rukeri*, *H. alvei*, *A. xylosoxidans* and *E. coli*) and four Gram-positive bacteria (*L. mesenteroides*, *L. planturum*, *P. acidilactici* and *B. subtilis*) were tested in this experiment. Each bacterium was cultured in SOB for a different period of time until the OD600 reached ~1. The CIM was then prepared with 1× RB_{Ba} and tested with EC1T/FS1 under the optimal reaction condition (50 mM HEPES, pH 7.5, 150 mM NaCl and 15 mM $BaCl_2$, room temperature, EC1T/FS1 = 50/1). Each reaction mixture was incubated at RT for 60 min, followed by 10% dPAGE analysis as described above.

3.14. Detection Sensitivity

First, a single colony of *E. coli* from an agar plate was taken, inoculated into 2 mL of SOB and grown for 14 h at 37 °C with shaking at 250 rpm. 10-fold serial dilution was then carried out as follows: 100 µL of the 14-h culture was mixed with 900 µL of fresh SOB. 100 µL of the diluted culture was again taken and mixed with 900 µL of fresh SOB. This process was repeated 7 times. 100 µL of the final dilution were plated onto a TSB agar plate (done in triplicate), which was incubated at 37°C for 15 h. Colonies in each plate were counted; the average number of colonies from the three plates was taken as the number of cells for this final dilution. This number was then used to calculate the number of cells for the other dilutions. 500 µL of each dilution was taken and centrifuged at 11,000 g for 5 min at RT. The cell pellet was re-suspended in 100 μL of 1× RB_{Ba} and used as the CEM-EC for this experiment (done in triplicate).

Cleavage reactions concerning EC1T/FS1 were set up and monitored as follows: 19.5 μL of water, 22.5 μL of 2× RB_{Ba}, 1 μL of 2.5 μM FS1, 1 μL of 125 μM EC1T were mixed in a quartz crystal cuvette, which was placed in a fluorimeter (Cary Eclipse Fluorescence Spectrophotometer; excitation wavelength = 488 nm and emission wavelength = 520 nm) set at RT. Fluorescence intensity was recorded every minute for 5 min; 5 μL of a relevant CIM-EC was then added into the cuvette and the solution was quickly mixed by pipetting the mixture up and down a few times. Following this step, the fluorescence intensity of the solution was recorded for 55 more minutes. All the reactions were conducted in 3 replicates and the average data are shown in Figure 6A. The final reaction mixture was also taken and analyzed by 10% dPAGE and data are shown in Figure 6B.

Cleavage reactions concerning RFD-EC1 were set up and monitored similarly: 20.5 μL of water, 22.5 μL of 2× RB_{Ba}, 1 μL of 2.5 μM RFD-EC1 was mixed in a cuvette. After reading fluorescence intensity for 5 min, 5 μL of a relevant CIM-EC was then added, followed by fluorescence intensity reading for 55 more minutes (Figure 6C). The final reaction mixture was also analyzed by 10% dPAGE (Figure 6D).

3.15. Single Cell Detection via Culturing

For isolating a single cell we followed our previously reported protocol [12]. Briefly, a glycerol stock containing 2 CFU/mL of *E. coli* was prepared. 100 μL of this stock was distributed to 10 culture tubes each with 2 mL of SOB. Since the concentration of the stock was 2 CFU/mL, only 2 out of the 10 tubes contained a single seeding cell (2 CFU/mL × 0.1 mL = 2). All the tubes were incubated at 37 °C with shaking at 250 rpm. At 2, 4, 6, 8 and 10 h, 200 μL of culture was harvested from each culture tube and CIMs were prepared (40 μL of 1× RB_{Ba} was used to dissolve the cell pellet). All the tubes were further incubated for 20 h to identify the two tubes containing *E. coli* cell (the culture in these tubes turned turbid while that in other 8 tubes stayed clear). Each CIM from *E. coli*-containing tubes was used to initiate the cleavage reaction by mixing 19.5 μL of water, 22.5 μL of 2× RB_{Ba}, 1 μL of 2.5 μM FS1, 1 μL of 125 μM EC1T, and 5 μL of a relevant CIM. The reaction and dPAGE analysis procedures were same as described above.

4. Conclusions

We recently described an RNA-cleaving fluorogenic DNAzyme, named RFD-EC1, which is active in the presence of the crude extracellular mixture (CEM) of the model Gram-negative bacterium *E. coli* [12–14]. RFD-EC1 was found to be highly active with CEM of *E. coli* but inactive with CEMs from a host of other Gram-negative and Gram-positive bacteria, and thus, RFD-EC1 can be used to develop a simple, "mix-and-read" fluorescence assay to achieve selective detection of *E. coli*. However, several parameters that are particularly relevant to the performance of this assay remained to be investigated. In this study we sought to establish a *trans*-acting DNA catalyst that cleaves an external substrate, optimize the reaction conditions that best support the catalytic activity of the DNAzyme, and determine the culturing conditions that enable the quickest detection of a single live bacterial cell.

The *trans*-acting DNAzyme was successfully established by segregating the substrate sequence domain from the sequence of the original DNA library. Also the two fixed sequence domains flanking the random-sequence domain could be removed without affecting the catalytic performance of the DNAzyme. The shortened, *trans*-acting DNAzyme, named EC1T, now contains 70 nucleotides.

Originally, the DNAzyme was isolated to cleave in the presence of the crude extracellular mixture (CEM) of *E. coli* and it has been determined that the target that activates the DNAzyme is a protein molecule based on the observation that the treatment of the CEM with proteases abolishes the DNAzyme activity [12]. Although the identity of this target is yet to be determined, we found that the target protein is much more abundant intracellularly and could be retrieved with a simple heating step (50 °C; 15 minutes). This led us to the use of the crude intracellular mixture (CIM) as the target of detection, translating into a better assay sensitivity.

Our results revealed that the nutritional factors in culture media played a vital role in growing the cells in faster rate (varying by as much as ~25-fold) with super Optimal Broth (SOB) which can substantially reduce the time required for single cell detection.

In order to establish an optimal reaction condition for EC1T, we examined the following reaction parameters: choice of divalent metal ions, reaction temperature and pH as well as the ratio between the substrate and the DNAzyme. Although EC1T was found to be active in the absence of any divalent ion, it exhibited much stronger activity in the presence of Ba^{2+}, Ca^{2+}, Mn^{2+} or Mg^{2+}. We chose Ba^{2+} as the divalent metal ion cofactor because this metal ion does not impose any fluorescence quenching effect. The DNAzyme was originally derived at room temperature (~23 °C) and a solution pH of 7.5 and therefore it was not surprising that EC1T exhibited the strongest activity at 23 °C and pH 7.5. We further found that when the concentration of FS1 was kept at 50 nM, 2.5 μM EC1T was required to reach the optimal cleavage activity. All the above optimization experiments led to the establishment of the optimal reaction condition for EC1T: 50 mM HEPES, 150 mM NaCl, 15 mM $BaCl_2$, pH 7.5, DNAzyme: substrate ratio = 50:1.

Under the above optimal reaction condition, the *trans*-acting system was able to detect 10^5 cells when the reaction was monitored in a fluorimeter. If the reaction mixture was analyzed by dPAGE (which separates the reaction product from the substrate), the system can detect 10^4 cells. When the original RFD-EC1 was used for the assay, the detection sensitivity was further improved: the fluorimeter method was able to detect 10^4 cells while the dPAGE method was able to detect as low as 10^3 cells. Importantly, the optimized assay did not compromise the specificity.

With a culturing step, the optimized assay is able to achieve the detection of *E. coli* from a single colony forming unit in 4–6 hours (dependent on the method of choice), which represents a significant deduction in time (12 h) required by the same probe under unoptimized conditions. Overall, we have significantly improved the performance of our DNAzyme probe and demonstrate the utility of such probes as simple biosensors to achieve sensitive and speedy detection of bacterial pathogens.

Acknowledgments

This work was supported by research grants from the Natural Sciences and Research Council of Canada (NSERC) and Sentinel Bioactive Paper Network. We would like to thank Gerard Wright, Brian Coombes and Russell Bishop for providing various bacterial cells.

References

1. World Health Organization (WHO). Fact sheet N^{0}237. Available online: https://apps.who.int/inf-fs/en/fact237.html (accessed on 16 August 2013).
2. Osterholm, M.T. Foodborne disease in 2011 — The rest of the story. *N. Engl. J. Med.* **2011**, *364*, 889–891.
3. Scallan, E.; Hoekstra, R.M.; Angulo, F.J.; Tauxe, R.V.; Widdowson, M.A.; Roy, S.L.; Griffin, P.M. Foodborne illness acquired in the United States-major pathogens. *Emerg. Infect. Dis.* **2011**, *17*, 7–15.
4. Newell, D.G.; Koopmans, M.; Verhoef, L.; Duizer, E.; Aidara-Kane, A.; Sprong, H.; Opsteegh, M.; Langelaar, M.; Threfall, J.; Scheutz, F.; *et al.* Food-borne diseases - the challenges of 20 years ago still persist while new ones continue to emerge. *Int. J. Food Microbiol.* **2010**, *139*, S3–S15.
5. Zourob, M.; Elwary, S.; Turner, A. *Principles of Bacterial Detection: Biosensors, Recognition Receptors and Microsystems*; Springer: New York, NY, USA, 2008.
6. Velusamy, V.; Arshak, K.; Korostynska, O.; Oliwa, K.; Adley, C. An overview of foodborne pathogen detection: In the perspective of biosensors. *Biotechnol. Adv.* **2010**, *28*, 232–254.
7. Lazcka, O.; Campo, F.J.D.; Muñoz, F.X. Pathogen detection: A perspective of traditional methods and biosensors. *Biosens. Bioelectron.* **2007**, *22*, 1205–1217.
8. Wright, A.C.; Danyluk, M.D.; Otwell, W.S. Pathogens in raw foods: What the salad bar can learn from the raw bar. *Curr. Opin. Biotechnol.* **2009**, *20*, 172–177.
9. Call, D.R. Challenges and opportunities for pathogen detections using DNA microarrays. *Crit. Rev. Microbiol.* **2005**, *31*, 91–99.
10. Yagi, K. Applications of whole-cell bacterial sensors in biotechnology and environmental science. *Appl. Microbiol. Biotechnol.* **2007**, *73*, 1251–1258.
11. Laberge, I.; Griffiths, M.W. Prevalence, detection and control of *Cryptosporidium parvum* in food. *Int. J. Food Microbiol.* **1996**, *32*, 1–26.
12. Ali, M.M.; Aguirre, S.D.; Lazim, H.; Li, Y. Fluorogenic DNAzyme probes as bacterial indicators. *Angew. Chem. Int. Ed.* **2011**, *50*, 3751–3754.
13. Aguirre, S.D.; Ali, M.M.; Li, Y. Detection of bacteria using fluorogenic DNAzymes. *J. Vis. Exp.* **2012**, e3961. doi: 10.3791/3961.
14. Li, Y. Advancements in using reporter DNAzymes for identifying pathogenic bacteria at speed and with convenience. *Future Microbiol.* **2011**, *6*, 973–976.
15. Mei, S.H.; Liu, Z.; Brennan, J.D.; Li, Y. An efficient RNA-cleaving DNA enzyme that synchronizes catalysis with fluorescence signaling. *J. Am. Chem. Soc.* **2003**, *125*, 412–420.
16. Liu, Z.; Mei, S.H.; Brennan, J.D.; Li, Y. Assemblage of signalling DNA enzymes with intriguing metal specificity and pH dependences. *J. Am. Chem. Soc.* **2003**, *125*, 7539–7545.
17. Shen, Y.; Brennan, J.D.; Li, Y. Characterizing the secondary structure and identifying functionally essential nucleotides of pH6DZ1, a fluorescence-signaling and RNA-cleaving deoxyribozyme. *Biochemistry* **2005**, *44*, 12066–12076.
18. Kandadai, S.A.; Li, Y. Characterization of a catalytically efficient acidic RNA-cleaving deoxyribozyme. *Nucleic Acids Res.* **2006**, *33*, 7164–7175.

19. Shen, Y.; Chiuman, W.; Brennan, J.D.; Li, Y. Catalysis and rational engineering of trans-acting pH6DZ1, an RNA-cleaving and fluorescence-signaling deoxyribozyme with a four-way junction structure. *ChemBioChem* **2006**, *7*, 1343–1348.
20. Ali, M.M.; Kandadai, S.A.; Li, Y. Characterization of pH3DZ1—An RNA-Cleaving deoxyribozyme with optimal activity at pH 3. *Can. J. Chem.* **2007**, *85*, 261–273.
21. Kandadai, S.A.; Mok, W.W.K.; Ali, M.M.; Li, Y. Characterization and optimization of an RNA-cleaving deoxyribozyme active at pH 5. *Biochemistry* **2009**, *48*, 7383–7391.
22. Tuerk, C.; Gold, L. Systematic evolution of ligands by exponential enrichment: RNA ligands to bacteriophage T4 DNA polymerase. *Science* **1990**, *249*, 505–510.
23. Ellington, A.D.; Szostak, J.W. In vitro selection of RNA molecules that bind specific ligands. *Nature* **1990**, *346*, 818–822.
24. Joyce, G.F. Forty years of in vitro evolution. *Angew. Chem. Int. Ed.* **2007**, *46*, 6420–6436.
25. Okumoto, Y.; Sugimoto, N. Effects of metal ions and catalytic loop sequences on the complex formation of a deoxyribozyme and its RNA substrate. *J. Inorg. Biochem.* **2000**, *82*, 189–195.
26. He, Q.C.; Zhou, J.M.; Zhou, D.M.; Nakamatsu, Y.; Baba, T.; Taira, K. Comparison of metal-ion-dependent cleavages of RNA by a DNA enzyme and a hammerheadribozyme. *Biomacromolecules* **2002**, *3*, 69–83.
27. Brown, A.K.; Li, J.; Pavot, C.M.; Lu, Y. A lead-dependent DNAzyme with a two-step mechanism. *Biochemistry* **2003**, *42*, 7152–7161.
28. Lan, T.; Lu, Y. Metal ion-dependent DNAzymes and their applications as biosensors. *Met. Ions Life Sci.* **2012**, *10*, 217–248.
29. Santoro, S.W.; Joyce, G.F. A general purpose RNA-cleaving DNA enzyme. *Proc. Natl. Acad. Sci. USA* **1997**, *94*, 4262–4266.
30. Li, J.; Zheng, W.; Kwon, A.H.; Lu, Y. In vitro selection and characterization of a highly efficient Zn(II)-dependent RNA-cleaving deoxyribozyme. *Nucleic Acids Res.* **2000**, *28*, 481–488.
31. Kim, H.K.; Liu, J.; Li, J.; Nagraj, N.; Li, M.; Pavot, C.M.; Lu, Y. Metal-dependent global folding and activity of the 8-17 DNAzyme studied byfluorescence resonance energy transfer. *J. Am. Chem. Soc.* **2007**, *129*, 6896–6902.
32. Rupcich, N.; Chiuman, W.; Nutiu, R.; Mei, S.; Flora, K.K.; Li, Y.; Brennan, J.D. Quenching of fluorophore-labeled DNA oligonucleotides by divalent metal ions: Implications for selection, design and applications of signaling aptamers and signaling deoxyribozymes. *J. Am. Chem. Soc.* **2006**, *128*, 780–790.

6

Microbial Enzymes: Tools for Biotechnological Processes

Jose L. Adrio [1] and Arnold L. Demain [2,*]

[1] Neol Biosolutions SA, BIC Granada, Granada 18016, Spain; E-Mail: jladrio@neolbio.com (J.L.A.)
[2] Research Institute for Scientists Emeriti (R.I.S.E.), Drew University, Madison, NJ 07940, USA

* Author to whom correspondence should be addressed; E-Mail: ademain@drew.edu

Abstract: Microbial enzymes are of great importance in the development of industrial bioprocesses. Current applications are focused on many different markets including pulp and paper, leather, detergents and textiles, pharmaceuticals, chemical, food and beverages, biofuels, animal feed and personal care, among others. Today there is a need for new, improved or/and more versatile enzymes in order to develop more novel, sustainable and economically competitive production processes. Microbial diversity and modern molecular techniques, such as metagenomics and genomics, are being used to discover new microbial enzymes whose catalytic properties can be improved/modified by different strategies based on rational, semi-rational and random directed evolution. Most industrial enzymes are recombinant forms produced in bacteria and fungi.

Keywords: biocatalysis; industrial enzymes; microbial hosts; protein engineering

1. Introduction

Microbial enzymes are known to play a crucial role as metabolic catalysts, leading to their use in various industries and applications. The end use market for industrial enzymes is extremely wide-spread with numerous industrial commercial applications [1]. Over 500 industrial products are being made using enzymes [2,3]. The demand for industrial enzymes is on a continuous rise driven by a growing need for sustainable solutions. Microbes have served and continue to serve as one of the largest and useful sources of many enzymes [4]. Many industrial processes, including chemical synthesis for production of chemicals and pharmaceuticals, have several disadvantages: low catalytic efficiency, lack of enantiomeric specificity for chiral synthesis, need for high temperature, low pH and high pressure. Also, the use of organic solvents leads to organic waste and pollutants. Enzymes are more useful for these applications as they work under mild reaction conditions (e.g., temperature, pH, atmospheric conditions), do not need protection of substrate functional groups, have a long half-life, a high stereo-selectivity yielding stereo- and regio-chemically-defined reaction products at an acceleration of 10^5 to 10^8-fold, and, in addition, they work on unnatural substrates [5]. Furthermore, enzymes can be selected genetically and chemically-modified to enhance their key properties: stability, substrate specificity and specific activity. There are drawbacks however, to the use of enzymes, e.g., certain enzymes require co-factors. However, various approaches such as cofactor recycling and use of whole cells can solve this problem. About 150 industrial processes use enzymes or whole microbial cell catalysts.

The global industrial enzymes market is very competitive with Novozymes being the largest player in the industry, followed by DSM, and DuPont (after it acquired a majority stake in Danisco and its Genencor division), among others. The companies mainly compete on the basis of product quality, performance, use of intellectual property rights, and the ability to innovate, among other such factors. North America and Europe are the largest consumers of industrial enzymes although the Asia Pacific region will undergo a rapid increase in enzyme demand in China, Japan and India, reflecting the size and strength of these country's economies.

2. Discovering Enzymes

Nature provides a vast amount of microbial enzyme resources. Our ability to tap into such immense biodiversity depends on the tools available to expand the search for new enzymes by (i) metagenome screening [6–10]); (ii) genome mining in more than 2,000 sequenced microbial genomes [11–14]; and (iii) exploring the diversity of extremophiles [15,16]).

2.1. Metagenomic Screening

Although numerous microbes inhabit the biosphere, less than 1% can be cultivated through standard laboratory techniques. Metagenomics has appeared as an alternative strategy to conventional microbe screening by preparing a genomic library from environmental DNA and systematically screening such a library for the open reading frames potentially encoding putative novel enzymes [10,17].

Metagenomic screening is mostly based on either function or sequence approaches [8,10]. Function-based screening is a straightforward way to isolate genes that show the desired function by direct phenotypical detection, heterologous complementation, and induced gene expression [18]. On the other hand, sequence-based screening is performed using either the polymerase chain reaction (PCR) or hybridization procedures. Usually, the common procedure is to use a set of degenerated primers that have been designed based on consensus amino acid sequences.

Studies from different habitats such as volcanic vents, arctic tundra, cow rumen [19], marine environments [20], and termite guts [21] have yielded microbial enzymes with potential for biocatalytic applications, such as lipase [22,23], oxidoreductase [24], amidase [25], amylase [26], nitrilase [27], beta-glucosidase [28,29]), decarboxylase [30], and epoxide hydrolase [31].

Function-based screening of metagenomic libraries is somehow problematic mainly due to insufficient, biased expression of foreign genes in *Escherichia coli* as this bacterium is usually used as surogate host [8]. To overcome such hurdles, alternative bacterial host and expression systems are currently being examined including *Streptomyces lividans*, *Pseudomonas putida* and *Rhizobium leguminosarum*, among others [32,33].

Hit rate (probability of identifying a certain gene) also depends on other factors such as size of the target gene and the assay method. Enzyme activities are usually assayed on agar plates supplemented with different substrates Sensitivity of agar plate-based screening can be improved by using cell lysates [34], screening for genes giving resistance to toxic compounds [35,36], or linking the target activity to the expression of a reporter gene such as green fluorescent protein (GFP) [37] or β-galactosidase [38]. Also, development of flow cytometry based screens as SIGEX are leading the way as they enable more rapid screening of large metagenomic libraries [39]. Genes obtained through sequence-based screening are limited to those having homology to the sequences used as probes.

2.2. Microbial Genomes

Recent success of genome sequencing programs has resulted in an explosion of information available from sequence databases, thus creating an opportunity to explore the possibility of finding new natural products (including enzymes) by database mining [40]. The next generation sequencing platforms (454 from Roche, Solexa from Illumina or SoLiD from ABI), hold promise to reduce time and cost of genome sequencing. Using these platforms, as well as cutting edge approaches such as resequencing, helps to complete multiple whole genomes in less than two weeks. So far, more than 2,000 genome sequences and draft assemblies are available in the NCBI database [41].

Two approaches are being followed to discover new enzymes [42]. On one hand, genome hunting is based on searching for open reading frames in the genome of a certain microorganism. Sequences that are annotated as putative enzymes are subjected to subsequent cloning, over-expression and activity screening. Another approach, called data mining, is based on homology alignment among all sequences deposited in databases. Using different bioinformatics tools (e.g., BLAST), a search for conserved regions between sequences yields homologous protein sequences that are identified as possible candidates for further characterization.

2.3. Extremophiles

Over 30 articles on extremophiles have been published in a special issue [43]. Due to their capability to survive under environments of extreme conditions, both physical as temperature (−2 to 12 °C, 60–110 °C), pressure or radiation, and geochemical such as salinity (2–5 NaCl) and pH (<2, >9), extremophiles are a very interesting source of enzymes with extreme stability under conditions regarded as incompatible with biological materials. Recent studies show that the diversity of organisms in these extreme environments could be even greater than was initially thought [16,44].

However, because the majority of these microorganisms have not yet been isolated in pure culture, useful characterization of their enzymes still remains quite difficult.

Thermophilic proteases, lipases as well as cellulases and amylases are being used in many different industrial applications [16,45]. Extreme thermophiles (growing at 60–80 °C) are widely distributed among several bacterial genera such as *Clostridium*, *Thermus*, *Thermotoga*, and *Bacillus*, whereas most of hyperthermophiles belong to Archaea such as *Pyrococcus*, *Thermococcus* or *Methanopyrus*, among others. The Taq DNA polymerase, isolated from the thermophilic *Thermus aquaticus*, had 2009 sales of $500 million [46].

Also, enzymes from psychrophiles have become quite interesting for many industrial applications, partly because of the ongoing efforts to reduce energy consumption. Therefore, enzymes like proteases, amylases or lipases have great commercial potential for development of detergents to reduce wear and tear of textile fibers. Polymer-degrading activities (e.g., cellulases, xylanases,) are quite interesting for the pulp and paper industry, as well as for saccharification of pre-treated lignocellulosic biomass for the production of second generation biofuels [47]. These cold-active enzymes also have potential for many other applications [16,48], such as extraction and clarification of fruit juices, improvement of bakery products, polishing and stone-washing of textiles, and bioremediation of waters contaminated with oils or hydrocarbons [49].

Enzymes from halophites have to cope with very high concentrations of salts (e.g., sodium or potassium chloride). Such proteins have adapted to these environments by acquiring a large number of negatively-charged amino acid residues of their surfaces to prevent precipitation. Such a property has been taken advantage of by use of non-aqueous media [50]. Xylanases, amylases, proteases and lipases from *Halobacterium*, *Halobacillus* and *Halothermothrix* have been produced and their potential has been reviewed [51,52]).

Microorganisms that can survive under extreme pH values could be good sources of thermoalkaliphilic enzymes, like proteases and lipases, particularly useful for applications as additives in laundry and dishwashing detergents [16,53].

3. Strategies to Improve Properties of Microbial Enzymes

The continuously expanding application of enzymes is creating a growing demand for biocatalysts that exhibit improved or new properties [46]. Although enzymes have favorable turnover numbers, they do not necessary fulfill all process requirements and need further fine tuning to achieve industrial scale production. Among those hurdles are: substrate/product inhibition, stability, narrow substrate specificity or enantioselectivity [54]. Genetic modification is very important and recombinant DNA techniques have increased production by 100-fold [55]. Development of new/improved biocatalysts is a challenging and complex task (Figure 1). There are two major ways in which enzymes can be modified to adapt their functions to applied ends: (i) rational redesign of existing biocatalysts and (ii) combinatorial methods which search for the desired functionality in libraries generated at random.

Figure 1. Discovery and development of biocatalysts.

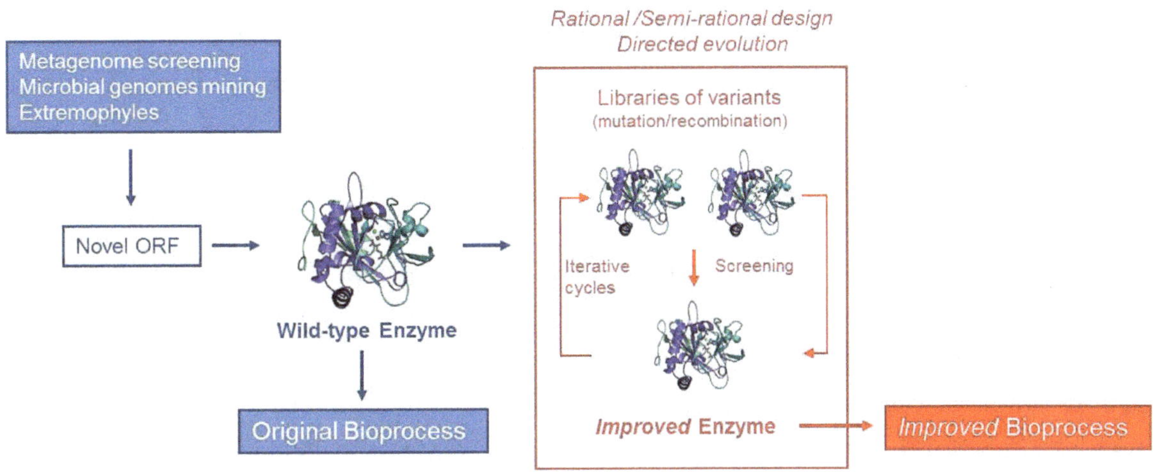

3.1. Rational Design

This approach includes site-directed mutagenesis to target amino acid substitutions, thus requiring knowledge of detailed information about the 3-dimensional structure and chemical mechanism of the enzymatic reaction, some of which may not be available. However, the increasing growth of databases containing protein structures and sequences is helping to overcome this lack of information. Comparison of the sequence of a new biocatalyst identified in a screening program with the thousands deposited in the databases can identify related proteins whose functions or/and structures are already known. Because new enzymes have evolved in nature by relatively minor modification of active-site structures, the goals of homology-driven experiments include engineering binding sites to fit different substrates as well as construction of new catalytic residues to modify functions and mechanisms [56]. A small number of variants are produced which are then screened. Although in many cases, results are poor compared to natural enzymes, there have been successes [57,58].

Computational protein design starts with the coordinates of a protein main chain and uses a force field to identify sequences and geometries of amino acids that are optimal for stabilizing the backbone geometry [59]. Because of the amazing number of possible sequences generated, the combination of predictive force fields and search algorithms is now being applied to functional protein design [60].

3.2. Directed Evolution

Combinatorial methods such as directed evolution create a large number of variants for screening for enantioselectivity, catalytic efficiency, catalytic rate, solubility, specificity and enzyme stability, but do not require extensive knowledge about the enzyme. Directed evolution is a fast and inexpensive way of finding variants of existing enzymes that work better than naturally occurring enzymes under specific conditions [3,61–63]. Directed evolution includes an entire range of molecular biological techniques that allow the achievement of genetic diversity mimicking mechanisms of evolution occurring in nature. It involves random mutagenesis of the protein-encoding gene by different techniques including the error-prone polymerase chain reaction (PCR) [64], repeated oligonucleotide directed mutagenesis [65], or chemical agents [66], among others. Error prone PCR accomplishes introduction of random point mutations in a population of enzymes. Such molecular breeding techniques (DNA shuffling, Molecular Breeding™) allow *in vitro* random homologous recombination, typically between parent genes with homology higher than 70% [67]. After cloning and expression, a large collection of enzyme variants (10^4–10^6) is typically generated and is subjected to screening or selection.

All the approaches mentioned above are not mutually exclusive as the fields of rational, semi-rational and random redesign of enzymes are moving closer. Thus, directed evolution techniques make use, where possible, of smaller enzyme variant libraries designed by rational or semi-rational methods to reduce the screening effort but without compromising the likelihood of finding better variants [61,63]. One option is to target the active site residues (about 10–15 amino acids) and those closest to it (another 20–30 amino acids) as mutations closer to these regions seem to be more beneficial [68]. Another strategy, called CASTing, is based on a combinatorial active site testing [69], in which libraries are generated from groups of two or three residues made from the active site residues. Hits obtained from these initial libraries are combined and new libraries are generated and screened in an iterative way. By changing the nucleotide components during PCR, the amino acid alphabet can be reduced yielding new proteins composed by only 12 amino acids [70]. Screening of such libraries can lead to remarkable improvements in activity by using saturation mutagenesis at homologous enzyme positions [71].

As the fitness of a gene will increase more rapidly in a breeding population with high genetic variability that is under the influence of selection, DNA Family Shuffling, an improvement in the breeding technique based on recombination of several homologous sequences, yielded an improvement in activity of about 30 to 80 times higher than when each gene was shuffled independently [72]. Further techniques can create shuffle exons or domains [73], loop regions [74], random truncations [75] or insertions and deletions of codons [76].

Random redesign techniques are being currently used to generate enzymes with improved properties, such as: activity and stability at different pH values and temperatures [77], increased or modified enantioselectivity [78], altered substrate specificity [79], stability in organic solvents [80], novel substrate specificity and activity [81] and increased biological activity of protein pharmaceuticals and biological molecules [82,83]. Directed evolution increased the activity of glyphosate-N-acetyltransferase by 10,000-fold and at the same time, its thermostability increased 5-fold [84]. Proteins from directed evolution work were already on the market in 2000 [85,86]. These were GFP of Clontech and Novo Nordisk's LipoPrime® lipase.

4. Production of Recombinant Proteins in Microbial Hosts

In 2005, there were 4,200 biotechnology companies worldwide. Total revenue amounted to $63 billion. Recombinant DNA technology has been remarkably advanced by their development of efficient and scale-up expression systems to produce enzymes from industrially-unknown microorganisms and other living organisms using industrial organisms such as *E. coli*, *Bacillus subtilis* and other species of *Bacillus*, *Ralstonia eutropha*, *Pseudomonas fluorescens*, *Saccharomyce cerevisiae*, *Pichia pastoris*, *Hansenula polymorpha*, and species of *Aspergillus and Trichoderma* [87]. Approximately 90% of industrial enzymes are recombinant versions.

E. coli has been extensively used as a recombinant host for many reasons including (i) ease of quickly and precisely modifying the genome; (ii) rapid growth to high cell densities; (iii) ease of culture in cheap media; and (iv) ease of reduction of protease activity. *E. coli* can accumulate heterologous proteins up to 50% of its dry cell weight [88]. However, it has several drawbacks, such as: inability for post-translational modifications, presence of toxic cell wall pyrogens or, sometimes, formation of inclusion bodies containing inactive, aggregated and insoluble heterologous protein. Despite this, high level production and excretion has been obtained with the following heterologous proteins: alkaline phosphatase (PhoA) at 5.2 g/L into the periplasm, and levan fructotransferase LFT at 4 g/L into the medium. In 1993, recombinant processes in *E. coli* were responsible for almost $5 billion worth of products (87). *R. eutropha* produced organophosphohydrolase at 10 g/L [89]. The *P. fluorescens* culture of the Pfenex company is capable of producing 20 g/L of protein [90].

S. cerevisiae offers certain advantages over bacteria as a cloning host [91]. This yeast can be grown rapidly in simple media and to a high cell density, can secrete heterologous proteins into the extracellular broth, and its genetics are more advanced than any other eukaryote. Glucose oxidase from *Aspergillus niger* has been produced at 9 g/L by this yeast (87). Despite these advantages, *S. cerevisiae* is sometimes regarded as a less than optimal host for large-scale production of mammalian proteins because of drawbacks such as hyperglycosylation, presence of α-1, 3-linked mannose residues that could cause an antigenic response in patients, and absence of strong and tightly-regulated promoters.

P. pastoris has become one of the most extensively used expression systems [87,92]. Over 700 proteins have been produced by this yeast [93]. Production of recombinant proteins reached 22 g/L for intracellular proteins [94] and 14.8 g/L for secreted proteins [95]. Claims have been made that *P. pastoris* can produce up to 30 g/L of recombinant proteins [96]. Among the advantages of this methylotrophic yeast over *S. cerevisiae* are (i) an efficient and tightly-regulated methanol promoter (AOX1) which yields alcohol oxidase at 30% of soluble protein; (ii) less extensive glycosylation, due to shorter chain lengths of N-linked high-mannose oligosaccharides, usually up to 20 residues lacking the terminal α-1,3-mannose linkages [97]; (iii) integration of multicopies of foreign DNA into chromosomal DNA yielding stable transformants; (iv) ability to secrete high levels of foreign proteins; and (v) high-density growth and straight-forward scale-up [98].

Heterologous gene expression in the methylotropic yeast *H. polymorpha* is similar to that of *P. pastoris*. The promoter of the methanol oxidase gene is used to express foreign genes. As with AOX1 in *P. pastoris*, the MOX gene in *H. polymorpha* is highly expressed and tightly-regulated, giving enzyme levels up to 37% of total cell protein [99]. One major difference is that expression of the MOX gene is significantly derepressed in the absence of glucose or during glucose limitation [100] and therefore tight regulation of the MOX promoter is lost in the high glucose conditions usually used for high-biomass fermentations. *H. polymorpha* can produce 1.4 g/L of secreted glucoamylase and 13.5 g/L of phytase.

The development of molecular techniques for production of recombinant heterologous proteins in filamentous fungi has been laborious and contrasts markedly with the success achieved in yeasts. The ability to introduce or delete genes remains difficult, although some advances in transformation have been reported, e.g., restriction enzyme-mediated integration [101] and *Agrobacterium tumefaciens*-Ti plasmid-mediated transformation [102]. Levels of production of non-fungal proteins have been low compared to those of homologous proteins. Different strategies have been developed to overcome these problems, including the construction of protease-deficient strains [103], introduction of a large number of gene copies [104], use of strong fungal promoters, efficient secretion signals [104,105], and gene fusions with a gene which encodes part of or an entire well-secreted protein [105]. Little research has been carried out on glycosylation in molds although hyper-glycosylation does not seem to occur and low-mannose side chains are formed. *Aspergillus awamori* produces 4.6 g/L of glucoamylase from *A. niger*. *Aspergillus oryzae* yields 3.3 g/L of *Mucor* rennin. *Acremonium chrysogenum* produces 4 g/L of *Fusarium* alkaline protease. *Trichoderma reesei* produces 35 g/L of recombinantproteins.A recent entry into the field is *Chrysosporium lucknowense*, which produces high levels of proteins (50–80 g/L), and from which low-viscosity and low-protease mutants have been obtained [106,107].

5. Enzymatic Biocatalysis

Biocatalysts have been widely applied in the brewing and food industries for a long time but ar finding new applications in many fields including industrial chemistry [108]. Biocatalysis involves the application of whole microbial cells, cell extracts, purified enzymes, immobilized cells, or immobilized enzymes as catalysts for any of the above mentioned processes [109,110]. As mentioned in the previous sections, its rapid development was due to the recent advances in large-scale genome sequencing, directed evolution, protein expression, metabolic engineering, high throughput screening, and structural biology [111,112].

5.1. Technical Applications

Enzymes are very important for processes of industrial, pharmaceutical and biotechnological significance [113]. The total market for industrial enzymes reached $3.3 billion in 2010 and it is estimated to reach a value of 4.4 billion by 2015 [114]. Of these, technical enzymes are typically used as bulk enzymes in the detergent, textile, pulp and paper industries, and in the biofuels industry, among others. Usage for leather and bioethanol is responsible for the highest sales figures. Technical enzymes had revenues of nearly $1.2 billion in 2011 which is expected to reach $1.5 billion in 2015 and $1.7 billion in 2016. The highest sales are expected to be in the biofuels (bioethanol) market [115]. Use of enzymes for foods and beverages is expected to reach $1.3 billion by 2015.

The use of enzymes as detergent additives represents a major application of industrial enzymes. Proteases, lipases, amylases, oxidases, peroxidases and cellulases are added to detergents where they catalyze the breakdown of chemical bonds on the addition of water. To be suitable, they must be active under thermophilic (60 °C) and alkalophilic (pH 9–11) conditions, as well as in the presence of the various components of washing powders.

Proteases constitute over 60% of the global market for enzymes. They are used to produce pharmaceuticals, foods, detergents, leather, silk and agrochemical products. In laundry detergents, they account for approximately 25% of the total worldwide sales of enzymes. The first detergent containing a bacterial protease ("Biotex") was introduced by Novo Industry A/S (now Novozymes) in 1956. It contained an alcalase produced by *Bacillus licheniformis*. In 1994, Novo Nordisk introduced Lipolase^{TM}, the first commercial recombinant lipase for use in a detergent, by cloning the *Humicola lanuginose* lipase into the *A. oryzae* genome. In 1995, Genencor International introduced two bacterial lipases, one from *Pseudomonas mendocina* (Lumafast^{TM}), and another from

Pseudomonas alcaligenes (Lipomax™). An enzyme added recently to detergents is Mannaway™, a *Bacillus* mannanase which removes food stains containing guar gum [116].

In the textile industry, enzymes are also increasingly being used to develop cleaner processes and reduce the use of raw materials and production of waste. The application of cellulases for denim finishing and laccases for decolorization of textile effluents and textile bleaching are the most recent commercial advances [117]. An alternative enzymatic process in the manufacturing of cotton has been recently developed based on a pectate lyase [118]. The process is performed at much lower temperatures and uses less water than the classical method.

Another application of enzymes with increasing importance is the use of lipases, xylanases and laccases in removing pitch (hydrophobic components of wood, mainly triglycerides and waxes) in the pulp industry [119]. A lipase from *Candida rugosa* is used by Nippon Paper Industries to remove up to 90% of these compounds [120]. The use of enzymes as alternatives to chemicals in leather processing has proved successful in improving leather quality and in reducing environmental pollution. Alkaline lipases from *Bacillus* strains, which grow under highly alkaline conditions, in combination with other alkaline or neutral proteases, are currently being used in this industry. Laccases oxidize phenolic and non-phenolic lignin-related compounds as well as environmental pollutants [121]. They are used to detoxify industrial effluents from the paper and pulp, textile, and petrochemical industries, as a medical diagnostic tool, for bioremediation of herbicides, pesticides, and explosives in soil, as a cleaning agent for water purification systems, as a catalyst in drug manufacture and as cosmetic ingredients.

Although cellulases have been widely used in textile applications for many years, these enzymes are gaining additional consideration in the enzyme market owing to their ability in the degradation of lignocellulosic feedstocks. The cost of cellulases is a key issue in achieving a low price conversion of lignocellulosic biomass into biofuels and other products [122–124]. Filamentous fungi can produce native cellulases at levels greater than 100 g/L [87]. Enzymes needed to hydrolyze cellulose include (1) endoglucanases, which break down cellulose chains in a random manner; (2) cellobiohydrolases, which liberate glucose dimers from both ends of cellulose chains; and (3) beta-glucosidases, which produce glucose from oligomer chains. *Hypocrea jecorina (Trichoderma reesei)* is the main industrial source of cellulases and hemicellulases used to depolymerize plant biomass to simple sugars [125–127]. The overall action of *T. reesei* on cellulosic biomass is limited by its low content of beta-glucosidase. The result is accumulation of cellobiose which limits further breakdown. The expression of the beta-glucosidase gene from *Pericona* sp in *T. reesei* resulted in an increased level of beta-glucosidase, thus increasing overall cellulase activity and action on biomass residues [128]. Cellulases are formed adaptively, and several positive (XYR1, ACE2, HAP2/3/5) and negative (ACE1, CRE1) components involved in this regulation are now known [127]. In addition, its complete genome sequence has been published [129], thus making the organism susceptible to targeted improvement by metabolic engineering. It has recently been reported that the extreme thermophilic bacterium *Caldicelluloseruptor bescil* produces a cellulase/hemicellulase system twice as active as that from *T. reesei* [130].

5.2. Enzymes in the Feed Industry

The global market for feed enzymes is a promising segment in the enzyme industry and is expected to reach about $730 million in 2015 [131]. Feed enzymes can increase the digestibility of nutrients, leading to greater efficiency in feed utilization. Also, they can degrade unacceptable components in feed, which are otherwise harmful or of little or no value [132]. Currently, feed enzymes commercially available are phytases, proteases, α-galactosidases, glucanases, xylanases, α-amylases, and polygalacturonases, mainly used for swine and poultry [133]. Developments of heat-stable enzymes, improved specific activity, some new non-starch polysaccharide-degrading enzymes, and rapid, economical and reliable assays for measuring enzyme activity have always been the focus and have been intensified recently. The use of enzymes as feed additives is restricted in many countries by local regulatory authorities [134] and applications may therefore vary from country to country.

5.3. Enzymes in Food Processing

Food and beverage enzymes constitute the largest segment of industrial enzymes with revenues of nearly $1.2 billion in 2011 which is expected to grow to $1.8 billion by 2016, at a composed annual growth rate of 10.4% [115].

Lipases are commonly used in the production of a variety of products, ranging from fruit juices, baked foods, and vegetable fermentations to dairy enrichment. Fats, oils and related compounds are the main targets of lipases in food technology. Accurate control of lipase concentration, pH, temperature and emulsion content is required to maximize the production of flavor and fragrance. The lipase mediation of carbohydrate esters of fatty acids offers a potential market for use as emulsifiers in foods, pharmaceuticals and cosmetics. There are three recombinant fungal lipases currently used in the food industry, one from *Rhizomucor miehi*, one from *Thermomyces lanuginosus* and another from *Fusarium oxysporum*; all being produced in *A. oryzae* [135,136].

The major application of proteases in the dairy industry is for the manufacture of cheese. Calf rennin had been preferred in cheese making due to its high specificity, but microbial proteases produced by GRAS microorganisms like *Mucor miehei*, *Bacillus subtilis*, *Mucor pusillus Lindt* and *Endothia parasitica* are gradually replacing it. The primary function of these enzymes in cheese making is to hydrolyze the specific peptide bond (Phe105-Met106) that generates para-k-casein and macropeptides [137]. Production of calf rennin (chymosin) in recombinant *A. niger var awamori* amounted to about 1 g/L after nitrosoguanidine mutagenesis and selection for 2-deoxyglucose resistance [138]. Further improvement was done by parasexual recombination resulting in a strain producing 1.5 g/L from parents producing 1.2 g/L [139]. Four recombinant proteases have been approved by FDA for cheese production [140,141].

Bacterial glucose isomerase, fungal α-amylase and glucoamylase are currently used to produce "high fructose corn syrup" from starch in a $1 billion business. Other enzymes useful in the food industry include invertase from *Kluyveromyces fragilis*, *Saccharomyces carlsbergensis* and *S. cerevisiae* for candy and jam manufacture, β-galactosidase (lactase) from *Kluyveromyces lactis*, *K. fragilis* or *Candida pseudotropicalis* for hydrolysis of lactose from milk or whey, and galactosidase from *S. carlsbergensis* for crystallization of beet sugar. "Fructose syrup", produced by xxylose isomerase on glucose is at a production level of 15 million tons per year [142].

In terms of government regulation, enzymes used in food can be divided into food additives and processing aids. Most food enzymes are considered as processing aids, with only a few used as additives, such as lysozyme and invertase [143]. The processing aids are used during the manufacturing process of foodstuffs, and do not have a technological function in the final food. All these materials are expected to be safe, under the guidance of good manufacturing practice (GMP). The key issue in evaluating safety of enzyme preparations is the safety assessment of the production strain. Only about nine recombinant microorganisms are considered "Generally Recognized As Safe" (GRAS) based on FDA regulations. These are from a relatively small number of bacterial and fungal species primarily *A. oryzae*, *A. niger*, *B. subtilis* and *B. licheniformis*. Olempska-Beer [144] has reviewed the microbial strains engineered for food enzyme production from a security point of view.

5.4. Enzymes in Chemical and Pharmaceutical Processes

Successful application of enzymatic processes in the chemical industry depends mainly on cost competitiveness with the existing and well-established chemical methods [145]. Lower energy demand, increased product titer, increased catalyst efficiency, less catalyst waste and byproducts, as well as lower volumes of wastewater streams, are the main advantages that biotechnological processes have as compared to well-established chemical processes. There are estimated to be only around 150 biocatalytic processes currently applied in industry [146]. However, new scientific developments in genomics, as well as in protein engineering, facilitate the tailoring of enzyme properties to increase that number significantly [147,148].

An enzymatic conversion was devised to produce the amino acid L-tyrosine [149]. Phenol, pyruvate, pyridoxal phosphate and ammonium chloride are converted to L-tyrosine using a thermostable and chemostable tyrosine phenol lyase obtained from *Symbiobacterium toebii*. The titer produced was 130 g/L after 30 h with continuous feeding of substrate.

Industrial scale biocatalysis is focused primarily on hydrolases, a few ketoreductases (KREDs), cofactor regeneration and protein stability in organic solvents. The numerous biocatalytic routes scaled up for pharmaceutical manufacturing have been recently reviewed by Bornscheuer *et al.* [112], showing the competitiveness of enzymes versus traditional chemical processes.

Enzymes are useful for preparing beta-lactam antibiotics such as semi-synthetic penicillins and cephalosporins [150]. Beta-lactams constitute 60%–65% of the total antibiotic market.

Preparation of chiral medicines, *i.e.*, the synthesis of complex chiral pharmaceutical intermediates efficiently and economically, is one of the most important applications in biocatalysis. Esterases, lipases, proteases and KREDs (ketoreductases) are widely applied in the preparation of chiral alcohols, carboxylic acids, amines or epoxides, among others [116,151,152].

The inherent inefficiency of kinetic resolution (maximum 50% yield) can be overcome by novel asymmetric reactions catalyzed by improved microbial enzymes which can provide a 100% yield [153]. Asymmetric reduction of tetrahydrothiophene-3-one with a wild-type reductase gave the desired alcohol ((R)-tetrahydrothiophene-3-ol), a key component in sulopenem, a potent antibacterial developed by Pfizer, but only in 80%–90% ee (enantiomeric excess). A combination of random mutagenesis, gene shuffling and ProSAR analysis was used to improve the enantio-selectivity of a ketoreductase towards tetrahydrothiophene-3-one. The best variant increased enantio-selectivity from 63% ee to 99% ee [154].

Atorvastatin, the active ingredient of Lipitor, a cholesterol-lowering drug that had global sales of US$12 billion in 2010, can be produced enzymatically. The process is based on three enzymatic activities: a ketone reductase, a glucose dehydrogenase and a halohydryn dehalogenase. Several iterative rounds of DNA shuffling for these three enzymes, with screening in the presence of iteratively higher concentrations of product, led to a 14-fold reduction in reaction time, a 7-fold increase in substrate loading, a 25-fold reduction in enzyme use, and a 50% improvement in isolated yield [155].

Kinetic resolution of racemic amines is a common method used in the synthesis of chiral amines. Acylation of a primary amine moiety by a lipase is used by BASF for the resolution of chiral primary amines in a multi-thousand ton scale [156]. Recently, asymmetric synthesis from the corresponding chiral ketones, using transaminases, is gaining attention [152]. Some (R)-selective transaminases have been recently discovered using *in silico* strategies for a sequence-based prediction of substrate specificity and enantio-preference [157]. Optically pure (S)-amines were obtained using a recombinant ω-transaminase with 99% ee and 97% yield [158]. These enantiopure amines may find use as inhibitors of monoamine oxidase in the treatment of diverse neurological disorders such as Parkinson's and Alzheimer's diseases.

An effective enzymatic process using enzyme evolution was developed by the biotechnology company Codexis, in cooperation with Pfizer, to produce 2-methyl pentanol, an important intermediate for manufacture of pharmaceuticals and liquid crystals [159]. Quite recently, protein engineering expanded the substrate range of transaminases to ketones. In an impressive piece of work developed by Merck and Codexis, the chemical manufacture of sitagliptin, the active ingredient in Januvia which is a leading drug for type 2 diabetes, was replaced by a new biocatalytic process. Several rounds of directed evolution were applied to create an engineered amine transaminase with a 40,000-fold increase in activity [160]. Such a process not only reduced total waste (by 19%), but also increased overall yield (by 13%) and productivity (by 53%). Codexis scientists also developed enzymatic processes for the production of montelukast (Singulair) and silopenem [161]. They also developed an improved LovD enzyme (an acyltransferase) for improved conversion of the cholesterol-lowering agent, lovastatin, to simvastatin [162].

Optically active carboxylic acids have been usually synthesized through different enzymatic routes catalyzed by lipases, nitrilases or hydroxynitrile lyases. Recent advances have improved the efficiency of such procedures. The synthesis of 2-arylpropanoic acids (e.g., ketoprofen, ibuprofen and naproxen) is mainly achieved through the kinetic resolution of racemic substrates by lipases from *Candida antarctica* or *Pseudomonas* sp. A process using a novel substrate, (R,S)-N-profenylazoles, instead of their correspondent esters, proved to be more efficient.

(R)-*o*-chloromandelic acid is a key intermediate for manufacture of Clopidogrel, a platelet aggregation inhibitor, with global sales of $10 billion per year. The asymmetric reduction of methyl o-chlorobenzoylformate with a versatile recombinant carbonyl reductase from *S. cerevisiae* expressed in *E. coli* yielded (R)-*o*-chloromandelate at a concentration of 178 g/L and 99% ee [163].

Application of enzymes in several industrial bioconversions has been broadened by the use of organic solvents replacing water [45,50], an important development in enzyme engineering. Many chemicals and polymers are insoluble in water and its presence leads to undesirable by-products and degradation of common organic reagents. Although switching from water to an organic solvent as the reaction medium might suggest that the enzyme would be denatured, many crystalline or lyophilized enzymes are actually stable and retain their activities in such anhydrous environments. Yeast lipases have been used to catalyze butanolysis in anhydrous solvents to obtain enantiopure 2-chloro- and 2-bromo-propionic acids that are used for the synthesis of herbicides and pharmaceuticals [151]. Another lipase is used in a stereoselective step, carried out in acetonitrile, for the acetylation of a symmetrical diol during the synthesis of an antifungal agent [164].

From a biotechnological perspective, there are many advantages of employing enzymes in organic, as opposed to aqueous, media [165], including higher substrate solubility, reversal of hydrolytic reactions, and modified enzyme specificity, which result in new enzyme activities. On the other hand, enzymes usually show lower catalytic activities in organic than in aqueous solution.

6. Conclusion and Future Perspectives

Microorganisms provide an impressive amount of catalysts with a wide range of applications across several industries such as household care, food, animal feed, technical industries, fine chemicals and pharma. The unique properties of enzymes such as high specificity, fast action and biodegradability allow enzyme-assisted processes in industry to run under milder reaction conditions, with improved yields and reduce waste generation.

However, naturally occurring enzymes are often not suitable for such biocatalytic processes without further tailoring or redesign of the enzyme itself in order to fine-tune substrate specificity activity or other key catalytic properties.

Recent advances in genomics, metagenomics, proteomics, efficient expression systems and emerging recombinant DNA techniques have facilitated the discovery of new microbial enzymes from nature (through genome and metagenome) or by creating (or evolving) enzymes with improved catalytic properties.

The ongoing progress and interest in enzymes provide further success in areas of industrial biocatalysis. The next years should see a lot of exciting developments in the area of biotransformations. There are many factors influencing the growing interest in biocatalysis which include enzyme promiscuity, robust computational methods combined with directed evolution and screening technologies to improve enzyme properties to meet process prospects, the application of one-pot multistep reactions using multifunctional catalysts and the *de novo* design and selection of catalytic proteins catalyzing any desired chemical reaction.

Many future investigations will use combinations of engineered and *de novo* designed enzymes coupled with chemistry to generate more (and most likely new) chemicals and materials from cheaper (and renewable) resources, which will consequently contribute to establishing a bio-based economy and achieving low carbon green growth.

References

1. Adrio, J.L.; Demain, A.L. Microbial Cells and Enzymes—A Century of Progress. In *Methods in Biotechnology. Microbial Enzymes and Biotransformations*; Barredo, J.L., Ed.; Humana Press: Totowa, NJ, USA, 2005; Volume 17, pp. 1–27.
2. Johannes, T.W.; Zhao, H. Directed evolution of enzymes and biosynthetic pathways. *Curr. Opin. Microbiol.* **2006**, *9*, 261–267.
3. Kumar, A.; Singh, S. Directed evolution: Tailoring biocatalysis for industrial application. *Crit. Rev. Biotechnol.* **2013**, *33*, 365–378.
4. Demain, A.L.; Adrio, J.L. Contributions of microorganisms to industrial biology. *Mol. Biotechnol.* **2008**, *38*, 41–45.
5. Johnson, E.A. Biotechnology of non-*Saccharomyces* yeasts- the ascomycetes. *Appl. Microbiol. Biotechnol.* **2013**, *97*, 503–517.
6. Rondon, M.R.; Goodman, R.M.; Handelsman, J. The Earth's bounty: Assessing and accessing soil microbial diversity. *Trends Biotechnol.* **1999**, *17*, 403–409.
7. Verenium, A Pioneer of 21st Century Bioscience, is Working to Transform Industries. Available online: http://www.diversa.com (access on 1 July 2013).
8. Uchiyama, T.; Miyazaki, K. Functional metagenomics for enzyme discovery: Challenges to efficient screening. *Curr. Opin. Biotechnol.* **2009**, *20*, 616–622.
9. Ferrer, M.; Beloqui, A.; Timmis, K.M.; Golyshin, P.N. Metagenomics for mining new genetic resources of microbial communities. *J. Mol. Microbiol. Biotechnol.* **2009**, *16*, 109–123.
10. Gilbert, J.A.; Dupont, C.L. Microbial metagenomics: Beyond the genome. *Annu. Rev. Mar. Sci.* **2011**, *3*, 347–371.
11. The Institute for Genomic Research (TIGR)-JCVI. http://cmr.jcvi.org/tigr (accessed on 1 July 2013).
12. The National Center for Biotechnology Information. Available online: http://www.ncbi.nlm.nih.gov (accessed on 1 July 2013).
13. Ahmed, N. A flood of microbial genomes—Do we need more? *PLoS One* **2009**, *4*, 1–5.
14. Kaul, P.; Asano, Y. Strategies for discovery and improvement of enzyme function: State of the art and opportunities. *Microbiol. Biotechnol.* **2012**, *5*, 18–33.
15. Schiraldini, C.; de Rosa, M. The production of biocatalysts and biomolecules from extremophiles. *Trends Biotechnol.* **2002**. *20*, 515–521.
16. Kumar, L.; Awasthi, G.; Singh, B. Extremophiles: A novel source of industrially important enzymes. *Biotechnology* **2011**, *10*, 1–15.
17. Steele, H.L.; Jaeger, K.E.; Daniel, R.; Streit, W.R. Advances in recovery of novel biocatalysts from metagenomes. *J. Mol. Microbiol. Biotechnol.* **2009**, *16*, 25–37.
18. Li, S.; Yang, X.; Yang, S.; Zhu, M.; Wang, X. Technology prospecting on enzymes: Application, marketing and engineering. *Comput. Struct. Biotechnol. J.* **2012**, *2*, 1–11.
19. Hess, M.; Sczyrba, A.; Egan, R.; Kim, T.W.; Chokhawala, H.; Schroth, G.; Luo, S.; Clark, D.S.; Chen, F.; Zhang, T.; *et al.* Metagenomic discovery of biomass-degrading genes and genomes from cow rumen. *Science* **2011**, *331*, 463–467.

20. Kennedy, J.; Marchesi, J.R.; Dobson, A.D. Marine metagenomics: Strategies for the discovery of novel enzymes with biotechnological applications from marine environments. *Microb. Cell Fact.* **2008**, *7*, 27–37.
21. Warnecke, F.; Luginbühl, P.; Ivanova, N.; Ghassemian, M.; Richardson, T.H.; Stege, J.T.; Cayouette, M.; McHardy, A.C.; Djordjevic, G.; Aboushadi, N.; *et al.* Metagenomic and functional analysis of hindgut microbiota of a wood-feeding higher termite. *Nature* **2007**, *450*, 560–565.
22. Jeon, J.H.; Kim, J.T.; Kim, Y.J.; Kim, H.K.; Lee, H.S.; Kang, S.G.; Kim, S.J.; Lee, J.H. Cloning and characterization of a new-cold active lipase from a deep-se sediment metagenome. *Appl. Microbiol. Biotechnol.* **2009**, *81*, 865–874.
23. Fernández-Álvaro, E.; Kourist, R.; Winter, J.; Böttcher, D.; Liebeton, K.; Naumer, C. Enantioselective kinetic resolution of phenylalkyl carboxylic acids using metagenome-derived esterases. *Microb. Biotechnol.* **2010**, *3*, 59–64.
24. Knietsch, A.; Waschkowitz, T.; Bowien, S.; Henne, A.; Daniel, R. Construction and screening of metagenomic libraries derived from enrichment cultures: Generation of a gene bank for genes conferring alcohol oxidoreductase activity on *Escherichia coli*. *Appl. Environ. Microbiol.* **2003**, *69*, 1408–1416.
25. Gabor, E.M.; de Vries, E.J.; Janssen, D.B. Construction, characterization, and use of small-insert gene banks of DNA isolated from soil and enrichment cultures for the recovery of novel amidases. *Environ. Microbiol.* **2004**, *6*, 948–958.
26. Rondon, M.R.; August, P.R.; Betterman, A.D.; Brady, S.F.; Grossman, T.H.; Liles, M.R. Cloning the soil metagenome: A strategy for accesing the genetic and functional diversity of uncultured microorganisms. *Appl. Environ. Microbiol.* **2000**, *66*, 2541–2547.
27. Bayer, S.; Birkemeyer, C.; Ballschmiter, M. A nitrilase from ametagenomic library acts regioselectively on aliphatic dinitriles. *Appl. Microbiol. Biotechnol.* **2011**, *89*, 91–98.
28. Wang, K.; Li, G.; Yu, S.Q.; Zhang, C.T.; Liu, Y.H. A novel metegenome-derived β-galactosidase: Gene cloning, overexpression, purification and characterization. *Appl. Microbiol. Biotechnol.* **2010**, *88*, 155–165.
29. Jiang, C.; Li, S.X.; Luo, F.F.; Jin, K.; Wang, Q.; Hao, Z.Y. Biochemical characterization of two novel β-glucosidase genes by metagenome expression cloning. *Bioresour. Technol.* **2011**, *102*, 3272–3278.
30. Jiang, C.; Shen, P.; Yan, B.; Wua, B. Biochemical characterization of a metagenome-derived decarboxylase. *Enzyme Microb. Technol.* **2009**, *45*, 58–63.
31. Kotik, M.; Stepanek, V.; Grulich, M.; Kyslik, P.; Archelas, A. Access to enantiopure aromatic epoxides and diols using epoxide hydrolases derived from total biofilter DNA. *J. Mol. Catal. B Enzym.* **2010**, *65*, 41–48.
32. Martínez, A.; Kolvek, S.J.; Yip, C.L.T.; Hopke, J.; Brown, K.A.; MacNeil, I.A.; Osburne, M.S. Genetically modified bacterial strains and novel bacterial artificial chromosome shuttle vectors for constructing environmental libraries and detecting heterologous naturals products in multiple expression hosts. *Appl. Environ. Microbiol.* **2004**, *70*, 2452–2463.

33. Wexler, M.; Bond, P.L.; Richardson, D.J.; Johnston, A.W. A wide host-range metagenomic library from a waste water treatment plants yields a novel alcohol/aldehyde dehydrogenase. *Environ. Microbiol.* **2005**, *7*, 1917–1926.
34. Suenaga, H.; Ohnuki, T.; Miyazaki, K. Functional screening of a metagenomic library for genes involved in microbial degradation of aromatic compounds. *Environ. Microbiol.* **2007**, *9*, 2289–2297.
35. Mirete, S.; de Figueras, C.G.; González-Pastor, J.E. Novel nickel resistance genes from the rhizosphere metagenome of plants adapted to acid mine drainage. *Appl. Environ. Microbiol.* **2007**, *73*, 6001–6011.
36. Mori, T.; Mizuta, S.; Suenaga, H.; Miyazaki, K. Metagenomic screening for bleomycin resistance genes. *Appl. Environ. Microbiol.* **2008**, *74*, 6803–6805.
37. Uchiyama, T.; Abe, T.; Ikemura, T.; Watanabe, K. Substrate-induced gene-expression screening of environmental metagenome libraries for isolation of catabolic genes. *Nat. Biotechnol.* **2005**, *23*, 88–93.
38. Schipper, C.; Hornung, C.; Bijtenhoorn, P.; Quistchau, M.; Grond, S.; Streit, W.R. Metagenome-derived clones encoding two novel lactonase family proteins involved in biofilm inhibition in *Pseudomonas aeruginosa*. *Appl. Environ. Microbiol.* **2009**, *75*, 224–233.
39. Uchiyama, T.; Watanabe, K. Substrate-induced gene expression (SIGEX) screening of metagenome libraries. *Nat. Protoc.* **2008**, *3*, 1202–1212.
40. Van Lanen, S.; Shen, B. Microbial genomics for the improvement of natural product discovery. *Curr. Opin. Microbiol.* **2006**, *9*, 252–260.
41. NCBI Microbial Genomes. Available online: http://www.ncbi.nlm.nhi.gov/genomes/microbial (accessed on 10 July 2013).
42. Luo, X.J.; Yu, H.L.; Xu, J.H. Genomic data mining: An efficient way to find new and better enzymes. *Enzyme Eng.* **2012**, *1*, 104–108.
43. Cavicchioli, R.; Amils, R.; Wagner, D.; McGenity, T. Life and applications of extremophiles. *Environ. Microbiol.* **2011**, *13*, 1903–1907.
44. Pikuta, E.V.; Hoover, R.B.; Tang, J. Microbial Extremophiles at the limit of life. *Crit. Rev. Microbiol.* **2007**, *33*, 183–209.
45. Atomi, H.; Sato, T.; Kanai, T. Application of hyperthermophiles and their enzymes. *Curr. Opin. Biotechnol.* **2011**, *22*, 618–626.
46. De Carvalho, C.C. Enzymatic and whole cell catalysis: Finding new strategies for old processes. *Biotechnol. Adv.* **2011**, *29*, 75–83.
47. Hess, M. Thermoacidophilic proteins for biofuels production. *Trends Microbiol.* **2008**, *16*, 414–419.
48. Chen, Z.W.; Liu, Y.Y.; Wu, J.F.; She, Q.; Jiang, C.Y.; Liu, S.J. Novel bacterial sulphur oxygenase reductases from bioreactors treating gold-bearing concentrates. *Appl. Microbiol. Biotechnol.* **2007**, *74*, 688–698.
49. Margesin, R.; Feller, G.; Gerday, C.; Rusell, N. Cold-Adapted Microorganisms: Adaptation Strategies and Biotechnological Potential. In *The Encyclopedia of Environmental Microbiology*; Bitton, G., Ed.; John Wiley and Sons: New York, NY, USA, 2002; pp. 871–885.

50. Klibanov, A.M. Improving enzymes by using them in organic solvents. *Nature* **2001**, *409*, 241–246.
51. Gomes, I.; Gomes, J.; Steiner, W. Highly thermostable amylase and pullulanase of the extreme thermophilic eubacterium *Rhodothermus marinus*: Production and partialcharacterization. *Bioresour. Technol.* **2003**, *90*, 207–214.
52. Van den Burg, B. Extremophiles as a source for novel enzymes. *Curr. Opin. Microbiol.* **2003**, *6*, 213–218.
53. Shukla, A.; Rana, A.; Kumar, L.; Singh, B.; Ghosh, D. Assessment of detergent activity of *Streptococcus* sp. AS02 protease isolated from soil of Sahastradhara, Doon Valley, Uttarakhand. *Asian J. Microbiol. Biotechnol. Environ. Sci.* **2009**, *11*, 587–591.
54. Marrs, B.; Delagrave, S.; Murphy, D. Novel approaches for discovering industrial enzymes. *Curr. Opin. Microbiol.* **1999**, *2*, 241–245.
55. Singhania, R.R.; Patel, A.K.; Pandy, A. *Industrial Biotechnology: Sustainable Growth and Economic Success*; Soetaert, W., Vandamme, E.J., Eds.; Wiley VCH Verlag GmbH: Weinheim, Germany, 2010; pp. 207–226.
56. Cedrone, F.; Menez, A.; Quemeneur, E. Tailoring new enzyme functions by rational redesign. *Curr. Opin. Struct. Biol.* **2000**, *10*, 405–410.
57. Beppu, T. Modification of milk-clotting aspartic proteinases by recombinant DNA techniques. *Ann. N. Y. Acad. Sci.* **1990**, *613*, 14–25.
58. Van den Burg, B.; de Kreij, A.; van der Veek, P.; Mansfeld, J.; Venema, G. Engineering an enzyme to resist boiling. *Proc. Natl. Acad. Sci. USA* **1998**, *95*, 2056–2060.
59. Bolon, D.N.; Voigt, C.A.; Mayo, S.L. *De novo* design of biocatalysts. *Curr. Opin. Struct. Biol.* **2002**, *6*, 125–129.
60. Shimaoka, M.; Shiftman, J.M.; Jing, H.; Tagaki, J.; Mayo, S.L.; Springer, T.A. Computational design of an integrin I domain stabilized in the open high affinity conformation. *Nat. Struct. Biol.* **2000**, *7*, 674–678.
61. Turner, N.J. Directed evolution drives the next generation of biocatalysts. *Nat. Chem. Biol.* **2009**, *5*, 567–573.
62. Schmidt, M.; Boettcher, D.; Bornscheuer, U.T. *Industrial Biotechnology: Sustainable Growth and Economic Success*; Soetaert, W., Vandamme, E.J., Eds.; Wiley VCH Verlag GmbH: Weinheim, Germany, 2010; pp. 155–187.
63. Dalby, P.A. Strategy and success for the directed evolution of enzymes. *Curr. Opin. Struct. Biol.* **2011**, *21*, 473–480.
64. Leung, D.W.; Chen, E.; Goeddel, D.V. A method for random mutagenesis of a defined DNA segment using a modified polymerase chain reaction. *Technique* **1989**, *1*, 11–15.
65. Reidhaar-Olson, J.; Bowie, J.; Breyer, R.M.; Hu, J.C.; Knight, K.L.; Lim, W.A.; Mossing, M.C.; Parsell, D.A.; Shoemaker, K.R.; Sauer, R.T. Random mutagenesis of protein sequences using oligonucleotide cassettes. *Methods Enzymol.* **1991**, *208*, 564–586.
66. Taguchi, S.; Ozaki, A.; Momose, H. Engineering of a cold-adapted protease by sequential random mutagenesis and a screening system. *Appl. Environ. Microbiol.* **1998**, *64*, 492–495.
67. Ness, J.E.; del Cardayre, S.B.; Minshull, J.; Stemmer, W.P. Molecular breeding: The natural approach to protein design. *Adv. Protein Chem.* **2000**, *55*, 261–292.

68. Morley, K.L.; Kazlauskas, R.J. Improving enzyme properties: When are closer mutations better? *Trends Biotechnol.* **2005**, *3*, 231–237.
69. Reetz, M.T.; Kahakeaw, D.; Lohmer, R. Addressing the numbers problem in directed evolution. *ChemBioChem* **2008**, *9*, 1797–1804.
70. Muñoz, E.; Deem, M.W. Amino acid alphabet size in protein evolution experiments: Better to search a small library thoroughly or a large library sparsely? *Protein Eng. Des. Sel.* **2008**, *1*, 311–317.
71. Reetz, M.T.; Wu, S. Greatly reduced amino acid alphabets in directed evolution: Making the right choice for saturation mutagenesis at homologous enzyme positions. *Chem. Commun.* **2008**, 5499–5501. doi:10.1039/b813388c.
72. Crameri, A.; Raillard, S.A.; Bermudez, E.; Stemmer, W.P. DNA shuffling of a family of genes from diverse species accelerates directed evolution. *Nature* **1998**, *391*, 288–291.
73. Kolkman, J.A.; Stemmer, W.P. Directed evolution of proteins by exon shuffling. *Nat. Biotechnol.* **2001**, *19*, 423–428.
74. Hackel, B.J.; Kapila, A.; Wittrup, K.D. Picomolar affinity fibronectin domain engineering utilizing loop length diversity, recursive mutagenesis, and loop shuffling. *J. Mol. Biol.* **2008**, *381*, 1238–1252.
75. Ostermeier, M.; Nixon, A.E.; Shim, J.H.; Benkovic, S.J. Combinatorial protein engineering by incremental truncation. *Proc. Natl. Acad. Sci. USA* **1999**, *96*, 3562–3567.
76. Fujii, R.; Kitakoa, M.; Hayashi, K. RAISE: A simple and novel method of generating random insertion and deletion mutations. *Nucleic Acid Res.* **2006**, *34*, e30.
77. Ness, J.E.; Welch, M.; Giver, L.; Bueno, M.; Cherry, J.R.; Borchert, T.V.; Stemmer, W.P.; Minshull, J. DNA shuffling of subgenomic sequences of subtilisin. *Nat. Biotechnol.* **1999**, *17*, 893–896.
78. Jaeger, K.E.; Reetz, M.T. Directed evolution of enantioselective enzymes for organic chemistry. *Curr. Opin. Chem. Biol.* **2000**, *4*, 68–73.
79. Suenaga, H.; Mitsokua, M.; Ura, Y.; Watanabe, T.; Furukawa, K. Directed evolution of biphenyl dioxygenase: Emergence of enhanced degradation capacity for benzene, toluene, and alkylbenzenes. *J. Bacteriol.* **2001**, *183*, 5441–5444.
80. Song, J.K.; Rhee, J.S. Enhancement of stability and activity of phospholipase A(1) in organic solvents by directed evolution. *Biochim. Biophys. Acta* **2001**, *1547*, 370–378.
81. Raillard, S.; Krebber, A.; Chen, Y.; Ness, J.E.; Bermudez, E.; Trinidad, R.; Fullem, R.; Davis, C.; Welch, M.; Seffernick, J.; *et al.* Novel enzyme activities and functional plasticity revealed by recombining highly homologous enzymes. *Chem. Biol.* **2001**, *8*, 891–898.
82. Kurtzman, A.L.; Govindarajan, S.; Vahle, K.; Jones, J.T.; Heinrichs, V.; Patten, P.A. Advances in directed protein evolution by recursive genetic recombination: Applications to therapeutic proteins. *Curr. Opin. Biotechnol.* **2001**, *12*, 361–370.
83. Patten, P.A.; Howard, R.J.; Stemmer, W.P. Applications of DNA shuffling to pharmaceuticals and vaccines. *Curr. Opin. Biotechnol.* **1997**, *8*, 724–733.
84. Siehl, D.L.; Castle, L.A.; Gorton, R.; Chen, Y.H.; Bertain, S.; Cho, H.J.; Keenan, R.; Liu, D.; Lassner, M.W. Evolution of a microbial acetyl transferase for modification of glyphosate: A novel tolerance strategy. *Pest Manag. Sci.* **2005**, *61*, 235–240.

85. Tobin, M.B.; Gustafsson, C.; Huisman, G.W. Directed evolution: The "rational" basis for "irrational" design. *Curr. Opin. Struct. Biol.* **2000**, *10*, 421–427.

86. Cherry, J.R.; Fidanstel, A.L. Directed evolution of industrial enzymes: An update. *Curr. Opin. Biotechnol.* **2003**, *14*, 438–443.

87. Demain, A.L.; Vaishnav, P. Production of recombinant proteins by microbes and higher organisms. *Biotechnol. Adv.* **2009**, *27*, 297–306.

88. Swartz, J.R. *Escherichia coli* Recombinant DNA Technology. In Escherichia coli *and Salmonella*: *Cellular and Molecular Biology*, 2nd ed.; Neidhardt, F.C., Ed.; American Society of Microbiology Press: Washington, DC, USA, 1996; pp. 1693–1771.

89. Barnard, G.C.; Henderson, G.E.; Srinivasan, S.; Gerngross, T.U. High level recombinant protein expression in *Ralstonia eutropha* using T7 RNA polymerase based amplification. *Protein Expr. Purif.* **2004**, *38*, 264–271.

90. Squires, C.H.; Lucy, P. Vendor voice: A new paradigm for bacterial strain engineering. *BioProcess Int.* **2008**, *6*, 22–27.

91. Romanos, M.A.; Scorer, C.A.; Clare, J.J. Foreign gene expression in yeast: A review. *Yeast* **1992**, *8*, 423–488.

92. Higgins, D.R.; Cregg, J.M. Introduction to *Pichia pastoris.* In *Pichia Protocols*; Higgins, D.R., Cregg, J.M., Eds.; Humana Press: Totowa, NJ, USA, 1998; pp. 1–15.

93. Zhang, A.L.; Luo, X.J.; Zhang, T.Y.; Pan, Y.N. Recent advances on the GAP promoter-derived expression system of *Pichia pastoris*. *Mol. Biol. Rep.* **2009**, *36*, 1611–1619.

94. Hasslacher, M.; Schall, M.; Hayn, M.; Bora, R.; Rumbold, K.; Luecki, J.; Grieng, H.; Kohlwein, S.D.; Schwab, H. High-level intracellular expression of hydroxynitrile lyase from the tropical rubber tree *Hevea brassiliensis* in microbial hosts. *Protein Expr. Purif.* **1997**, *11*, 61–71.

95. Werton, M.W.T.; van den Bosch, T.J.; Wind, R.D.; Mooibroek, H.; de Wolf, F.A. High-yield secretion of recombinant gelatins by *Pichia pastoris*. *Yeast* **1999**, *15*, 1087–1096.

96. Morrow, K.J. Improving protein production processes. *Gen. Eng. Biotechnol. News* **2007**, *27*, 50–54.

97. Bretthauer, R.K.; Castellino, F.J. Glycosylation of *Pichia pastoris*-derived proteins. *Biotechnol. Appl. Biochem.* **1999**, *30*, 193–200.

98. Romanos, M.A. Advances in the use of *Pichia pastoris* for high-level expression. *Curr. Opin. Biotechnol.* **1995**, *6*, 527–533.

99. Giuseppin, M.; van Eijk, H.M.; Bes, B.C. Molecular regulation of methanol oxidase activity in continuous cultures of *Hansenula polymorpha*. *Biotechnol. Bioeng.* **1988**, *32*, 577–583.

100. Egli, T.; van Dijken, J.P.; Veenhuis, M.; Harder, W.; Feichter, A. Methanol metabolism in yeasts: Regulation of the synthesis of catabolite enzymes. *Arch. Microbiol.* **1980**, ***124***, 115–121.

101. Shuster, J.R.; Connelley, M.B. Promoter-tagged restriction enzyme-mediated insertion mutagenesis in *Aspergillus niger*. *Mol. Gen. Genet.* **1999**, *262*, 27–34.

102. Gouka, R.J.; Gerk, C.; Hooykaas, P.J.J.; Bundock, P.; Musters, W.; Verrips, C.T.; de Groot, M.J.A. Transformation of *Aspergillus awamori* by *Agrobacterium tumefaciens*-mediated homologous recombination. *Nat. Biotechnol.* **1999**, *6*, 598–601.

103. Van der Hombergh, J.P.; van de Vondervoort, P.J.; van der Heijden, N.C.; Visser, J. New protease mutants in *Aspergillus niger* result in strongly reduced *in vitro* degradation of target proteins; genetical and biochemical characterization of seven complementation groups. *Curr. Genet.* **1997**, *28*, 299–308.
104. Gouka, R.J.; Punt, P.J.; van den Hondel, C.A.M.J.J. Efficient production of secreted proteins by *Aspergillus*: Progress, limitations and prospects. *Appl. Microbiol. Biotechnol.* **1997**, *47*, 1–11.
105. Moralejo, F.J.; Cardoza, R.E.; Gutierrez, S.; Martín, J.F. Thaumatin production in *Aspergillus awamori* by use of expression cassettes with strong fungal promoters and high gene dosage. *Appl. Environ. Microbiol.* **1999**, *65*, 1168–1174.
106. Gusakov, A.V.; Salanovich, T.N.; Antonov, A.I.; Ustinov, B.B.; Okunev, O.N.; Burlingame, R.; Emalfarb, M.; Baez, M.; Sinitsyn, A.P. Design of highly efficient cellulase mixtures for enzymatic hydrolysis of cellulose. *Biotechnol. Bioeng.* **2007**, *97*, 1028–1038.
107. Verdoes, J.C.; Punt, P.J.; Burlingame, R.; Bartels, J.; van Dijk, R.; Slump, E.; Meens, M.; Joosten, R.; Emalfarb, M. A dedicated vector for efficient library construction and high throughput screening in the hyphal fungus *Chrysosporium lucknowense*. *Ind. Biotechnol.* **2007**, *3*, 48–57.
108. Wohlgemuth, R. Biocatalysis—Key to sustainable industrial chemistry. *Curr. Opin. Biotechnol.* **2010**, *21*, 713–724.
109. Gong, J.S.; Lu, Z.M.; Li, H.; Shi, J.S.; Zhou, Z.M.; Xu, Z.H. Nitrilases in nitrile biocatalysis: Recent progress and forthcoming research. *Microb. Cell Fact.* **2012**, *11*, 142–160.
110. Schmid, A.; Dordick, J.S.; Hauer, B.; Kiener, A.; Wubbolts, M.; Witholt, B. Industrial biocatalysis today and tomorrow. *Nature* **2001**, *409*, 258–268.
111. Ran, N.; Zhao, L.; Chen, Z.; Tao, J. Recent applications of biocatalysis in developing green chemistry for chemical synthesis at the industrial scale. *Green Chem.* **2008**, *10*, 361–372.
112. Bornscheuer, U.T.; Huisman, G.W.; Kazlauskas, R.J.; Lutz, S.; Moore, J.C.; Robins, K. Engineering the third wave of biocatalysis. *Nature* **2012**, *485*, 185–194.
113. Sanchez, S.; Demain, A.L. Enzymes and bioconversions of industrial, pharmaceutical, and biotechnological significance. *Org. Process Res. Dev.* **2011**, *15*, 224–230.
114. BBC Research. *Report BIO030 F Enzymes in Industrial Applications: Global Markets*; BBC Research: Wellesley, MA, USA, **2011**.
115. World Enzymes. Freedonia Group: Cleveland, OH, USA, **2011**.
116. Kirk, O.; Borchert, T.V.; Fuglsang, C.C. Industrial enzyme applications. *Curr. Opin. Biotechnol.* **2002**, *13*, 345–351.
117. Araujo, R.; Casal, M.; Cavaco-Paulo, A. Application of enzymes for textiles fibers processing. *Biocatal. Biotechnol.* **2008**, *26*, 332–349.
118. Tzanov, T.; Calafell, M.; Guebitz, G.M.; Cavaco-Paulo, A. Bio-preparation of cotton fabrics. *Enzyme Microb. Technol.* **2001**, *29*, 357–362.
119. Gutiérrez, A.; del Río, J.C.; Martínez, A.T. Microbial and enzymatic control of pitch in the pulp and paper industry. *Appl. Microbiol. Biotechnol.* **2009**, *82*, 1005–1018.
120. Farrell, R.L.; Hata, K.; Wall, M.B. Solving pitch problems in pulp and paper processes by the use of enzymes or fungi. *Adv. Biochem. Eng. Biotechnol.* **1997**, *57*, 197–212.

121. Rodriguez-Couto, S.; Toca-Herrera, J.L. Industrial and biotechnological applications of laccases: A review. *Biotechnol. Adv.* **2006**, *24*, 500–513.
122. Rubin, E.M. Genomics of cellulosic biofuels. *Nature* **2008**, *454*, 841–845.
123. Wilson, B.D. Cellulases and biofuels. *Curr. Opin. Biotechnol.* **2009**, *20*, 295–299.
124. Zhang, Y.H.P. What is vital (and not vital) to advance economically-competitive biofuels production. *Process Biochem.* **2011**, *46*, 2091–2110.
125. Zhang, Y.H.P.; Himmmel, M.E.; Mielenz, J.R. Outlook for cellulose improvement: Screening and selection strategies. *Biotechnol. Adv.* **2006**, *24*, 452–481.
126. Kumar, R.; Singh, S.; Singh, O.V. Bioconversion of lignocellulosic biomass: Biochemical and molecular perspectives. *J. Ind. Microbiol. Biotechnol.* **2008**, *35*, 377–391.
127. Kubicek, C.P.; Mikus, M.; Schuster, A.; Schmoll, M.; Seiboth, B. Metabolic engineering strategies for the improvement of cellulase production by *Hypocrea jecorina*. *Biotechnol. Biofuels* **2009**, *2*, 19–31.
128. Dashtban, M.; Qin, W. Overexpression of an exotic thermotolerant β-glucosidase in *Trichoderma reesei* and its significant increase in cellulolytic activity and saccharification of barley straw. *Microb. Cell Fact.* **2012**, *11*, 63–78.
129. Martinez, D.; Berka, R.M.; Henrissat, B.; Saloheimo, M.; Arvas, M.; Baker, S.E.; Chapman, J.; Chertkov, O.; Coutinho, P.M.; Cullen, D.; *et al.* Genome sequence analysis of the cellulolytic fungus *Trichoderma reesei* (syn. *Hypocrea jecorina*) reveals a surprisingly limited inventory of carbohydrate active enzymes. *Nat. Biotechnol.* **2008**, *26*, 553–560.
130. Kanafusa-Shinkai, S.; Wakayama, J.; Tsukamoto, K.; Hayashi, N.; Miyazaki, Y.; Ohmori, H.; Tajima, K.; Yokoyama, H. Degradation of microcrystalline cellulose and non-pretreated plant biomass by a cell-free extracellular cellulose/hemicellulase system from the extreme thermophilic bacterium *Caldicellulosiruptor bescii*. *J. Biosci. Bioeng.* **2013**, *115*, 64–70.
131. *Feed Enzymes: The Global Scenario*; Frost & Sullivan: London, UK, **2007**.
132. Choct, M. Enzymes for the feed industry: Past, present and future. *World Poult. Sci.* **2006**, *62*, 5–15.
133. Selle, P.H.; Ravindran, V. Microbial phytase in poultry nutrition. *Anim. Feed Sci. Technol.* **2007**, *135*, 1–41.
134. Pariza, M.W.; Cook, M. Determining the safety of enzymes used in animal feed. *Regul. Toxicol. Pharmcol.* **2010**, *56*, 332–342.
135. Mendez, C.; Salas, J.A. Altering the glycosylation pattern of bioactive compounds. *Trends Biotechnol.* **2001**, *19*, 449–456.
136. Okanishi, M.; Suzuki, N.; Furita, T. Variety of hybrid characters among recombinants obtained by interspecific protoplast fusion in streptomycetes. *Biosci. Biotechnol. Biochem.* **1996**, *6*, 1233–1238.
137. Rao, M.B.; Tanksale, A.M.; Ghatge, M.S.; Deshpande, V.V. Molecular and biotechnological aspects of microbial proteases. *Microbiol. Mol. Biol. Rev.* **1998**, *62*, 597–635.
138. Dunn-Coleman, N.S.; Bloebaum, P.; Berka, R.; Bodie, E.; Robinson, N.; Armstrong, G.; Ward, M.; Przetak, M.; Carter, G.L.; LaCost, R.; *et al.* Commercial levels of chymosin production by *Aspergillus*. *Biotechnology* **1991**, *9*, 976–981.

139. Bodie, E.A.; Armstrong, G.L.; Dunn-Coleman, N.S. Strain improvement of chymosin-producing strains of *Aspergillus niger var awamori* using parasexual recombination. *Enzyme Microb. Technol.* **1994**, *16*, 376–382.
140. Pariza, M.W.; Johnson, E.A. Evaluating the safety of microbial enzyme preparations used in food processing: Update for a new century. *Regul. Toxicol. Pharmacol.* **2001**, *33*, 173–186.
141. US FDA. Center for Food Safety and Applied Nutrition. http://vm.cfsan.fda.gov (accessed on 1 July 2013).
142. Vandamme, E.J.; Cerdobbel, A.; Soetaert, W. Biocatalysis on the rise. Part 1. Principles. *Chem. Today* **2005**, *23*, 57–61.
143. *Collection of Information on Enzymes*; Austrian Federal Evironment Agency: Vienna, Austria, **2002**.
144. Olempska-Beer, Z.S.; Merker, R.I.; Ditto, M.D.; DiNovi, M.J. Food-processing enzymes from recombinant microorganisms—A review. *Regul. Toxicol. Pharmcol.* **2006**, *45*, 144–158.
145. Tufvesson, P.; Lima-Ramos, J.; Nordblad, M.; Woodley, J.M. Guidelines and cost analysis for catalyst production in biocatalytic processes. *Org. Process Res. Dev.* **2011**, *15*, 266–274.
146. Panke, S.; Wubbolts, M. Advances in biocatalytic synthesis of pharmaceutical intermediates. *Curr. Opin. Chem. Biol.* **2005**, *9*, 188–194.
147. Jäckel, C.; Hilvert, D. Biocatalysts by evolution. *Curr. Opin. Biotechnol.* **2010**, *21*, 753–759.
148. Lutz, S. Reengineering enzymes. *Science* **2010**, *329*, 285–287.
149. Kim, D.Y.; Rha, E.; Choi, S.-L.; Song, J.J.; Hong, S.-P.; Sung, M.-H.; Lee, S.-G. Development of bioreactor system for L-tyrosine synthesis using thermostable tyrosine phenol-lyase. *J. Microb. Biotechnol.* **2007**, *17*, 116–122.
150. Volpato, G.; Rodrigues, R.C.; Fernandez-Lafuente, R. Use of enzymes in the production of semi-synthetic penicillins and cephalosporins: Drawbacks and perspectives. *Curr. Med. Chem.* **2010**, *17*, 3855–3873.
151. Kirchner, G.; Scollar, M.P.; Klibanov, A. Resolution of racemic mixtures via lipase catalysis in organic solvents. *J. Am. Chem. Soc.* **1995**, *107*, 7072–7076.
152. Zheng, G.W.; Xu, J.H. New opportunities for biocatalysis: Driving the synthesis of chiral chemicals. *Curr. Opin. Biotechnol.* **2011**, *22*, 784–792.
153. Wohlgemuth, R. Asymmetric biocatalysis with microbial enzymes and cells. *Curr. Opin. Microbiol.* **2010**, *13*, 283–292.
154. Liang, J.; Mundorff, E.; Voladri, R.; Jenne, S.; Gilson, L.; Conway, A.; Krebber, A.; Wong, J.; Huisman, G.; Truesdell, S.; *et al.* Highly Enantioselective Reduction of a small heterocyclic ketone: Biocatalytic reduction of tetrahydrothiophene-3-one to the corresponding (*R*)-alcohol. *Org. Process Res. Dev.* **2010**, *14*, 185–192.
155. Ma, S.K.; Gruber, J.; Davis, C.; Newman, L.; Gray, D.; Wang, A.; Grate, J.; Huisman, G.W.; Sheldon, R.A. A green-by-design biocatalytic process for atorvastatin intermediate. *Green Chem.* **2010**, *12*, 81–86.
156. Sheldom, R.A. E factors, green chemistry and catalysis: An odyssey. *Chem. Commun.* **2008**, 3352–3365. doi:10.1039/b803584a.
157. Höhne, M.; Schätzle, S.; Jochens, H.; Robins, K.; Bornscheuer, U.T. Rational assignment of key motifs for function guides in silico enzyme identification. *Nat. Chem. Biol.* **2010**, *6*, 807–813.

158. Koszelewski, D.; Göritzer, M.; Clay, D.; Seisser, B.; Kroutil, W. Synthesis of optically active amines employing recombinant ω-transaminases in *E. coli* cells. *ChemCatChem* **2010**, *2*, 73–77.
159. Gooding, O.; Voladri, R.; Bautista, A.; Hopkins, T.; Huisman, G.; Jenne, S.; Ma, S.; Mundorff, E.C.; Savile, M.M.; Truesdell, S.J. Development of a practical biocatalytic process for (R)-2-methylpentanol. *Org. Process Res. Dev.* **2010**, *14*, 119–126.
160. Savile, C.K.; Janey, J.M.; Mundorff, E.C.; Moore, J.C.; Tam, S.; Jarvis, W.R.; Colbeck, J.C.; Krebber, A.; Fleits, F.J.; Brands, J.; *et al.* Biocatalytic asymmetric synthesis of chiral amines from ketones applied to sitagliptin manufacture. *Science* **2010**, *329*, 305–309.
161. Liang, J.; Lalonde, J.; Borup, B.; Mitchell, V.; Mundorff, E.; Trinh, N.; Kochreckar, D.A.; Cherat, R.N.; Ganesh, P.G. Development of a biocatalytic process as an alternative to the (-)-DIP-Cl-mediated asymmetric reduction of a key intermediate of Montelukast. *Org. Process Res. Dev.* **2010**, *14*, 193–198.
162. Ritter, S. Green chemistry awards. Honors: Presidential challenge highlights innovations that promote sustainability. *Chem. Eng. News* **2012**, *90*, 11.
163. Ema, T.; Sayaka, I.; Nobuyasu, O.; Sakai, T. Highly efficient chemoenzymatic synthesis of methyl (*R*)-*o*-chloromandelate, a key intermediate for clopidogrel, via asymmetric reduction with recombinant *Escherichia coli. Adv. Synth. Catal.* **2008**, *350*, 2039–2044.
164. Zaks, A.; Dodds, D.R. Application of biocatalysis and biotransformations to the synthesis of pharmaceuticals. *Drug Discov. Today* **1997**, *2*, 513–531.
165. Lee, M.Y.; Dordick, J.S. Enzyme activation for nonaqueous media. *Curr. Opin. Biotechnol.* **2002**, *13*, 376–384.

7

Microbial Enzymes with Special Characteristics for Biotechnological Applications

Poonam Singh Nigam

Biomedical Science Research Institute, University of Ulster, Coleraine BT52 1SA, UK;
E-Mail: p.singh@ulster.ac.uk

Abstract: This article overviews the enzymes produced by microorganisms, which have been extensively studied worldwide for their isolation, purification and characterization of their specific properties. Researchers have isolated specific microorganisms from extreme sources under extreme culture conditions, with the objective that such isolated microbes would possess the capability to bio-synthesize special enzymes. Various Bio-industries require enzymes possessing special characteristics for their applications in processing of substrates and raw materials. The microbial enzymes act as bio-catalysts to perform reactions in bio-processes in an economical and environmentally-friendly way as opposed to the use of chemical catalysts. The special characteristics of enzymes are exploited for their commercial interest and industrial applications, which include: thermotolerance, thermophilic nature, tolerance to a varied range of pH, stability of enzyme activity over a range of temperature and pH, and other harsh reaction conditions. Such enzymes have proven their utility in bio-industries such as food, leather, textiles, animal feed, and in bio-conversions and bio-remediations.

Keywords: microbial-enzymes; thermophilic; alkalophilic; thermostable; protease; keratinase; amylase; xylanase; laccase

---

1. Enzymes from Microbial Sources

Enzymes are the bio-catalysts playing an important role in all stages of metabolism and biochemical reactions. Certain enzymes are of special interest and are utilized as organic catalysts in numerous processes on an industrial scale. Microbial enzymes are known to be superior enzymes obtained from different microorganisms, particularly for applications in industries on commercial scales. Though the enzymes were discovered from microorganisms in the 20th century, studies on their isolation, characterization of properties, production on bench-scale to pilot-scale and their application in bio-industry have continuously progressed, and the knowledge has regularly been updated. Many enzymes from microbial sources are already being used in various commercial processes. Selected microorganisms including bacteria, fungi and yeasts have been globally studied for the bio-synthesis of economically viable preparations of various enzymes for commercial applications [1].

In conventional catalytic reactions using biocatalysts the use of enzymes, either in free or in immobilized forms, is dependent on the specificity of enzyme. In recent advances of biotechnology, according to the requirements of a process, various enzymes have been and are being designed or purposely engineered. Various established classes of enzymes are specific to perform specialized catalytic reactions and have established their uses in selected bio-processes. A large number of new enzymes have been designed with the input of protein-engineering, biochemical-reaction engineering and metagenomics. Various molecular techniques have also been applied to improve the quality and performance of microbial enzymes for their wider applications in many industries [2]. As a result, many added-value products are being synthesized in global market with the use of established bioprocess-technology employing purposely engineered biocatalyst-enzymes.

Most of the commercially applicable proteases are alkaline and are bio-synthesized mainly by bacteria such as *Pseudomonas*, *Bacillus*, and *Clostridium*, and some fungi are also reported to produce these enzymes [3]. The xylanases with significant applications in bio-industries are produced by the fungal species belonging to genera *Trichoderma*, *Penicillium* and *Aspergillus*; the xylanases produced by these microorganisms have been found to possess high activity over a wide range of temperatures (40–60 °C) [4].

2. Enzymes with Special Characteristics

Special characteristics of microbial enzymes include their capability and appreciable activity under abnormal conditions, mainly of temperature and pH. Hence, certain microbial enzymes are categorized as thermophilic, acidophilic or alkalophilic. Microorganisms with systems of thermostable enzymes that can function at higher than normal reaction temperatures would decrease the possibility of microbial contamination in large scale industrial reactions of prolonged durations [5–7]. The quality of thermostability in enzymes promotes the breakdown and digestion of raw materials; also the higher reaction temperature enhances the penetration of enzymes [8]. The complete saccharification and hydrolysis of polysaccharides containing agricultural residues requires a longer reaction time, which is often associated with the contamination risks over a period of time. Therefore, the hydrolytic enzymes are well sought after, being active at higher temperatures as well as retaining stability over a prolonged period of processing at a range of temperatures. The high temperature enzymes also help in enhancing the mass-transfer and reduction of the substrate viscosity [9,10] during the progress of hydrolysis of substrates or raw materials in industrial processes. Thermophilic xylanase are considered to be of commercial interest in many industries particularly in the mashing process of brewing. The thermostable plant xerophytic isoforms of laccase enzyme are considered to be useful for their applications in textile, dyeing, pulping and bioremediation [1,4].

3. Enzymes with Special Characteristics in Biotechnology

3.1. Protease

Though the hydrolytic enzymes belong to the largest group of enzymes and are the most commercially-applicable enzymes, among the enzymes within this group the microbial proteases have been extensively studied [11–16]. Proteases prepared from microbial systems are of three types: acidic, neutral and alkaline. Alkaline proteases are efficient under alkaline pH conditions and consist of a serine residue at their active site [15]. Alkaline serine proteases have the largest applications in bio-industry. Alkaline proteases are of particular interest being more suitable for a wide range of applications, since these possess high activity and stability in abnormal conditions of extreme physiological parameters. Alkaline proteases have shown their capability to work under high pH, temperature and in presence of inhibitory compounds [15–18].

Vijayalakshmi *et al.* [16] have optimized and characterized the cultural conditions for the production of alkalophilic as well as a thermophilic extracellular protease enzyme from *Bacillus*. This bacteria named *Bacillus* RV.B2.90 was found to be capable of producing an enzyme preparation possessing special characteristics such as being highly alkalophilic, moderately halophilic, thermophilic, and exhibiting the quality of a thermostable protease enzyme. Alkaline proteases possess the property of a great stability in their enzyme activity when used in detergents [16,18,19]. The alkaline protease produced from *Bacilli* and proteases from other microorganisms have found more applications overall in bio-industries such as: washing powders, tannery, food-industry, leather processing, pharmaceuticals, for studies in molecular biology and in peptide synthesis [1,3].

3.2. Keratinases

Keratin is an insoluble and fibrous structural protein that is a constituent of feathers and wool. The protein is abundantly available as a by-product from keratinous wastes, representing a valuable source of proteins and amino acids that could be useful for animal feeds or as a source of nitrogen for plants [20]. However, the keratin-containing substrates and materials have high mechanical stability and hence are difficult to be degraded by common proteases. Keratinases are specific proteolytic enzymes which are capable of degrading insoluble keratins. The importance of these enzymes is being increasingly recognized in fields as diverse as animal feed production, textile processing, detergent formulation, leather manufacture, and medicine. Proteolytic enzymes with specialized keratinase activity are required to degrade keratins and for this purpose the keratinases have been isolated and purified from certain bacteria, actinomycetes, and fungi [20,21].

Keratinases have been classified as serine- or metallo-proteases. Cloning and expression of keratinase genes in a variety of expression systems have also been reported [22]. A higher operation temperature is required in the degradation of materials like feathers and wool, which would be possible using a thermostable keratinase. This aspect is of added advantage in achieving a higher reactivity due to lower diffusional restrictions and hence a higher reaction rate would be established. The enhanced stability of keratinase would increase the overall process yield due to the increased solubility of keratin and favorable equilibrium displacement in endothermic reactions.

Baihong *et al.* [23] have reported the enhanced thermostability of a preparation of keratinase by computational design and empirical mutation. The quadruple mutant of *Bacillus subtilis* has been characterised to exhibit the synergistic and additive effects at 60 °C with an increase of 8.6-fold in the t1/2 value. The N122Y substitution also led to an approximately 5.6-fold increase in catalytic efficiency compared to that of the wild-type keratinase.

An alkalophilic strain of *Streptomyces albidoflavus* has been reported to produce extracellular proteases [24]. This particular type of protease was capable of hydrolyzing keratin. The biosynthesis of this specific enzyme was optimized in submerged batch cultures at highly alkaline pH 10.5 and the enzyme yield was stimulated by using an inducer substrate containing keratin in the form of white chicken feathers. An enhanced (six-fold) protease production could be achieved with modified composition of culture-medium containing inducer at the concentration of 0.8% in the fermentation medium. The novelty of this crude enzyme has been reported to be its activity and stability in neutral and alkaline conditions. The maximum activity has been obtained at pH 9.0 and in the temperature range of 60–70 °C. This type of protease (keratinase- hydrolyzing keratins) is of particular significance for its application in industries since the crude enzyme showed its tolerance to the detergents and solvents tested [24]. Liu *et al.* [25] have studied the expression of extreme alkaline, oxidation-resistant keratinase from *Bacillus licheniformis* into the recombinant *Bacillus subtilis* WB600 expression system. The alkaline keratinase was characterized for its application in the processing of wool fibers.

3.3. Amylase

Amylases are significant enzymes for their specific use in the industrial starch conversion process [26]. Amylolytic enzymes act on starch and related oligo- and polysaccharides [27]. The global research on starch hydrolyzing enzymes based on the DNA sequence, structural analysis and catalytic mechanism has led to the concept of one enzyme family—the alpha amylase. The amylolytic and related enzymes have been classified as glycoside hydrolases. The enzymes have been produced by a wide range of microorganisms and substrates [28–30] and categorized as exo-, endo-, de-branching and cyclodextrin producing enzyme. The application of these enzymes has been established in starch liquefaction, paper, food, sugar and pharmaceutical industries. In the food industry amylolytic enzymes have a large scale of applications, such as the production of glucose syrups, high fructose corn syrups, maltose syrup, reduction of viscosity of sugar syrups, reduction of turbidity to produce clarified fruit juice for longer shelf-life, solubilisation and saccharification of starch in the brewing industry [31]. The baking industry uses amylases to delay the staling of bread and other baked products; the paper industry uses amylases for the reduction of starch viscosity to achieve the appropriate coating of paper. Amylase enzyme is used in the textile industry for warp sizing of textile fibers, and used as a digestive aid in the pharmaceutical industry [28].

Li *et al.* [32] have recently isolated, characterized and cloned a themotolerant isoamylase. For this purpose the enzyme was bio-synthesized using a thermophilic bacterium *Bacillus* sp. This novel enzyme has been reported to display its optimal activity at a remarkably high temperature of 70 °C, as well as being active in the alkaline range. This thermophilic enzyme has also been found to be thermo-stable between 30 and 70 °C, and its activity has been reported to be stable within a pH range of 5.5 to 9.0.

Gurumurthy *et al.* [33] completed the molecular characterization of an extremely thermostable alpha-amylase for industrial applications. This novel enzyme was produced by a bacterium *Geobacillus* sp which was isolated from the thermal water of a geothermal spring. This isolated bacterium showed the characteristics of thermo-tolerance and alkali-resistance. A purified preparation of amylase suitable for application was obtained using a DEAE-cellulose column and Sephadex G-150 gel filtration chromatography. The enzyme is a novel alpha-amylase due to its optimum activity at a very high temperature of 90 °C and an alkaline pH 8.0. However, this purified preparation enzyme was found to be stable only for 10 min at 90 °C.

3.4. Xylanase

Hemicellulose is one of main constituents of agricultural residues and plants along with cellulose, lignin and pectin [34]. Xylan is the major component of hemicellulose consisting of β-1,4-linked D-xylopyranosyl residues. The hydrolysis of xylan in plant materials is achieved by the use of a mixture of hydrolytic enzymes including endo-β-1,4-xylanase and β-D-xylosidase [35]. The importance of xylanase has tremendously increased due to its biotechnological applications for pentose production, fruit-juice clarification, improving rumen digestion and the bioconversion of lignocellulosic agricultural residues to fuels and chemicals [34]. Collins *et al.* [36] have extensively studied the xylanase enzyme and its families as well as the special xylanases possessing extremophilic characteristics. Xylanases have established their uses in the food, pulp, paper and textile industries, agri-industrial residues utilization, and ethanol and animal feed production [37,38].

The enzyme used for the purpose of bio-bleaching of wood pulp should be active in the conditions of alkaline pH, high temperature and at the same time it is desirable that this enzyme is stable at high reaction temperatures. Xylanase preparations used for wood processing in the paper industry should be free of cellulose activity. Cellulase-free xylanase preparations have applications in the paper industry to provide brightness to the paper due to their preferential solubilisation of xylans in plant materials and selective removal of hemicelluloses from the kraft-pulp. Kohli *et al.* [39] have studied the production of a c cellulase free extracellular endo-1,4-β-xylanase at a higher temperature of 50 °C and at pH 8.5 employing a selected microorganism: *Thermoactinomyces thalophilus*. The enzyme preparation was found to be thermostable at 65 °C, retaining its activity at 50% after 125 min of incubation at 65 °C. The crude enzyme preparation showed no cellulase activity and the optimum temperature and pH for maximum xylanase activity was found to be 65 °C and 8.5–9.0, respectively. A thermotolerant and alkalotolerant xylanase has been reported to be produced by *Bacillus* sp [40]. To make the applications of xylanase viable on commercial scales, heterologous systems of *Escherichis coli*, *Pichia pastoris* and *Bacillus* sp have been used to express xylanase activity [41,42]. The thermophilic microorganism *Humicola* spp. has been studied for its capability of bio-synthesising an alkali-tolerant β-mannase xylanase [43]. Acidophilic xylanases stable under acidic conditions of reaction are reported to be produced by an acidophilic fungus *Bispora* [44], in contrast a xylanase active under conditions of alkaline pH has been studied by Mamo *et al.* [45] for the mechanism of their high pH catalytic adaptation.

Recently three novel xylanases thermophilic in nature (XynA,B,C) have been characterized by Yanlong *et al.* [46], these were produced by *Humicola* sp. for their potential applications in the brewing industry. One xylanase gene, XynA, has been found to adapt to alkaline conditions and stability at higher temperatures. This XynA also possessed higher catalytic efficiency and specificity for a range of substrates. Yanlong *et al.* [46] have reported the application of three xylanases, XynA-C, in simulated mashing conditions in the brewing industry and found the better performance of 37% on filtration acceleration and 13% reduction in viscosity of substrate in comparison to the performance of a commercial trade enzyme, Ultraflo, a product from Novozyme.

3.5. Laccase/Ligninase

Ligninolytic enzymes are applicable in the hydrolysis of lignocellulosic agricultural residues, particularly for the degradation of the complex and recalcitrant constituent lignin. This group of enzymes is a mixture of synergistic enzymes, hence they are highly versatile in nature and can be used in a range of industrial processes [47–49]. The complex enzyme system consists of three oxidative enzymes: lignin peroxidase (LiP), manganese peroxidase (MnP) and laccase. These enzymes have established their applications in bio-remediation, pollution control and in the treatment of industrial effluents containing recalcitrant and hazardous chemicals such as textile dyes, phenols and other xenobiotics [50–53].

The paper and pulp industry requires a step of separation and degradation of lignin from plant material, where the pretreatment of wood pulp using ligninolytic enzymes is important for a milder and cleaner strategy of lignin removal compared to chemical bleaching. Bleach enhancement of mixed wood pulp has been achieved using co-culture strategies, through the combined activity of xylanase and laccase [54]. The ligninolytic enzyme system is used in bio-bleaching of craft pulp and in other industries such as for the stabilization of wine and fruit juices, denim washing [49], the cosmetic industry and biosensors [1,34]. Fungi are the most potent producers of lignin degrading enzymes. White rot fungi have been specifically studied for the production of these enzymes by Robinson *et al.* [50–52]. For the economical production of ligninolytic enzymes, agricultural residues have been used as the substrate in microbial production of lignin degrading enzymes [34].

Thermophilic laccase enzyme is of particular use in the pulping industry. Recently, Gali and Kotteazeth [55] reported the biophysical characterization of thermophilic laccase isoforms. These were initially isolated from the xerophytic plant species *Cereus pterogonus* and *Opuntia vulgaris* and showed thermophilic property [56–58]. In order to prepare laccase enzymes with special characteristics, several studies have been conducted to provide a scientific basis for the employment of laccases in biotechnological processes [59–62]. Forms of laccase with unusual properties have been isolated from the basidiomycetes culture of *Steccherinum ochraceum* [63], *Polyporus versicolor* [64] and a microbial consortium [65].

3.6. Cellulase

Cellulase enzymes are the third most important enzyme for industrial uses: world-wide research has been focused on the commercial potential of cellulolytic enzymes for the commercial production of glucose feedstock from the agricultural cellulosic materials [1]. The significance of cellulose hydrolyzing thermophilic enzymes in various industries includes the production of bio-ethanol and value-added organic compounds from renewable agricultural residues [66]. Cellulose is the most abundant natural resource available globally for bioconversion into numerous products in bio-industry on a commercial scale. For efficient bioconversion a strategy of efficient saccharification using cellulolytic enzymes is required. Hardiman *et al.* [66] used the approach of thermophilc directed evolution of a thermophilic β-glucosidase.

Cellulase is complex of three important enzymes which work synergistically owing to the crystalline and amorphous complex structure of cellulose. These enzymes, acting synergistically, hydrolyse cellulose to cello-biose, glucose and oligo-saccharides. Endoglucanase enzyme is the first one acting on amorphous cellulose fibers, attacking the glucose-polymer chain randomly, which releases small fibers consisting of free-reducing and non-reducing ends. The free-ends of the chain are then exposed to the activity of exoglucanase enzyme, which produces cellobiose. The third component of cellulase is β-glucosidase, which hydrolyses the cellobiose, producing the glucose as the final product of cellulose sacharification.

Thermostability is an important technical property for cellulases: since the saccharification of cellulose is faster at higher temperatures, the stability of enzyme activity is necessary to be maintained for the completion of the process. Though the enzymes have been prepared using thermophilic microorganisms, these enzyme preparations are not necessarily heat-stable. The activity profile for the thermal activation and stability of cellulases derived from two *Basidiomycetes* cultures was studied by Nigam and Prabhu [67]. The results proved that the prior heat-treatment of enzyme preparation caused activation of exo- and endo-glucanase activities, and improved the stability of enzymes over a period of reaction time. Therefore, the efficiency of cellulolytic enzymes may be increased by heat-treatment, by incubating buffered enzyme preparations without cellulose or substrates prior to the saccharification process [67].

Cellulolytic enzymes have been produced by a range of microorganisms including bacteria and fungi. The studies have been performed for the biosynthesis of a high-activity preparation in high yields [68–70]. Researchers have cultivated microorganisms to achieve cellulases of desired quality under submerged and solid state fermentation conditions for the economical production of enzyme using waste agricultural residues [1].

3.7. Miscellaneous Enzymes in Biotechnology

Various enzymes other than those described above have a significant place in the list of microbial enzymes, which have established their applications in bio-industries. Lipases have been widely studied for their properties and utilization in many industries [71–75]. Pectinases have established their role in the fruit and juice industries [76]. Certain enzymes are specifically required in pharmaceutical industry for diagnostic kits and analytical assays [77–80].

Bornscheuer *et al.* [81] have currently mentioned that in all the research and developments so far in the field of biocatalysis, the researchers have contributed in three waves of outcomes. These innovations have played an important role in the establishment of current commercially successful level of bio-industries. As a result recent bioprocess-technology is capable of meeting future challenges and the requirements of conventional and modern industries, for example Trincone [82] has reviewed the options for unique enzymatic preparation of glycosides. Earlier enzymatic process were performed within the limitations of an enzyme, whereas currently with the knowledge of modern techniques, the enzyme can be engineered to be a suitable biocatalyst to meet the process requirement. Riva [83] has identified the scope of a long-term research in biocatalysis, since there are underlying problems in the shift from classical processes to bio-based processes for commercial market.

Table 1 summarizes some enzymes produced by microorganisms possessing special characteristics useful in various bio-processes. There is a tremendous scope for research and development to meet the challenges of third generation biorefeneries [83], for the production of numerous chemicals and bio-products from renewable biomasses [34]; or by the new glycoside hydrolases [82]; or new enzymes found in marine environments [84]. Although the research for the hemicellulases as important biorefining enzymes has not well established, biocatalysis for xylan processing is slowly progressing and a wide range of hemicellulases have been isolated and characterized [85]. Specifically about the biobased glycosynthesis, Trincone [82] has mentioned that the new prospects are open for the use of pentose sugars as main building blocks for engineered pentosides to be used as non-ionic surfactants or as the ingredients for prebiotic food and feed preparations.

Table 1. A summarized overview of some microbial enzymes with special characteristics of industrial importance.

Enzyme	Properties	Producer Microbes-	Applications	Reference
PROTEASE (Proteolytic activity)	Acidic, Neutral, Alkaline, Thermophilic, Active in presence of inhibitory compounds	*Bacilli*; *Pseudomonas*; *Clostridium*; *Rhizopus*; *Penicillium*; *Aspergillus*	Washing Powders; Detergents; Tannery; Food Industry; Leather processing; Pharmaceuticals; Molecular Biology; Peptide synthesis	[1,3,11–19,34]
KERATINASE (Keratin-hydrolysing activity)	Specific Proteolytic Activity for Insoluble & Fibrous Proteins in furs, feathers, wool, hair; Thermophilic; Alkalophilic; Oxidation-Resistant	Bacteria; Actinomycetes; Fungi	Animal Feed Production; Textile Processing; Detergent Formulation; Leather Manufacturing; Medicine	[1,20–25,34]
AMYLASE (Starch-hydrolyzing activity)	Thermotolerant, Thermostable, Alkali-resistant-Exo-, endo-, de-branching, cyclodextrin-producing enzymes	*Bacillus* sp.; *Geobacillus*	Starch industry (for liquefaction); Paper, Food industry (Glucose & Maltose syrups, High Fructose Corn syrups, clarified fruit-juices); Pharmaceutical industries (Digestive aid); Brewing Industry (Starch-processing); Textile industry (Warp-sizing of fibers); Baking industry (delayed staling)	[1,26–34]

Table 1. *Cont.*

Enzyme	Properties	Producer Microbes-	Applications	Reference
XYLANASE (Xylan–Pentose polymer hydrolyzing activity)	Extremophilic characteristics–Alkalophilic, Thermophilic & Thermostable	*Thermoactinomyces thalophilus*; *Bacillus* sp.; *Humicola insolens. Bispora* (acidophilic fungus)	Pentose production - Bioconversion of hemicellulose for fuel & Chemicals; Fruit-juice clarification; Improving rumen digestion; Paper industry- selective removals of xylans from kraft-pulp; Brewing industry	[1,34–46]
LIGNINASE (Ligninolytic-Complex-enzyme)	Oxidative properties in Lignin peroxidase, Manganese peroxidase & Laccase; Thermophilic	Basidiomycetes strains—*Steccherinum ochraceum*, *Polyporus versicolor*, *Panus tigrinus*	Denim washing; Bio-sensors; Bio-bleaching of Kraft-pulp; Bioremediation; Pollution-control; Treatment of recalcitrant chemicals in Textile and Industrial effluents	[1,34,47–65]
CELLULASE (Cellulolytic-complex enzyme)	Saccahrification of crystalline & amorphous cellulose; Thermophilic; Thermostable	Basidiomycetes strains *Polyporus* sp.; *Pleurotus* sp.; *Trichoderma* sp.; *Aspergillus* sp.	Glucose feedstock from cellulose; Bio-refinery; Bio-ethanol; Paper-pulp industry	[1,34,66–70]
LIPASE (Lipolytic activity)	Fat- splitting; Stereoselectivity; Racemic-Resolution activity; Solvents-resistant; Thermotolerant	Yeasts and Fungal strains- *Candida* sp., *Aspergillus* sp, *Penicillium* sp., *Rhizopus*, *Mucor*	Detergents; Dairy Industry- oils, fats, Butter, Cream, Fat-Spreads; Feed supplement; Therapeutic agent	[1,34,71–75]

4. Conclusions

Biotechnology is utilizing a wide range of enzymes synthesized on a commercial scale employing purposely screened microorganisms. Selected microorganisms have been characterized, purposely designed and optimized to produce a high-quality enzyme preparation on large scales for industrial applications. Different industries require enzymes for different purposes; hence microbial enzymes have been studied for their special characteristics applicable in various bio-processes. Recent molecular biology techniques have allowed to tailor a specific microorganism, to produce not only the high yields of an enzyme, but also enzyme with desired special characteristics such as thermostability, tolerance at high temperature and its stability in acidic or alkaline environment, and retaining the enzyme activity under severe reaction conditions such as in presence of other metals and compounds.

References

1. Pandey, A.; Selvakumar, P.; Soccol, C.R.; Nigam, P. Solid-state fermentation for the production of industrial enzymes. *Curr. Sci.* **1999**, *77*, 149–162.
2. Chirumamilla, R.R.; Muralidhar, R.; Marchant, R.; Nigam, P. Improving the quality of industrially important enzymes by directed evolution. *Mol. Cell. Biochem.* **2001**, *224*, 159–168.
3. Kumar, C.G.; Takagi, H. Microbial alkaline proteases: From a bioindustrial viewpoint. *Biotechnol. Adv.* **1999**, *17*, 561–594.
4. Ahmed, S.; Riaz, S.; Jamil, A. Molecular cloning of fungal xylanases: An overview. *Appl. Microbiol. Biotechnol.* **2009**, *84*, 19–35.
5. Wang, X.; Li, D.; Watanabe, T.; Shigemori, Y.; Mikawa, T.; Okajima, T.; Mao, L.Q.; Ohsaka, T. A glucose/o-2 biofuel cell using recombinant thermophilic enzymes. *Int. J. Electrochem. Sci.* **2012**, *7*, 1071–1078.
6. Banat, I.M.; Nigam, P.; Marchant, R. Isolation of a thermotolerant, fermentative yeasts growing at 52 °C and producing ethanol at 45 °C & 50 °C. *World J. Microbiol. Biotechnol.* **1992**, *8*, 259-263.
7. Wati, L.; Dhamija, S.S.; Singh, D.; Nigam, P.; Marchant, R. Characterisation of genetic control of thermotolerance in mutants of *Saccharomyces cerevisiae*. *Genet. Eng. Biotechnol.* **1996**, *16*, 19-26.
8. Zhang, S.B.; Wu, Z.L. Identification of amino acid residuesresponsible for increased thermostability of feruloyl esterase A from *Aspergillus niger* using the PoPMuSiC algorithm. *Bioresour. Technol.* **2011**, *102*, 2093–2096.
9. Berka, R.M.; Grigoriev, I.V.; Otillar, R.; Salamov, A.; Grimwood, J.; Reid, I.; Ishmael, N.; John, T.; Darmond, C.; Moisan, M.C.; *et al.* Comparative genomic analysis of the thermophilic biomass-degrading fungi. *Myceliophthora thermophila* and *Thielavia terrestris*. *Nat. Biotechnol.* **2011**, *29*, 922–927.
10. Cai, H.; Shi, P.; Bai, Y.; Huang, H.; Yuan, T.; Yang, P.; Luo, H.; Meng, K.; Yao, B. A novel thermoacidophilic family 10 xylanase from *Penicillium pinophilum* C1. *Process Biochem.* **2011**, *46*, 2341–2346.
11. Mukherjee, A.K.; Adhikari, H.; Rai, S.K. Production of alkaline protease by a thermophilic *Bacillus subtilis* under solid-state fermentation (SSF) condition using *Imperata* cylindrical grass and potato peel as low-cost medium: Characterization and application of enzyme in detergent formulation. *J. Biochem. Eng.* **2008**, *39*, 353–361.
12. Rahman, R.N.Z.R.A.; Basri, M.; Salleh, A.B. Thermostable alkaline protease from *Bacillus stearothermophilus* F1; Nutritional factors affecting protease production. *Ann. Microbiol.* **2003**, *53*, 199–210.
13. Chudasama, C.J.; Jani, S.A.; Jajda, H.M.; Pate, H.N. Optimization and production of alkaline protease from *Bacillus thuringiensis* CC7. *J. Cell Tissue Res.* **2010**, *10*, 2257–2262.
14. Genckal, H.; Tari, C. Alkaline protease production from alkalophilic *Bacillus* sp. isolated from natural habitats. *Enzym. Microb. Technol.* **2006**, *39*, 703–710.
15. Gupta, R.; Beg, Q.K.; Lorenz, P. Bacterial alkaline proteases: Molecular approaches and Industrial Applications. *Appl. Microbiol. Biotechnol.* **2002**, *59*, 15–32.

16. Vijayalakshmi, S.; Venkat Kumar, S.; Thankamani, V. Optimization and cultural characterization of *Bacillus* RV.B2.90 producing alkalophilic thermophilic protease. *Res. J. Biotechnol.* **2011**, *6*, 26–32.
17. Gupta, A.; Joseph, B.; Mani, A.; Thomas, G. Biosynthesis and properties of an extracellular thermostable serine alkaline protease from *Virgibacillus pantothenticus*. *World J. Microbiol. Biotechnol.* **2008**, *24*, 237–243.
18. Johnvesly, B.; Naik, G.K. Studies on production of thermostable alkaline protease from thermophilic and alkaliphilic *Bacillus* sp. JB-99 in a chemical defined medium. *Process Biochem.* **2001**, *37*, 139–144.
19. Hadj-Ali, N.E.; Rym, A.; Basma, G.F.; Alya, S.K.; Safia, K.; Moncef, N. Biochemical and molecular characterization of a detergent stable alkaline serineprotease from a newly isolated *Bacillus licheniformis* NH1. *Enzym. Microb. Technol.* **2007**, *40*, 515–523.
20. Gushterova, A.; Vasileva-Tonkova, E.; Dimova, E.; Nedkov, P.; Haertle, T. Keratinase production by newly isolated Antarctic actinomycete strains. *World J. Microbiol. Biotechnol.* **2005**, *21*, 831–834.
21. Brandelli, A.; Daroit, D.J.; Riffel, A. Biochemical features of microbial keratinases and their production and applications. *Appl. Microbiol. Biotechnol.* **2010**, *85*, 1735–1750.
22. Gupta, R.; Sharma, R.; Beg, Q.K. Revisiting microbial keratinases: Next generation proteases for sustainable biotechnology. *Crit. Rev. Biotechnol.* **2013**, *33*, 216–228.
23. Baihong, L.; Juan, Z.; Zhen, F.; Lei, G.; Xiangru, L.; Guocheng, D.; Jian, C. Enhanced thermostability of keratinase by computational design and empirical mutation. *J. Ind. Microbiol. Biotechnol.* **2013**, *40*, 697–704.
24. Indhuja, S.; Shiburaj, S.; Pradeep, N.S.; Thankamani, V.; Abraham, T.K. Extracellular keratinolytic proteases from an Alkalophilic *Streptomyces albidoflavus* TBG-S13A5: Enhanced production and characterization. *J. Pure Appl. Microbiol.* **2012**, *6*, 1599–1607.
25. Liu, B.; Zhang, J.; Li, B.; Liao, X.; Du, G.; Chen, J. Expression and characterization of extreme alkaline, oxidation-resistant keratinase from *Bacillus licheniformis* in recombinant *Bacillus subtilis* WB600 expression system and its application in wool fiber processing. *World J. Microbiol. Biotechnol.* **2013**, *29*, 825–832.
26. Nigam, P.; Singh, D. Enzyme and microbial systems involved in starch processing. *Enzym. Microb. Technol.* **1995**, *17*, 770–778.
27. Pandey, A.; Soccol, C.R.; Nigam, P. Biotechnological potential of agro-industrial residues, II-Cassava Bagasse. *Bioresour. Technol.* **2000**, *74*, 81–87.
28. Sivaramakrishnan, S.; Gangadharan, D.; Nampoothiri, K.M.; Soccol, C.R.; Pandey, A. α-amylases from microbial sources – An overview on recent developments. *Food Technol. Biotechnol.* **2006**, *44*, 173–184.
29. Kumar, J.; Dahiya, J.S.; Singh, D.; Nigam, P. Production of endo-1, 4- β-glucanase by a biocontrol fungus *Cladorrhinum foecundissimum*. *Bioresour. Technol.* **2000**, *75*, 95–97.
30. Singh, D.; Dahiya, J.S.; Nigam, P. Simultaneous raw starch hydrolysis and ethanol fermentation by glucoamylase from *Rhizoctonia solani* and *Saccharomyces cerevisiae*. *J. Basic Microbiol.* **1995**, *35*, 117–121.

31. Pandey, A.; Nigam, P.; Soccol, C.R.; Soccol, V.T.; Singh, D.; Mohan, R. Advances in Microbial Amylases. *Biotechnol. Appl. Biochem.* **2000**, *31*, 135–152.
32. Li, Y.; Niu, D.; Zhang, L.; Wang, Z.; Shi, G. Purification, characterization and cloning of a thermotolerant isoamylase produced from Bacillus sp. CICIM 304. *J. Ind. Microbiol. Biotechnol.* **2013**, *40*, 437–446.
33. Gurumurthy, D.M.; Neelagund, S.E. Molecular characterization of industrially viable extreme thermostable novel alpha-amylase of geobacillus sp Iso5 Isolated from geothermal spring. *J. Pure Appl. Microbiol.* **2012**, *6*, 1759–1773.
34. Nigam, P., Pandey, A., Eds.; *Biotechnology for Agro-Industrial Residues Utilisation*; Publisher Springer Science Business Media B.V. 2009; pp. 1–466.
35. Polizeli, M.L.; Rizzatti, A.C.; Monti, R.; Terenzi, H.F.; Jorge, J.A.; Amorim, D.S. Xylanases from fungi: Properties and industrial applications. *Appl. Microbiol. Biotechnol.* **2005**, *67*, 577–591.
36. Collins, T.; Gerday, C.; Feller, G. Xylanases, xylanase families and extremophilic xylanases. *FEMS Microbiol. Rev.* **2005**, *29*, 3–23.
37. Srinivasan, M.C.; Rele, M.V. Cellulase free xylanase from microorganisms and their applications to pulp and paper biotechnology: An overview. *Indian J. Microbiol.* **1995**, *35*, 93–101.
38. Garg, A.P.; Roberts, J.C.; McCarthy, A. Bleach boosting effect of cellulase free xylanase of *Streptomyces thermoviolaceus* and its comparison with two commercial enzyme preparations on birchwood Kraft pulp. *Enzym. Microb. Biotechnol.* **1998**, *22*, 594–598.
39. Kohli, U.; Nigam, P.; Singh, D.; Chaudhary, K. Thermostable, alkalophilic and cellulase free xylanase production by *Thermonoactinomyces* thalophilus subgroup C. *Enzym. Microb. Technol.* **2001**, *28*, 606–610.
40. Marques, S.; Alves, L.; Ribeiro, S.; Girio, F.M.; Amaralcollaco, M.T. Characterisation of a thermotolerant and alkalotolerant xylanase from a *Bacillus* sp. *Appl. Biochem. Biotechnol. A* **1998**, *73*, 159–172.
41. Jhamb, K.; Sahoo, D.K. Production of soluble recombinant proteins in *Escherichia coli*: Effects of process conditions and chaperone co-expression on cell growth and production of xylanase. *Bioresour. Technol.* **2012**, *123*, 135–143.
42. Prade, R.A. Xylanases: From biology to biotechnology. *Biotechnol. Genet. Eng. Rev.* **1996**, *13*, 100–131.
43. Luo, H.; Wang, K.; Huang, H.; Shi, P.; Yang, P.; Yao, B. Gene cloning, expression and biochemical characterization of an alkali-tolerant b-mannanase from *Humicola insolens* Y1. *J. Ind. Microbiol. Biotechnol.* **2012**, *39*, 547–555.
44. Luo, H.; Li, J.; Yang, J.; Wang, H.; Yang, Y.; Huang, H.; Shi, P.; Yuan, T.; Fan, Y.; Yao, B. A thermophilic and acid stable family-10 xylanase from the acidophilic fungus *Bispora* sp. MEY-1. *Extremophiles* **2009**, *13*, 849–857.
45. Mamo, G.; Thunnissen, M.; Hatti-Kaul, R.; Mattiasson, B. An alkaline active xylanase: Insights into mechanisms of high pH catalytic adaptation. *Biochimie* **2009**, *91*, 1187–1196.
46. Du, Y.; Shi, P.; Huang, H.; Zhang, X.; Luo, H.; Wang, Y.; Yao, B. Characterization of three novel thermophilic xylanases from Humicola insolens Y1 with application potentials in the brewing industry. *Bioresour. Technol.* **2013**, *130*, 161–167.

47. Nigam, P.; Pandey, A.; Prabhu, K.A. Cellulase and ligninase production by Basidiomycetes culture in solid-state fermentation. *Biol. Wastes* **1987a**, *20*, 1–9.
48. Nigam, P.; Pandey, A.; Prabhu, K.A. Ligninolytic activity of two Basidiomycetes moulds in the decomposition of bagasse. *Biol. Wastes* **1987b**, *21*, 1–10.
49. Dahiya, J.S.; Singh, D.; Nigam, P. Characterisation of laccase produced by *Coniotherium minitans*. *J. Basic Microbiol.* **1998**, *38*, 349–359.
50. Robinson, T.; Chandran, B.; Nigam, P. Studies on the production of enzymes by white-rot fungi for the decolourisation of textile dyes. *Enzym. Microb. Technol.* **2001**, *29*, 575–579.
51. Robinson, T.; Chandran, B.; Nigam, P. Studies on the decolourisation of an artificial effluent through lignolytic enzyme production by white-rot fungi in N-rich and N-limited media. *Appl. Microbiol. Biotechnol.* **2001b**, *57*, 810–813.
52. Robinson, T.; Nigam, P. Remediation of textile dye-waste water using a white rot fungus *Bjerkandera adusta* through solid state fermentation (SSF). *Appl. Biochem. Biotechnol.* **2008**, *151*, 618-628.
53. Dahiya, J.; Singh, D.; Nigam, P. Decolourisation of synthetic and spentwash-melanoidins using the white-rot fungus *Phanerochaete chrysosporium* JAG-40. *Bioresour. Technol.* **2001**, *78*, 95–98.
54. Dwivedi, P.; Vivikanand, V.; Pareek, N.; Sharma, A.; Singh, R.P. Bleach enhancement of mixed wood pulp by xylanase-laccase concoction derived through co-culture strategy. *Appl. Biochem. Biotechnol.* **2010**, *160*, 255–268.
55. Gali, N.K.; Kotteazeth, S. Biophysical characterization of thermophilic laccase from the xerophytes: *Cereus pterogonus* and *Opuntia vulgaris*. *Cellulose* **2013**, *20*, 115–125.
56. Gali, N.K.; Kotteazeth, S. Isolation, purification and characterization of thermophilic laccase from xerophyte *Cereus pterogonus*. *Chem. Nat. Compd.* **2012**, *48*, 451–456.
57. Kumar, G.N.; Srikumar, K. Thermophilic laccase from xerophyte species *Opuntia vulgaris*. *Biomed. Chromatogr.* **2011**, *25*, 707–711.
58. Kumar, G.N.; Srikumar, K. Characterization of xerophytic thermophilic laccase exhibiting metal ion-dependent dye decolorization potential. *Appl. Biochem. Biotechnol.* **2012**, *167*, 662–676.
59. Quaratino, D.; Federici, F.; Petruccioli, M.; Fenice, M.; D'Annibale, A. Production, purification and partial characterisation of a novel laccase from the white-rot fungus *Panus tigrinus* CBS 577.79. *Anton. Leeuw. Int J.G.* **2007**, *91*, 57–69.
60. Uthandi, S.; Saad, B.; Humbard, M.A.; Maupin-Furlow, J.A. LccA, an archaeal laccase secreted as a highly stable glycoprotein into the extracellular medium by *Haloferax volcanii*. *Appl. Environ. Microbiol.* **2010**, *76*, 733–743.
61. Papinutti, L.; Dimitriu, P.; Forchiassin, F. Stabilization studies of *Fomes sclerodermeus* laccases. *Bioresour. Technol.* **2008**, *99*, 419–424.
62. Mishra, A.; Kumar, S. Kinetic studies of laccase enzyme of *Coriolus versicolor* MTCC 138 in an inexpensive culture medium. *Biochem. Eng. J.* **2009**, *46*, 252–256.
63. Chernykh, A.; Myasoedova, N.; Kolomytseva, M.; Ferraroni, M.; Briganti, F.; Scozzafava, A.; Golovleva, L. Laccase isoforms with unusual properties from the basidiomycete Steccherinum ochraceum strain 1833. *J. Appl. Microbiol.* **2008**, *105*, 2065–2075.
64. Nigam, P.; Prabhu, K.A. The effects of some added carbohydrates on cellulases and ligninase and decomposition of bagasse. *Agric. Wastes* **1986**, *17*, 293–299.

65. Wongwilaiwalin, S.; Rattanachomsri, U.; Laothanachareon, T.; Eurwilaichitr, L.; Igarashi, Y.; Champreda, V. Analysis of a thermophilic lignocellulose degrading microbial consortium and multi-species lignocellulolytic enzyme system. *Enzym. Microb. Technol.* **2010**, *47*, 283–290.
66. Hardiman, E.; Gibbs, M.; Reeves, R.; Bergquist, P. Directed Evolution of a thermophilic beta-glucosidase for Cellulosic Bioethanol Production. *Appl. Biochem. Biotechnol.* **2010**, *161*, 301–312.
67. Nigam, P.; Prabhu, K.A. Thermal activation and stability of cellulases derived from two Basidiomycetes. *Biotechnol. Lett.* **1988**, *10*, 919–920.
68. Nigam, P.; Prabhu, K.A. Effect of cultural factors on cellulase biosynthesis in submerged bagasse fermentation by basidiomycetes cultures. *J. Basic Microbiol.* **1991**, *31*, 285–292.
69. Nigam, P.; Prabhu, K.A. Isolation and recovery of cellulase and ligninase from crude enzymes produced by two basidiomycetes cultures in submerged bagasse fermentation. *Sharkara* **1988**, *27*, 40–46.
70. Nigam, P.; Prabhu, K.A. Microbial degradation of bagasse: Isolation and cellulolytic properties of Basidiomycetes Spp. from biomanure from a biogas plant. *Agric. Wastes* **1985**, *12*, 273–285.
71. Reddivari, M.; Chirumamilla, R.; Nigam, P. Understanding lipase stereoselectivity. *World J. Microbiol. Biotechnol.* **2002**, *18*, 81–97.
72. Muralidhar, R.; Chirumamilla, R.R.; Nigam, P. Resolution of proglumide using lipase from *Candida cylindraceae. Bioorg. Med. Chem.* **2002**, *10*, 1471–1475.
73. Muralidhar, R.; Chirumamilla, R.R.; Marchant, R.; Nigam, P. A response surface approach for the comparison of lipase production by *Candida cylindraceae* using two different carbon sources. *Biochem. Eng. J.* **2001**, *9*, 17–23.
74. Pandey, A.; Benzamin, S.; Soccol, C.R.; Nigam, P.; Krieger, N.; Soccol, V.T. The realm of microbial lipases in biotechnology. *Biotechnol. Appl. Biochem.* **1999**, *29*, 119–131.
75. Muralidhar, R.; Chirumamilla, C.; Marchant, R.; Nigam, P. Lipases in racemic resolutions. *J. Chem. Technol. Biotechnol.* **2001**, *76*, 3–8.
76. Sunnotel, O.; Nigam, P. Pectinolytic activity of bacteria isolated from soil and two fungal strains during submerged fermentation. *World J. Microbiol. Biotechnol.* **2002**, *18*, 835–839.
77. Zhou, D.M.; Nigam, P.; Marchant, R.; Jones, J. Production of salicylate hydroxylase from Pseudomonas putida UUC-1 and its application in the construction of biosensor. *J. Chem. Technol. Biotechnol.* **1995**, *64*, 331–338.
78. Banat, I.M.; Marchant, A.; Nigam, P.; Gaston, S.J.S.; Kelly, B.; Marchant, R. Production, partial characterization and potential diagnostic use of salicylate hydroxylase from Pseudomonas putida UUC-1. *Enzym. Microb. Technol.* **1994**, *16*, 665–670.
79. Nigam, P.; Marchant, R. Production of enzyme dihydrofolate reductase by methotrexate-resistant bacteria isolated from soil. *J. Chem. Technol. Biotechnol.* **1993**, *56*, 35–40.
80. Nigam, P.; Banat, I.M.; Kelly, B.; Marchant, R. Dihydrofolate reductase synthesis in continuous culture using methotrexate-resistant *Escherichia coli. Enzym. Microb. Technol.* **1993**, *15*, 652–656.
81. Bornscheuer, U.T.; Huisman, G.W.; Kazlausaks, R.J.; Lutz, S.; Moore, J.C.; Robins, K. Engineering the third wave of biocatalysis. *Nature* **2012**, *485*, 185–194.
82. Trincone, A. Angling for uniqueness in enzymatic preparation of glycosides. *Biomolecules* **2013**, *3*, 334–350.

83. Riva, S. 1983–2013: The long wave of biocatalysis. *Trends Biotechnol.* **2013**, *31*, 120–121.
84. Trincone, A. Potential biocatalysts originating from sea environments. *J. Mol. Catal. B-Enzym.* **2010**, *66*, 241–256.
85. Dumon, C.; Songa, L.; Bozonneta, S.; Fauréa, R.; O'Donohue, M.J. Progress and future prospects for pentose-specific biocatalysts in biorefining. *Proc. Biochem.* **2012**, *47*, 346–357.

8

Arming Technology in Yeast—Novel Strategy for Whole-Cell Biocatalyst and Protein Engineering

Kouichi Kuroda and Mitsuyoshi Ueda *

Division of Applied Life Sciences, Graduate School of Agriculture, Kyoto University, Sakyo-ku, Kyoto 606-8502, Japan; E-Mail: k_kuro@kais.kyoto-u.ac.jp

* Author to whom correspondence should be addressed; E-Mail: miueda@kais.kyoto-u.ac.jp

---

Abstract: Cell surface display of proteins/peptides, in contrast to the conventional intracellular expression, has many attractive features. This arming technology is especially effective when yeasts are used as a host, because eukaryotic modifications that are often required for functional use can be added to the surface-displayed proteins/peptides. A part of various cell wall or plasma membrane proteins can be genetically fused to the proteins/peptides of interest to be displayed. This technology, leading to the generation of so-called "arming technology", can be employed for basic and applied research purposes. In this article, we describe various strategies for the construction of arming yeasts, and outline the diverse applications of this technology to industrial processes such as biofuel and chemical productions, pollutant removal, and health-related processes, including oral vaccines. In addition, arming technology is suitable for protein engineering and directed evolution through high-throughput screening that is made possible by the feature that proteins/peptides displayed on cell surface can be directly analyzed using intact cells without concentration and purification. Actually, novel proteins/peptides with improved or developed functions have been created, and development of diagnostic/therapeutic antibodies are likely to benefit from this powerful approach.

Keywords: cell surface engineering; arming yeast; whole-cell biocatalyst; molecular display; arming technology; cell surface display; protein engineering; directed evolution; combinatorial bioengineering; single cell analysis

---

1. Introduction

Arming technology in yeast using the cell surface display system is an innovative technology for the construction of whole-cell biocatalysts and protein engineering through high-throughput screening of protein libraries. Although intracellular expression and extracellular secretion have been employed in conventional strategies for molecular breeding, the cell surface (including the cell wall and plasma membrane) is an attractive location for heterologous gene expression. Proteins localized at the cell surface play an important role in signal transduction, recognition and transport of environmental substances, morphology formation, and various other reactions. In arming technology, functional heterologous proteins/peptides are genetically immobilized on the cell surface by fusion with the domain for cell wall- or plasma membrane-anchoring domains. Cell surface-engineered yeasts constructed using the arming technology, have been termed "arming yeasts" [1–3]. Arming technology has advantages that are not found in conventional intracellular expression and extracellular secretion approaches. Immobilization on the cell surface itself retains the features of immobilized enzymes, and increases the thermal stability of the displayed proteins/peptides [4,5]. Preparation of protein/peptide-displaying cells can be achieved only by cell cultivation, in which the two processes of production of proteins/peptides and immobilization on cell surface are simultaneously undertaken, leading to cost effectiveness. Furthermore, the displayed proteins/peptides can interact with environmental substances with high molecular mass that cannot be imported into a cell. Surface-engineered cells prepared by cell cultivation are ready to use as microparticles covered by proteins/peptides, whereas troublesome protein purification and concentration procedures are required for the analysis and subsequent application of proteins produced by intracellular expression and extracellular secretion. Therefore, arming technology is also suitable for high-throughput screening of proteins/peptides from the mutant library at the single-cell level, and amino acid sequences of the screened proteins/peptides can be easily determined by DNA sequencing of the introduced plasmid [6–10]. Therefore, arming technology has been applied to wide range of targets such as whole-cell biocatalysts, bioremediation, biosensors, live vaccines, high-throughput screening of ligand peptides for receptors, and protein engineering [2,3,11–14].

The cell surface display system has also been established in gram-negative bacteria, gram-positive bacteria, and phages [11,15–18]. So far, many proteins/peptides have been successfully displayed with maintenance of their functions. In contrast to bacteria and phages, yeasts are equipped with the quality control and modification systems of eukaryotic secretory pathways. Therefore, in the case of target proteins that have a high molecular mass or require glycosylation modification, yeasts are suitable hosts for cell surface display. In addition, simultaneous display of multiple kinds of proteins/peptides on the same cell surface can be performed in yeasts by using different auxotrophic markers, leading to the enhanced potential of surface-engineered yeasts. Recently, the cell surface display system has also been developed in *Aspergillus oryzae* as a eukaryote except yeasts [19,20]. Here, we review the advances of molecular breeding of novel cells and protein engineering by arming technology in yeasts such as *Saccharomyces cerevisiae*, *Pichia pastoris*, and *Yarrowia lipolytica*.

2. Cell Surface Display System in Yeast

Among yeasts, arming technology was first developed in *S. cerevisiae*. Subsequently, based on the success of this approach using full length or cell wall-anchoring domains of cell wall proteins from *S. cerevisiae*, similar display systems have been transferred into other yeasts, including *P. pastoris* and *Y. lipolytica*. The cell surface display systems (Figure 1) are classified into two systems. One is the *N*-terminus free display in which target proteins/peptides are produced as fusions with the secretion signal sequence at the *N*-terminus and the cell wall-anchoring domain at the *C*-terminus. The other is the *C*-terminus free display, in which secretion signal sequence, cell wall-anchoring domain, and target proteins/peptides are fused in this order. The effect of the orientation on the display efficiency and on the functional properties depends on the kinds of target proteins/peptides to be displayed. Therefore, the fusion order itself, namely whether the *N*- or *C*-terminus of target proteins/peptides is fused with cell wall-anchoring domain, is important. In addition, the length of spacer peptides between target proteins/peptides and the cell wall-anchoring domain is also important. By optimizing the spacer length, there are cases in which the function of displayed proteins/peptides has been improved [21,22].

Figure 1. Cell surface display system in *S. cerevisiae* (**a**) α-agglutinin-based display system; (**b**) **a**-agglutinin-based display system; (**c**) Flo1p-based display system; (**d**) Membrane display system by anchoring domain of Yps1p.

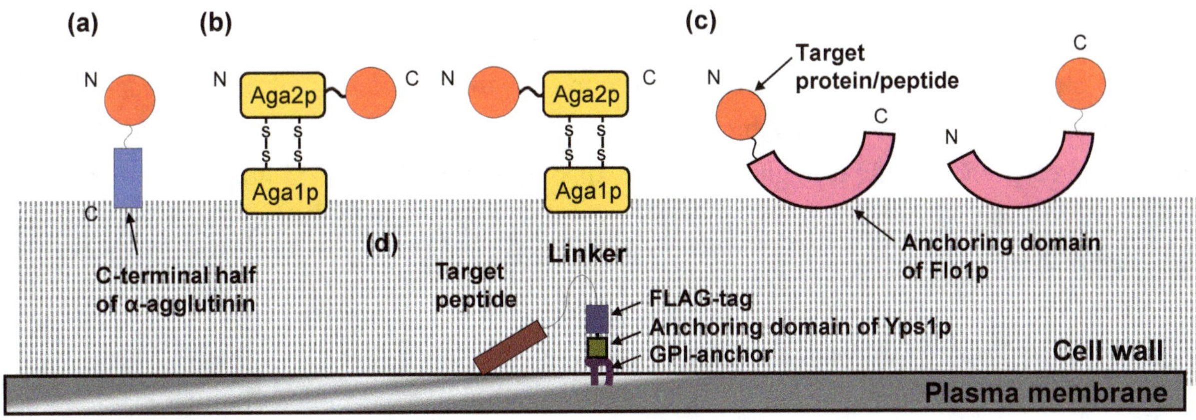

2.1. Cell Surface Display System in *S. cerevisiae*

S. cerevisiae has the "generally regarded as safe" (GRAS) approval of the Food and Drug Administration, and thus is suitable for industrial use. Among the cell wall proteins that are linked to the glucan layer by a covalent bond in yeast, α-agglutinin and **a**-agglutinin are representatives of cell wall-anchoring proteins used for cell surface display of target proteins/peptides. These proteins are involved in sexual adhesion of mating type α and **a** cells, and have characteristics specific to cell wall proteins. Specifically, they possess an *N*-terminal secretion signal sequence for transportation to the cell surface and *C*-terminal glycosylphosphatidylinositol (GPI) anchor attachment signal sequence for transient anchoring in the plasma membrane [23]. GPI-anchored proteins in the plasma membrane are released via cleavage by phosphatidylinositol-specific phospholipase C (PI-PLC), and then bind to the cell wall with a covalent bond by cell wall-anchoring domain. For the *N*-terminus free display system, α-agglutinin is the most frequently used approach (Figure 1a). Many proteins/peptides, including those with relatively large molecular masses and glycosylation requirements, have been successfully displayed on the yeast cell surface using the cell wall-anchoring domain of α-agglutinin [3,13,14,24]. In the α-agglutinin-based display system, target proteins/peptides are fused to the secretion signal sequence at the *N*-terminus and to the cell wall-anchoring domain (including the GPI anchor attachment signal sequence) at the *C*-terminus. In addition, Flo1p, which is a GPI-anchored cell wall protein involved in flocculation, is available for the *N*-terminal free display in a similar manner (Figure 1c) [25]. Other GPI-anchored cell wall proteins (Cwp1p, Cwp2p, Tip1p, Sed1p, YCR89w, and Tir1p) have the potential as cell wall-anchoring domains, although they are less frequently used because of the past successful performance of α- and a-agglutinin-based display system [26].

On the other hand, a-agglutinin has been used in the *C*-terminus free display system (Figure 1b). a-Agglutinin consists of Aga1p and Aga2p subunits [7]. The secreted Aga2p subunit is linked to the Aga1p subunit via two disulfide bonds, which are incorporated in the cell wall. Therefore, in the a-agglutinin-based display system, target proteins/peptides are fused with the *C*-terminus of Aga2p. In addition, this system has also been used in the *N*-terminal free display by fusing target proteins/peptides with the *N*-terminus of Aga2p [27]. Flo1p is also available for the *C*-terminus free display, in which truncated Flo1p without GPI anchor attachment signal sequence is used as an adhesive region (Figure 1c) [28]. Pir proteins (Pir1–4p) have been used as anchoring proteins in the *C*-terminus free display, and this permits extraction of displayed proteins/peptides from the cell wall by alkali treatment [29].

Several strategies have been employed to improve the efficiency of cell surface display in *S. cerevisiae*. Vector engineering by high copy number plasmid and the improvement of host strain enhanced the efficiency in the α-agglutinin-based display system [30]. Additionally, screening from a cDNA library identified five genes whose overexpression improved the efficiency of the **a**-agglutinin-based display system [31].

2.2. Cell Surface Display System in P. pastoris

The methylotrophic strain *P. pastoris* can grow on an economical carbon source and allows high-density culture. Therefore, it is also a suitable host for use in large-scale fermentation cultures of surface-engineered cells. In *P. pastoris*, both the *N*- and *C*-terminus-free display systems have been established using cell wall proteins (α-agglutinin, **a**-agglutinin, Flo1p, Pir1p, Sed1p, and Tip1p) from *S. cerevisiae* in the same strategy as outlined above [32–34].

2.3. Cell Surface Display System in Y. lipolytica

The oleaginous yeast *Y. lipolytica* is a heterothallic and dimorphic yeast, and has the high potential for secreting heterologous proteins, which is a preferred feature in industrial uses. Several cell wall proteins in *Y. lipolytica* have been identified. YlCWP1 and YlPIR1 from *Y. lipolytica*, which are GPI-anchored cell wall proteins, allow *N*-terminus free display and *C*-terminus free display, respectively [35].

2.4. Membrane Display System in S. cerevisiae

Almost all displayed proteins/peptides in yeasts are localized to the cell wall as described above. However, the display of proteins/peptides on the plasma membrane is desirable when there is a requirement for interaction with membrane proteins such as receptors. Membrane display of proteins/peptides is performed using the anchoring domain of Yps1p, a GPI-anchored plasma membrane protein. The fusion of this domain with the *C*-terminus of the peptide ligand allows the display on the plasma membrane and activation of either endogenous or heterologous G protein-coupled receptors (GPCRs) in *S. cerevisiae* [36,37].

3. Whole-Cell Biocatalyst by Arming Technology

3.1. Biofuel Production

Development of whole-cell biocatalysts to produce biofuel from biomass such as grain or cellulosic biomass has attracted attention as a means of creating a sustainable society based on biomass resources. Especially, consolidated bioprocessing (CBP) of lignocellulose to ethanol is an ideal system combining all processes such as enzyme production, hydrolytic degradation, and fermentation of sugar [38]. The arming technology has been applied to yeasts for the construction of whole-cell biocatalysts that can perform saccharification and fermentation. Starch and cellulose are major components in grain and cellulosic biomass, respectively. Therefore, the cell surface display of enzymes for hydrolytic degradation of these components was attempted in order to achieve CBP (Figure 2). As a starch-degrading enzyme, an exotype glucoamylase from *Rhizopus oryzae* was displayed on *S. cerevisiae* using the α-agglutinin-based display system, which allowed the direct production of ethanol from starch through the saccharification of starch on the cell surface and the subsequent fermentation of released glucose [39]. In addition to glucoamylase, an endotype α-amylase from *Bacillus stearothermophilus* was co-displayed with glucoamylase, leading to an improved production efficiency of ethanol from starch [40]. In the same way, α-amylase from *Streptococcus bovis* was also displayed using the Flo1p-based display system on glucoamylase-displaying yeast constructed by the α-agglutinin-based display system [41]. The arming yeasts described above show improved ethanol production and faster growth in the medium including starch as the sole carbon source.

Figure 2. Whole-cell biocatalyst constructed by arming technology and their applications.

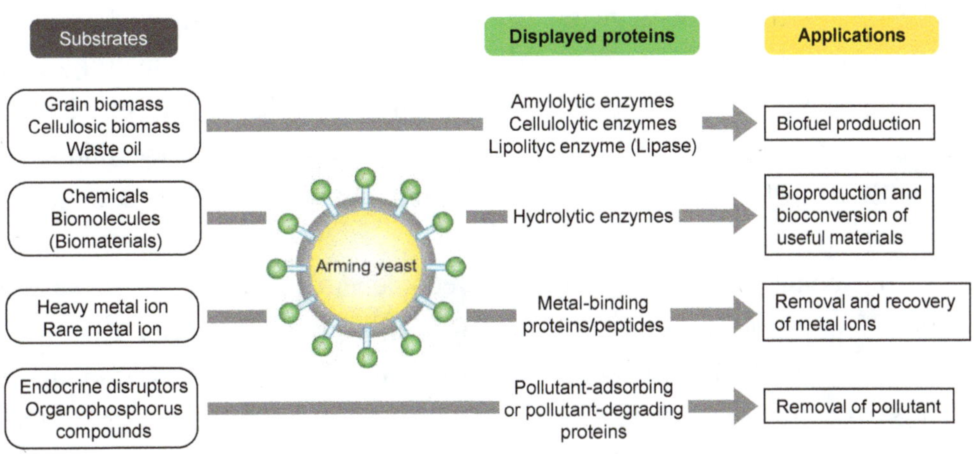

For degradation of cellulose, various enzymes including endoglucanases (EGs), cellobiohydrolases (CBHs), and β-glucosidases (BGLs) are necessary. Therefore, the potential of co-display on the same cell surface is an important feature in yeast, which is suitable for the reaction by multiple enzymes. Actually, the co-display of carboxymethylcellulase (CMCase) and BGL1 from *Aspergillus aculeatus* on *S. cerevisiae* enabled assimilation of cellobiose or oligosaccharide and growth in medium containing these materials as the sole carbon source [42]. EG acts randomly against the amorphous region of the cellulose chain to produce reducing and nonreducing ends, and CBH releases cellobiose from both ends. Therefore, efficient degradation of cellulose to cellobiose and cellooligosaccharides is achieved by the endo-exo synergism of EG and CBH. Finally, BGL hydrolyzes the generated cello-oligosaccharides to glucose. The arming yeast constructed by co-display of endoglucanase II (EG II) and cellobiohydrolase II (CBH II) from *Trichoderma reesei* and BGL1 from *A. aculeatus* can produce ethanol directly from phosphoric-acid-swollen cellulose due to the combined activities of these three cell surface enzymes [43]. In sake yeast, ethanol production from β-glucan can be achieved by cell surface display of EG and BGL from *A. oryzae* [44]. Furthermore, in arming yeast co-displaying EG II, CBH II, and BGL1, four non-conserved amino acids in the carbohydrate-binding module (CBM) of EG II were comprehensively mutated. In the case of co-displaying multiple kinds of enzymes on the same cell surface, the control or design of the display ratio of enzymes is a next challenge in arming technology. Recently, some approaches to this challenge were performed for optimized synergistic effects of displayed enzymes [45,46]. The optimal combination of yeasts displaying EG II with mutated CBMs was determined by inoculating a yeast library into selection liquid medium that included newspaper as the sole carbon source. The selected yeast mixture showed the improved ethanol production from newspaper, suggesting that this strategy is useful for the selection of the optimal combination of CBMs for each type of biomass [47].

The cellulosome, which is a multi-enzyme complex composed of scaffolding proteins and various cellulosomal enzymes, is produced by some clostridia and plays an important role in efficient degradation of polysaccharides in the plant cell wall. The complex is constructed through the interaction between a cohesin module in the scaffolding protein and a dockerin module in cellulosomal enzymes. Recently, reconstruction of mini-cellulosome on the yeast cell surface has been attempted. The engineered scaffolding protein consisting of three cohesin domains derived from *Clostridium thermocellum*, *Clostridium cellulolyticum*, and *Ruminococcus flavefaciens* was displayed on the *S. cerevisiae* cell surface using the **a**-agglutinin-based display system. The incubation of the scaffolding protein-displaying yeast with three recombinant cellulases (EG, CBH, and BGL) fused with a dockerin domain produced by *Escherichia coli* led to the synergistic hydrolytic degradation of cellulose [48]. Furthermore, a yeast consortium system has been developed, in which four kinds of engineered yeasts are co-cultivated [49]. In this system, construction of the mini-cellulosome was achieved by co-cultivation of a yeast displaying scaffolding protein and yeasts secreting three kinds of dockerin-fused enzymes.

Hemicellulose, including xylan, is also a major component of cellulosic biomass, Furthermore, ethanol production from hemicellulose is important for full utilization of cellulosic biomass. Xylan is hydrolyzed to xylo-oligosaccharides by endo-β-xylanase, and the produced xylo-oligosaccharides are hydrolyzed to D-xylose by β-xylosidase. *S. cerevisiae* co-displaying xylanase II (XYN II) from *T. reesei* and β-xylosidase (XylA) from *A. oryzae* was constructed using the α-agglutinin-based display system, and shown to hydrolyze xylan to xylose [50]. In a further development for ethanol production from xylan, xylose reductase (XR) and xylitol dehydrogenase (XDH) from *Pichia stipitis* and xylulokinase (XK) from *S. cerevisiae* were produced in the XYN II- and XylA-co-displaying yeast. The arming yeast catalyzes simultaneous saccharification and fermentation of xylan [50,51]. In addition to the two-step isomerization of xylose into xylulose, xylose isomerase (XI) is an alternative enzyme that shows great promise, as it is not associated with a cofactor imbalance. Recently, XI from *Clostridium cellulovorans* was successfully displayed and retained activity on *S. cerevisiae*. The constructed XI-displaying yeast could grow in medium containing xylose as the sole carbon source and directly produce ethanol from xylose [52].

Lignin is also a major component of cellulosic biomass, and inhibits cellulose degradation by cellulases, because of its physiological recalcitrancy and its masking the cellulose fibers. Therefore, the removal of lignin from cellulosic biomass is required for efficient degradation. Laccase from white-rot fungus, which participates in several biological pathways including lignin degradation, was displayed on *S. cerevisiae* via the α-agglutinin-based display system. By pretreatment of hydrothermally processed rice straw with laccase-displaying yeast, ethanol production by yeast co-displaying EG II, CBH II, and BGL1 was improved [53].

3.2. Bioproduction of Chemicals

Lipase is one of the most applied enzymes in cell surface display on yeast because lipases are used in wide range of industries and the display of lipase is an attractive approach for creating a whole-cell biocatalyst (Figure 2). *R. oryzae* lipase (ROL) and *Candida antarctica* lipase B (CALB) have been extensively studied and are one of the most widely used lipases. ROL was displayed on *S. cerevisiae* by the α-agglutinin-based display system, and the activity of displayed ROL was improved by inserting a spacer peptide with a Gly/Ser repeat between ROL and the cell wall-anchoring domain of α-agglutinin [21,22]. ROL displayed on *S. cerevisiae* showed higher activity than commercially available free and immobilized lipases in organic solvents because the displayed lipase is stabilized by the cell wall [54]. ROL-displaying *S. cerevisiae*, constructed by the Flo1p-based display system, could synthesize methylesters from triglyceride and methanol [28]. In addition, yeast displaying ROL could be used for the optical resolution of (*R,S*)-1-benzyloxy-3-chloro-2-propyl monosuccinate [55]. CALB has also been displayed on *S. cerevisiae* in several studies. Mutated CALB displayed on yeast using the α-agglutinin-based display system has higher thermal stability [56]. Furthermore, CALB-displaying yeasts constructed using the α-agglutinin- or Flo1p-display systems have been used for several ester syntheses with reduced amount of by-products [57–60]. With regard to lipases, several cell surface display strategies have been attempted in yeasts other than *S. cerevisiae*. In *P. pastoris*, CALB was displayed on the cell surface using the *S. cerevisiae* α-agglutinin-based display system. Compared to CALB-displaying *S. cerevisiae*, the engineered *P. pastoris* showed higher activity, and the ability to synthesize ethyl hexanoate was enhanced [61]. CALB was also displayed on *P. pastoris* via the Sed1p-based display system, and exhibited improved thermal stability [4]. ROL, *Y. lipolytica* lipases, and *Pseudomonas fluorescens* lipase were displayed on *P. pastoris* using the Flo1p-based display system [32,62,63]. Furthermore, *Y. lipolytica* lipase was displayed on *Y. lipolytica* by using the Flo1p-based display system [35].

Isoflavone aglycones, which are bioactive and easily adsorbed by human cells, are hydrolysates of isoflavone glycosides by β-glucosidase. Three kinds of β-glucosidases were independently displayed on *S. cerevisiae* using the α-agglutinin-based display system. Among them, BGL1-displaying yeast could convert isoflavone glycosides into isoflavone aglycones most efficiently [64]. Carnosine and chitosan oligosaccharides also show bioactivities such as antioxidant, antiglycation, cytoplasmic buffering, antitumor, and anticancer properties. The synthesis of carnosine from β-alanine and histidine was achieved through a reverse reaction catalyzed by human carnosinase (CN1)-displaying *S. cerevisiae* in organic solvents and ionic liquids [65]. Cell surface display of chitosanase from *Paenibacillus fukuinensis* enabled the production of chitosan oligosaccharides from chitosan [66]. In *Y. lipolytica*, alginate lyase from *Vibrio* sp. was displayed by fusion with the cell wall-anchoring domain of YlCWP1. The *Y. lipolytica* displaying alginate lyase could hydrolyze poly-β-D-mannuronate, poly-α-L-guluronate, and sodium alginate to produce oligosaccharides [67].

3.3. Bioadsorption

Arming technology has been applied to bioadsorption of toxic metal ions and rare-metal ions, leading to bioremediation and resource recovery (Figure 2) [12,13]. The cell surface display of metal-binding proteins/peptides enables the enhanced adsorption and recovery of metal ions on the cell surface. Cell surface adsorption has advantages that are lacking in intracellular accumulation; (i) no requirement for cell disruption for recovery of the adsorbed metal ions; (ii) repeated use of arming yeasts for the further adsorption of metal ions; (iii) rapid and selective adsorption of target metal ions. For the construction of bioadsorbents for bioremediation, toxic metal-binding proteins/peptides such as hexa-His, NP peptides (harboring the CXXEE metal fixation motif of the bacterial Pb^{2+}-transporting P1-type ATPases), and metallothionein were displayed on *S. cerevisiae* using the α-agglutinin-based display system [68–71]. The bioadsorbents constructed by arming technology showed enhanced adsorption of heavy metal ions. Furthermore, cell surface adsorption of metal ions improved cellular tolerance to heavy metal ions [69,71].

Bioadsorption to the cell surface has also been used for adsorption and recovery of rare metal ions. Some of these ions are essential trace metals that play an important role in living cells. For the specific binding and recovery of target metal ions, metal-responsive transcription factors that can bind and dissociate metal ions are displayed and repurposed as cell surface metal-binding proteins. Uptake of molybdenum, one of the rare metals, is regulated by transcription factor ModE that binds molybdate and controls the expression of downstream operon in *E. coli* [72]. Therefore, the full length or *C*-terminal domain of ModE from *E. coli* has been displayed on *S. cerevisiae* via the α-agglutinin-based display system. Molybdate adsorption was achieved using yeast displaying the *C*-terminal domain of ModE, and more than 50% of the molybdate adsorbed on the cell surface could be recovered by papain treatment [73]. Furthermore, a single amino acid mutation (T163Y) of the metal-binding pocket of ModE was efficient to convert it to a selective binder of tungstate, in contrast to the wild-type protein, which binds both tungstate and molybdate. Arming yeast displaying this mutant ModE was shown to selectively uptake tungstate [74].

3.4. Bioremediation

To remove or degrade environmental pollutants other than heavy metal ions, arming yeasts for bioremediation have been constructed (Figure 2). For example, to antagonize endocrine disruptors, which are environmental pollutants that perturb natural endocrine function, the ligand-binding domain of the rat estrogen receptor (ERLBD) has been displayed on *S. cerevisiae* using the α-agglutinin-based display system. The ERLBD-displaying yeast was able to bind estrogen-like compounds with an affinity comparable to the native receptor, suggesting the possible application in screening, adsorption, and removal of endocrine disruptor-like chemicals from the environment [75].

Organophosphorus compounds (OPs) are one of the most widely used pesticides, but discharge to the environment is a problem to be solved due to significant threat to public health. Therefore, to degrade OPs using a cell-based system, organophosphorus hydrolase (OPH) from *Flavobacterium* spp. was displayed on *S. cerevisiae* using the α-agglutinin- and Flo1p-based display systems. OPH-displaying yeasts showed hydrolase activity against paraoxon; the activity in the Flo1p-based display system was higher than that in the α-agglutinin-based display system because the active center is located near the *C*-terminal of OPH [76,77]. Furthermore, the development of a biosensor for the sensitive and rapid detection of OPs was achieved by detecting the protons generated by OPs hydrolysis. In this system, additional display of EGFP was performed in OPH-displaying yeast to evaluate the proton generation by the change in fluorescence intensity of EGFP. Together with OPH-dependent hydrolysis, the fluorescence intensity of EGFP was decreased [78]. In another system for OP sensing, the concentration of *p*-nitrophenol produced by OP hydrolysis in OPH-displaying yeast was examined [79].

3.5. Other Applications

Other applications of arming technology include biosensors, oral vaccines, antibodies, and stress tolerance. For the non-invasive sensing of environmental changes or monitoring of heterologous protein production, the cell surface display of fluorescent proteins on *S. cerevisiae* was used as a reporter system. The strategy for biosensor construction is that different fluorescent protein variants under the control of different promoters are displayed on the cell surface. GFP display has been placed under the control of the *GAPDH* promoter that is induced in the presence of glucose, whereas BFP was displayed under the control of the *UPR-ICL* promoter from *Candida tropicalis*, which is activated upon the exhaustion of glucose. Therefore, the concentration of intra- or extracellular glucose could be estimated by measuring the fluorescence intensities of GFP and BFP [80]. In addition, monitoring of intra- and extracellular concentrations of phosphate and ammonium ions has been performed by arming yeasts in which the *PHO5* and *MEP2* promoters regulate the genes for cell surface display of ECFP and EYFP, respectively [81]. By using the same *GAL1* promoter for the production of heterologous proteins and cell surface display of EGFP, protein production could be monitored by measuring the fluorescence intensity of EGFP [82].

Construction of an oral vaccine was attempted by displaying antigen on the cell surface of *S. cerevisiae*. 380R antigen from the red sea bream iridovirus (RSIV) was displayed for oral vaccination of cultured marine fish using the α-agglutinin-based display system [83]. In *P. pastoris*, hemagglutinin protein from a highly pathogenic avian influenza (HPAI), subtype H5N1, was displayed with the α-agglutinin-based display system. Oral vaccination of chickens with the hemagglutinin-displaying *P. pastoris* caused the production of virus neutralizing antibodies in the serum [84]. For diagnostics and therapeutics, antibodies and related molecules are attractive targets to be displayed. As a feature of cell surface display system in yeast, produced proteins undergo post-translational modification and efficient disulfide isomerization in a manner that is mechanistically similar to that in mammalian cells. Therefore, hetero-oligomeric proteins such as Fab fragments of catalytic antibodies have been successfully displayed on *S. cerevisiae*, in which the light chain (Lc fragment) of a Fab fragment was displayed on cell surface and the heavy chain (Fd fragment) of a Fab fragment was produced by secretion. As one example, the arming yeast displaying the Fab fragment of hydrolytic antibody 6D9 could catalyze the hydrolysis of a chloramphenicol monoester derivative and showed high stability in binding with a transition-state analog (TSA) [85,86]. In a similar strategy, an engineered Fab fragment, in which mutations were introduced into the Lc fragment to form a catalytic triad, was displayed on *S. cerevisiae*, leading to a higher catalytic activity compared to the wild type Fab fragment [87]. The ZZ domain of protein A from *Staphylococcus aureus* is a repeat of the Z domain that binds to the Fc fragment of human or rabbit IgG. Arming yeast displaying the ZZ domain could be applied to the detection of IgG by enzyme-linked immunosorbent assay (ELISA) and repeated affinity purification of IgG from serum [88].

Arming technology has been also used in molecular breeding of stress-tolerant *S. cerevisiae*. The yeast cell surface was modified by displaying combinatorial random peptides, and acid-tolerant yeast from a yeast library displaying random peptides was successfully selected by culture in acidic conditions. As a result, Scr35 peptide (25 a.a.) was found to enhance the acid tolerance of yeast when displayed on the cell surface [89]. In addition, a combinatorial random protein library was constructed from cDNA of *S. cerevisiae*, and this library was displayed on the yeast cell surface. From the yeast library, nonane-tolerant yeast has been obtained by screening on nonane-overlaid culture medium [90,91]. Cell surface display of leucine-rich peptides has also allowed the creation of yeasts with increased resistance to salt, ethanol, and acetonitrile [92].

4. Protein Engineering and Directed Evolution by Arming Technology

4.1. Engineering of Enzymes

The advantages of arming technology, including direct analysis of target proteins/peptides using intact cells without concentration and purification, are suitable for high-throughput screening of protein/peptide libraries containing random or comprehensive mutations. Therefore, protein engineering has been performed by display of randomly and/or comprehensively mutated proteins/peptides and subsequent high-throughput screening (Figure 3). A combinatorial library of the lid domain of ROL was displayed on the *S. cerevisiae* cell surface. Using a halo assay on soybean oil-containing plates and a assay using fluorescent substrates, lipase mutants with a high substrate specificity toward short-chain substrates and a unique oxyanion hole have been identified from the library [93]. Neurolysin is a metalloendopeptidase that cleaves the bioactive peptide neurotensin. A mutant library of neurolysin constructed by semirational mutagenesis was displayed, and screening was performed using two fluorescence-quenching peptides, the matrix metalloproteinases-2/9- (MMPs-2/9) and MMP-3-specific substrates. As a result, the Y610L mutant of neurolysin was found to show the altered substrate specificity [94]. In addition, the protease inhibitory activity of matrix metalloproteinase-2/9 was improved by arming technology and an automatic single-cell pickup system [6]. Luciferase is also an interesting target for protein engineering, as it plays an important role in the pyrosequencing system of next-generation DNA sequencers. A mutant luciferase library was constructed by arming technology, and interesting mutant luciferases with improved specific activity and dATP discrimination were obtained through only three step-wise screenings [95]. The further technological innovation of the system for single cell analysis and isolation would lead to more efficient protein engineering and directed evolution by arming technology.

Figure 3. Creation of novel proteins/peptides with improved or developed function by arming technology.

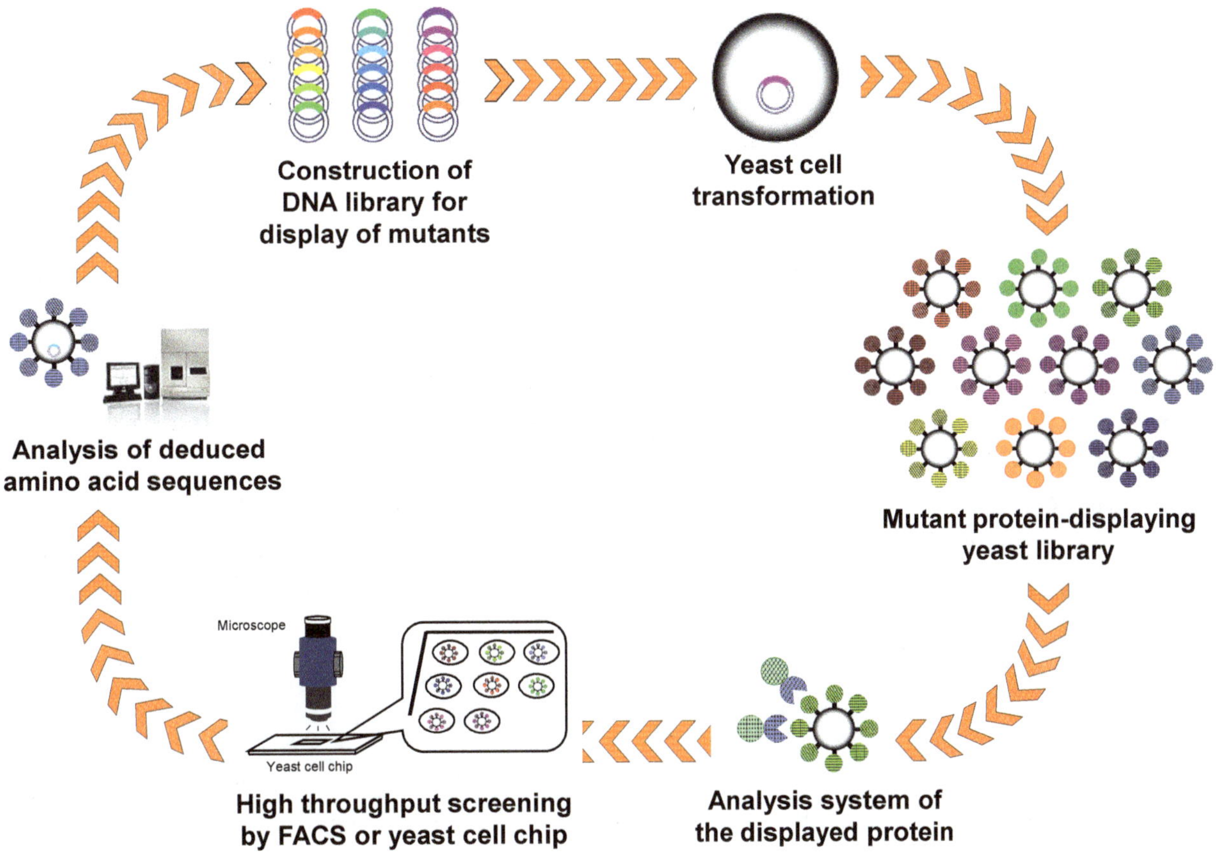

4.2. Antibody Engineering

Improvement of the affinity of antibodies for target biomolecules is an important and essential challenge in diagnostic and therapeutic development. Single-chain variable fragments (scFv) are those in which the light and heavy chains are linked by a flexible linker. A scFv library was displayed on *S. cerevisiae* by using the **a**-agglutinin-based display system, and the display efficiency was monitored using *N*-terminal or *C*-terminal epitope tags. scFv clones with improved properties could be efficiently isolated from the library through the incubation of scFv-displaying yeast library with fluorescently labeled antigen and the subsequent screening by fluorescently activated cell sorting (FACS) [7]. Furthermore, antibodies to botulinum neurotoxins, carcinoembryonic antigen, CD3 diphtheria toxin, lysozyme, and streptavidin have been improved or developed using the similar strategies based on arming technology [96–101]. Thus, arming technology is an attractive methodology for the isolation of antibodies with high affinity and other specific biological functions.

4.3. Ligand Screening for Activation of GPCRs

G protein-coupled receptors (GPCRs) are seven transmembrane-domain proteins with the ability to mediate rapid responses to extracellular signals, and they play an important role in many aspects of cellular physiology. Therefore, the design and identification of peptide-ligands for GPCRs are an attractive research area contributing to diagnostics. In the study of ligand-GPCR interactions and identification of ligands against orphan GPCRs, display of peptide ligands specifically on the plasma membrane is important, because ligands displayed only on the cell wall would be unable to access GPCRs on the plasma membrane. Hara *et al.* [36] have developed a display system of peptide-ligand on the yeast plasma membrane for activation of GPCR signaling. First, this system was applied to activation of Ste2 signaling in *S. cerevisiae*, where activation was detected using a Ste2-responsive *FUS1* promoter driving EGFP reporter expression. The α-factor displayed on the plasma membrane was shown to functionally activate the pheromone response pathway. Furthermore, this system has been applied to the activation of human GPCR signaling as well as the yeast pheromone response pathway using chimeric Gα protein. Somatostatin is a naturally occurring gastrointestinal hormone that regulates various endocrine and exocrine processes. After the construction of yeast producing human somatostatin receptor subtype-2 (SSTR2) and chimeric Gα protein, somatostatin was displayed on plasma membrane. The somatostatin displayed on the plasma membrane could activate human SSTR2 in *S. cerevisiae* [37]. This technological platform, namely “PepDisplay” is useful for identification of novel peptide-ligands for heterologously produced GPCRs by membrane display of peptides with random and/or comprehensive sequences and screening based on the activation of reporter genes.

5. Conclusions

Arming technology is well established and has been applied to a wide range of research areas, owing to the advantages that are absent in the conventional expression systems. In the construction of whole-cell biocatalysts, cell surface display of enzymes enables bioconversion of the substrate with high-molecular mass that can not enter in the cells, and multistep and synergistic reactions can be achieved on the cell surface by co-displaying multiple kinds of enzymes on the cell surface. Bioadsorption on the cell surface is advantageous in that it does not affect intracellular biological mechanisms, and the adsorbed materials such as metal ions can be easily concentrated and rapidly recovered. Furthermore, arming technology is an innovative molecular tool for protein engineering and directed evolution. The fact that proteins/peptides displayed on the cell surface can be directly, speedily, and conveniently analyzed using intact cells without concentration and purification is suitable for high-throughput screening of a protein/peptide library carrying random and/or comprehensive mutations. Therefore, arming technology is a powerful technology that will facilitate the development of a wide range of biotechnologies and contribute to the production of various kinds of materials for a sustainable society in the future.

References

1. Anonymous. Arming yeast with cell-surface catalysts. *Chem. Eng. News* **1997**, *75*, 32.
2. Ueda, M.; Tanaka, A. Genetic immobilization of proteins on the yeast cell surface. *Biotechnol. Adv.* **2000**, *18*, 121–140.
3. Ueda, M.; Tanaka, A. Cell surface engineering of yeast: Construction of arming yeast with biocatalyst. *J. Biosci. Bioeng.* **2000**, *90*, 125–136.
4. Su, G.D.; Zhang, X.; Lin, Y. Surface display of active lipase in *Pichia pastoris* using Sed1 as an anchor protein. *Biotechnol. Lett.* **2010**, *32*, 1131–1136.
5. Chen, Y.P.; Hwang, I.E.; Lin, C.J.; Wang, H.J.; Tseng, C.P. Enhancing the stability of xylanase from *Cellulomonas fimi* by cell-surface display on *Escherichia coli*. *J. Appl. Microbiol.* **2012**, *112*, 455–463.
6. Aoki, W.; Yoshino, Y.; Morisaka, H.; Tsunetomo, K.; Koyo, H.; Kamiya, S.; Kawata, N.; Kuroda, K.; Ueda, M. High-throughput screening of improved protease inhibitors using a yeast cell surface display system and a yeast cell chip. *J. Biosci. Bioeng.* **2011**, *111*, 16–18.
7. Boder, E.T.; Wittrup, K.D. Yeast surface display for screening combinatorial polypeptide libraries. *Nat. Biotechnol.* **1997**, *15*, 553–557.
8. Chen, W.; Georgiou, G. Cell-Surface display of heterologous proteins: From high-throughput screening to environmental applications. *Biotechnol. Bioeng.* **2002**, *79*, 496–503.
9. Fukuda, T.; Kato-Murai, M.; Kadonosono, T.; Sahara, H.; Hata, Y.; Suye, S.; Ueda, M. Enhancement of substrate recognition ability by combinatorial mutation of β-glucosidase displayed on the yeast cell surface. *Appl. Microbiol. Biotechnol.* **2007**, *76*, 1027–1033.
10. Yeung, Y.A.; Wittrup, K.D. Quantitative screening of yeast surface-displayed polypeptide libraries by magnetic bead capture. *Biotechnol. Prog.* **2002**, *18*, 212–220.
11. Georgiou, G.; Poetschke, H.L.; Stathopoulos, C.; Francisco, J.A. Practical applications of engineering gram-negative bacterial cell surfaces. *Trends Biotechnol.* **1993**, *11*, 6–10.
12. Kuroda, K.; Ueda, M. Engineering of microorganisms towards recovery of rare metal ions. *Appl. Microbiol. Biotechnol.* **2010**, *87*, 53–60.
13. Kuroda, K.; Ueda, M. Molecular design of the microbial cell surface toward the recovery of metal ions. *Curr. Opin. Biotechnol.* **2011**, *22*, 427–433.
14. Kuroda, K.; Ueda, M. Cell surface engineering of yeast for applications in white biotechnology. *Biotechnol. Lett.* **2011**, *33*, 1–9.
15. Samuelson, P.; Gunneriusson, E.; Nygren, P.A.; Ståhl, S. Display of proteins on bacteria. *J. Biotechnol.* **2002**, *96*, 129–154.
16. Ståhl, S.; Uhlén, M. Bacterial surface display: Trends and progress. *Trends Biotechnol.* **1997**, *15*, 185–192.
17. Smith, G.P. Filamentous fusion phage: Novel expression vectors that display cloned antigens on the virion surface. *Science* **1985**, *228*, 1315–1317.

18. Smith, G.P. Surface presentation of protein epitopes using bacteriophage expression systems. *Curr. Opin. Biotechnol.* **1991**, *2*, 668–673.
19. Tabuchi, S.; Ito, J.; Adachi, T.; Ishida, H.; Hata, Y.; Okazaki, F.; Tanaka, T.; Ogino, C.; Kondo, A. Display of both N- and C-terminal target fusion proteins on the *Aspergillus oryzae* cell surface using a chitin-binding module. *Appl. Microbiol. Biotechnol.* **2010**, *87*, 1783–1789.
20. Adachi, T.; Ito, J.; Kawata, K.; Kaya, M.; Ishida, H.; Sahara, H.; Hata, Y.; Ogino, C.; Fukuda, H.; Kondo, A. Construction of an *Aspergillus oryzae* cell-surface display system using a putative GPI-anchored protein. *Appl. Microbiol. Biotechnol.* **2008**, *81*, 711–719.
21. Breinig, F.; Schmitt, M.J. Spacer-elongated cell wall fusion proteins improve cell surface expression in the yeast *Saccharomyces cerevisiae*. *Appl. Microbiol. Biotechnol.* **2002**, *58*, 637–644.
22. Washida, M.; Takahashi, S.; Ueda, M.; Tanaka, A. Spacer-mediated display of active lipase on the yeast cell surface. *Appl. Microbiol. Biotechnol.* **2001**, *56*, 681–686.
23. Lipke, P.N.; Kurjan, J. Sexual agglutination in budding yeasts: Structure, function, and regulation of adhesion glycoproteins. *Microbiol. Rev.* **1992**, *56*, 180–194.
24. Kuroda, K.; Ueda, M. Generation of arming yeasts with active proeins and peptides via cell surface display system—Cell surface engineering, bio-arming technology. *Methods Mol. Biol.* 2013, in press.
25. Sato, N.; Matsumoto, T.; Ueda, M.; Tanaka, A.; Fukuda, H.; Kondo, A. Long anchor using Flo1 protein enhances reactivity of cell surface-displayed glucoamylase to polymer substrates. *Appl. Microbiol. Biotechnol.* **2002**, *60*, 469–474.
26. Van der Vaart, J.M.; te Biesebeke, R.; Chapman, J.W.; Toschka, H.Y.; Klis, F.M.; Verrips, C.T. Comparison of cell wall proteins of *Saccharomyces cerevisiae* as anchors for cell surface expression of heterologous proteins. *Appl. Environ. Microbiol.* **1997**, *63*, 615–620.
27. Wang, Z.; Mathias, A.; Stavrou, S.; Neville, D.M., Jr. A new yeast display vector permitting free scFv amino termini can augment ligand binding affinities. *Protein Eng. Des. Sel.* **2005**, *18*, 337–343.
28. Matsumoto, T.; Fukuda, H.; Ueda, M.; Tanaka, A.; Kondo, A. Construction of yeast strains with high cell surface lipase activity by using novel display systems based on the Flo1p flocculation functional domain. *Appl. Environ. Microbiol.* **2002**, *68*, 4517–4522.
29. Abe, H.; Ohba, M.; Shimma, Y.; Jigami, Y. Yeast cells harboring human α-1,3-fucosyltransferase at the cell surface engineered using Pir, a cell wall-anchored protein. *FEMS Yeast Res.* **2004**, *4*, 417–425.
30. Kuroda, K.; Matsui, K.; Higuchi, S.; Kotaka, A.; Sahara, H.; Hata, Y.; Ueda, M. Enhancement of display efficiency in yeast display system by vector engineering and gene disruption. *Appl. Microbiol. Biotechnol.* **2009**, *82*, 713–719.
31. Wentz, A.E.; Shusta, E.V. A novel high-throughput screen reveals yeast genes that increase secretion of heterologous proteins. *Appl. Environ. Microbiol.* **2007**, *73*, 1189–1198.
32. Tanino, T.; Fukuda, H.; Kondo, A. Construction of a *Pichia pastoris* cell-surface display system using Flo1p anchor system. *Biotechnol. Prog.* **2006**, *22*, 989–993.
33. Wang, Q.; Li, L.; Chen, M.; Qi, Q.; Wang, P.G. Construction of a novel system for cell surface display of heterologous proteins on *Pichia pastoris*. *Biotechnol. Lett.* **2007**, *29*, 1561–1566.
34. Wang, Q.; Li, L.; Chen, M.; Qi, Q.; Wang, P.G. Construction of a novel *Pichia pastoris* cell-surface display system based on the cell wall protein Pir1. *Curr. Microbiol.* **2008**, *56*, 352–357.

35. Yuzbasheva, E.Y.; Yuzbashev, T.V.; Laptev, I.A.; Konstantinova, T.K.; Sineoky, S.P. Efficient cell surface display of Lip2 lipase using C-domains of glycosylphosphatidylinositol-anchored cell wall proteins of *Yarrowia lipolytica*. *Appl. Microbiol. Biotechnol.* **2011**, *91*, 645–654.
36. Hara, K.; Ono, T.; Kuroda, K.; Ueda, M. Membrane-displayed peptide ligand activates the pheromone response pathway in *Saccharomyces cerevisiae*. *J. Biochem.* **2012**, *151*, 551–557.
37. Hara, K.; Shigemori, T.; Kuroda, K.; Ueda, M. Membrane-displayed somatostatin activates somatostatin receptor subtype-2 heterologously produced in *Saccharomyces cerevisiae*. *AMB Express* **2012**, *2*, e63.
38. Lynd, L.R.; van Zyl, W.H.; McBride, J.E.; Laser, M. Consolidated bioprocessing of cellulosic biomass: An update. *Curr. Opin. Biotechnol.* **2005**, *16*, 577–583.
39. Murai, T.; Ueda, M.; Yamamura, M.; Atomi, H.; Shibasaki, Y.; Kamasawa, N.; Osumi, M.; Amachi, T.; Tanaka, A. Construction of a starch-utilizing yeast by cell surface engineering. *Appl. Environ. Microbiol.* **1997**, *63*, 1362–1366.
40. Murai, T.; Ueda, M.; Shibasaki, Y.; Kamasawa, N.; Osumi, M.; Imanaka, T.; Tanaka, A. Development of an arming yeast strain for efficient utilization of starch by co-display of sequential amylolytic enzymes on the cell surface. *Appl. Microbiol. Biotechnol.* **1999**, *51*, 65–70.
41. Shigechi, H.; Koh, J.; Fujita, Y.; Matsumoto, T.; Bito, Y.; Ueda, M.; Satoh, E.; Fukuda, H.; Kondo, A. Direct production of ethanol from raw corn starch via fermentation by use of a novel surface-engineered yeast strain codisplaying glucoamylase and α-amylase. *Appl. Environ. Microbiol.* **2004**, *70*, 5037–5040.
42. Murai, T.; Ueda, M.; Kawaguchi, T.; Arai, M.; Tanaka, A. Assimilation of cellooligosaccharides by a cell surface-engineered yeast expressing β-glucosidase and carboxymethylcellulase from *Aspergillus aculeatus*. *Appl. Environ. Microbiol.* **1998**, *64*, 4857–4861.
43. Fujita, Y.; Ito, J.; Ueda, M.; Fukuda, H.; Kondo, A. Synergistic saccharification, and direct fermentation to ethanol, of amorphous cellulose by use of an engineered yeast strain codisplaying three types of cellulolytic enzyme. *Appl. Environ. Microbiol.* **2004**, *70*, 1207–1212.
44. Kotaka, A.; Bando, H.; Kaya, M.; Kato-Murai, M.; Kuroda, K.; Sahara, H.; Hata, Y.; Kondo, A.; Ueda, M. Direct ethanol production from barley β-glucan by sake yeast displaying *Aspergillus oryzae* β-glucosidase and endoglucanase. *J. Biosci. Bioeng.* **2008**, *105*, 622–627.
45. Ito, J.; Kosugi, A.; Tanaka, T.; Kuroda, K.; Shibasaki, S.; Ogino, C.; Ueda, M.; Fukuda, H.; Doi, R.H.; Kondo, A. Regulation of the display ratio of enzymes on the *Saccharomyces cerevisiae* cell surface by the immunoglobulin G and cellulosomal enzyme binding domains. *Appl. Environ. Microbiol.* **2009**, *75*, 4149–4154.
46. Yamada, R.; Taniguchi, N.; Tanaka, T.; Ogino, C.; Fukuda, H.; Kondo, A. Cocktail δ-integration: A novel method to construct cellulolytic enzyme expression ratio-optimized yeast strains. *Microb. Cell Fact.* **2010**, *9*, e32.
47. Nakanishi, A.; Bae, J.; Kuroda, K.; Ueda, M. Construction of a novel selection system for endoglucanases exhibiting carbohydrate-binding modules optimized for biomass using yeast cell-surface engineering. *AMB Express* **2012**, *2*, e56.
48. Tsai, S.L.; Oh, J.; Singh, S.; Chen, R.; Chen, W. Functional assembly of minicellulosomes on the *Saccharomyces cerevisiae* cell surface for cellulose hydrolysis and ethanol production. *Appl. Environ. Microbiol.* **2009**, *75*, 6087–6093.

49. Goyal, G.; Tsai, S.L.; Madan, B.; DaSilva, N.A.; Chen, W. Simultaneous cell growth and ethanol production from cellulose by an engineered yeast consortium displaying a functional mini-cellulosome. *Microb. Cell Fact.* **2011**, *10*, e89.
50. Katahira, S.; Fujita, Y.; Mizuike, A.; Fukuda, H.; Kondo, A. Construction of a xylan-fermenting yeast strain through codisplay of xylanolytic enzymes on the surface of xylose-utilizing *Saccharomyces cerevisiae* cells. *Appl. Environ. Microbiol.* **2004**, *70*, 5407–5414.
51. Katahira, S.; Mizuike, A.; Fukuda, H.; Kondo, A. Ethanol fermentation from lignocellulosic hydrolysate by a recombinant xylose- and cellooligosaccharide-assimilating yeast strain. *Appl. Microbiol. Biotechnol.* **2006**, *72*, 1136–1143.
52. Ota, M.; Sakuragi, H.; Morisaka, H.; Kuroda, K.; Miyake, H.; Tamaru, Y.; Ueda, M. Display of *Clostridium cellulovorans* xylose isomerase on the cell surface of *Saccharomyces cerevisiae* and its direct application to xylose fermentation. *Biotechnol. Prog.* **2013**, *29*, 346–351.
53. Nakanishi, A.; Bae, J.G.; Fukai, K.; Tokumoto, N.; Kuroda, K.; Ogawa, J.; Nakatani, M.; Shimizu, S.; Ueda, M. Effect of pretreatment of hydrothermally processed rice straw with laccase-displaying yeast on ethanol fermentation. *Appl. Microbiol. Biotechnol.* **2012**, *94*, 939–948.
54. Shiraga, S.; Kawakami, M.; Ishiguro, M.; Ueda, M. Enhanced reactivity of *Rhizopus oryzae* lipase displayed on yeast cell surfaces in organic solvents: Potential as a whole-cell biocatalyst in organic solvents. *Appl. Environ. Microbiol.* **2005**, *71*, 4335–4338.
55. Nakamura, Y.; Matsumoto, T.; Nomoto, F.; Ueda, M.; Fukuda, H.; Kondo, A. Enhancement of activity of lipase-displaying yeast cells and their application to optical resolution of (*R*,*S*)-1-benzyloxy-3-chloro-2-propyl monosuccinate. *Biotechnol. Prog.* **2006**, *22*, 998–1002.
56. Kato, M.; Fuchimoto, J.; Tanino, T.; Kondo, A.; Fukuda, H.; Ueda, M. Preparation of a whole-cell biocatalyst of mutated *Candida antarctica* lipase B (mCALB) by a yeast molecular display system and its practical properties. *Appl. Microbiol. Biotechnol.* **2007**, *75*, 549–555.
57. Inaba, C.; Maekawa, K.; Morisaka, H.; Kuroda, K.; Ueda, M. Efficient synthesis of enantiomeric ethyl lactate by *Candida antarctica* lipase B (CALB)-displaying yeasts. *Appl. Microbiol. Biotechnol.* **2009**, *83*, 859–864.
58. Tanino, T.; Aoki, T.; Chung, W.Y.; Watanabe, Y.; Ogino, C.; Fukuda, H.; Kondo, A. Improvement of a *Candida antarctica* lipase B-displaying yeast whole-cell biocatalyst and its application to the polyester synthesis reaction. *Appl. Microbiol. Biotechnol.* **2009**, *82*, 59–66.
59. Tanino, T.; Ohno, T.; Aoki, T.; Fukuda, H.; Kondo, A. Development of yeast cells displaying *Candida antarctica* lipase B and their application to ester synthesis reaction. *Appl. Microbiol. Biotechnol.* **2007**, *75*, 1319–1325.
60. Han, S.Y.; Pan, Z.Y.; Huang, D.F.; Ueda, M.; Wang, X.N.; Lin, Y. Highly efficient synthesis of ethyl hexanoate catalyzed by CALB-displaying *Saccharomyces cerevisiae* whole-cells in non-aqueous phase. *J. Mol. Catal. B* **2009**, *59*, 168–172.
61. Su, G.D.; Huang, D.F.; Han, S.Y.; Zheng, S.P.; Lin, Y. Display of *Candida antarctica* lipase B on *Pichia pastoris* and its application to flavor ester synthesis. *Appl. Microbiol. Biotechnol.* **2010**, *86*, 1493–1501.
62. Jiang, Z.; Gao, B.; Ren, R.; Tao, X.; Ma, Y.; Wei, D. Efficient display of active lipase LipB52 with a *Pichia pastoris* cell surface display system and comparison with the LipB52 displayed on *Saccharomyces cerevisiae* cell surface. *BMC Biotechnol.* **2008**, *8*, e4.

63. Jiang, Z.B.; Song, H.T.; Gupta, N.; Ma, L.X.; Wu, Z.B. Cell surface display of functionally active lipases from *Yarrowia lipolytica* in *Pichia pastoris*. *Protein Expr. Purif.* **2007**, *56*, 35–39.
64. Kaya, M.; Ito, J.; Kotaka, A.; Matsumura, K.; Bando, H.; Sahara, H.; Ogino, C.; Shibasaki, S.; Kuroda, K.; Ueda, M.; *et al.* Isoflavone aglycones production from isoflavone glycosides by display of β-glucosidase from *Aspergillus oryzae* on yeast cell surface. *Appl. Microbiol. Biotechnol.* **2008**, *79*, 51–60.
65. Inaba, C.; Higuchi, S.; Morisaka, H.; Kuroda, K.; Ueda, M. Synthesis of functional dipeptide carnosine from nonprotected amino acids using carnosinase-displaying yeast cells. *Appl. Microbiol. Biotechnol.* **2010**, *86*, 1895–1902.
66. Fukuda, T.; Isogawa, D.; Takagi, M.; Kato-Murai, M.; Kimoto, H.; Kusaoke, H.; Ueda, M.; Suye, S. Yeast cell-surface expression of chitosanase from *Paenibacillus fukuinensis*. *Biosci. Biotechnol. Biochem.* **2007**, *71*, 2845–2847.
67. Liu, G.; Yue, L.; Chi, Z.; Yu, W.; Chi, Z.; Madzak, C. The surface display of the alginate lyase on the cells of *Yarrowia lipolytica* for hydrolysis of alginate. *Mar. Biotechnol. (NY)* **2009**, *11*, 619–626.
68. Kotrba, P.; Ruml, T. Surface display of metal fixation motifs of bacterial P1-type ATPases specifically promotes biosorption of Pb^{2+} by *Saccharomyces cerevisiae*. *Appl. Environ. Microbiol.* **2010**, *76*, 2615–2622.
69. Kuroda, K.; Shibasaki, S.; Ueda, M.; Tanaka, A. Cell surface-engineered yeast displaying a histidine oligopeptide (hexa-His) has enhanced adsorption of and tolerance to heavy metal ions. *Appl. Microbiol. Biotechnol.* **2001**, *57*, 697–701.
70. Kuroda, K.; Ueda, M. Bioadsorption of cadmium ion by cell surface-engineered yeasts displaying metallothionein and hexa-His. *Appl. Microbiol. Biotechnol.* **2003**, *63*, 182–186.
71. Kuroda, K.; Ueda, M. Effective display of metallothionein tandem repeats on the bioadsorption of cadmium ion. *Appl. Microbiol. Biotechnol.* **2006**, *70*, 458–463.
72. Grunden, A.M.; Ray, R.M.; Rosentel, J.K.; Healy, F.G.; Shanmugam, K.T. Repression of the *Escherichia coli* modABCD (molybdate transport) operon by ModE. *J. Bacteriol.* **1996**, *178*, 735–744.
73. Nishitani, T.; Shimada, M.; Kuroda, K.; Ueda, M. Molecular design of yeast cell surface for adsorption and recovery of molybdenum, one of rare metals. *Appl. Microbiol. Biotechnol.* **2010**, *86*, 641–648.
74. Kuroda, K.; Nishitani, T.; Ueda, M. Specific adsorption of tungstate by cell surface display of the newly designed ModE mutant. *Appl. Microbiol. Biotechnol.* **2012**, *96*, 153–159.
75. Yasui, M.; Shibasaki, S.; Kuroda, K.; Ueda, M.; Kawada, N.; Nishikawa, J.; Nishihara, T.; Tanaka, A. An arming yeast with the ability to entrap fluorescent 17β-estradiol on the cell surface. *Appl. Microbiol. Biotechnol.* **2002**, *59*, 329–331.
76. Fukuda, T.; Tsuchiyama, K.; Makishima, H.; Takayama, K.; Mulchandani, A.; Kuroda, K.; Ueda, M.; Suye, S. Improvement in organophosphorus hydrolase activity of cell surface-engineered yeast strain using Flo1p anchor system. *Biotechnol. Lett.* **2010**, *32*, 655–659.
77. Takayama, K.; Suye, S.; Kuroda, K.; Ueda, M.; Kitaguchi, T.; Tsuchiyama, K.; Fukuda, T.; Chen, W.; Mulchandani, A. Surface display of organophosphorus hydrolase on *Saccharomyces cerevisiae*. *Biotechnol. Prog.* **2006**, *22*, 939–943.

78. Fukuda, T.; Tsuchiya, K.; Makishima, H.; Tsuchiyama, K.; Mulchandani, A.; Kuroda, K.; Ueda, M.; Suye, S. Organophosphorus compound detection on a cell chip with yeast coexpressing hydrolase and eGFP. *Biotechnol. J.* **2010**, *5*, 515–519.
79. Takayama, K.; Suye, S.; Tanaka, Y.; Mulchandani, A.; Kuroda, K.; Ueda, M. Estimation of enzyme kinetic parameters of cell surface-displayed organophosphorus hydrolase and construction of a biosensing system for organophosphorus compounds. *Anal. Sci.* **2011**, *27*, 823–826.
80. Shibasaki, S.; Ueda, M.; Ye, K.; Shimizu, K.; Kamasawa, N.; Osumi, M.; Tanaka, A. Creation of cell surface-engineered yeast that display different fluorescent proteins in response to the glucose concentration. *Appl. Microbiol. Biotechnol.* **2001**, *57*, 528–533.
81. Shibasaki, S.; Ninomiya, Y.; Ueda, M.; Iwahashi, M.; Katsuragi, T.; Tani, Y.; Harashima, S.; Tanaka, A. Intelligent yeast strains with the ability to self-monitor the concentrations of intra- and extracellular phosphate or ammonium ion by emission of fluorescence from the cell surface. *Appl. Microbiol. Biotechnol.* **2001**, *57*, 702–707.
82. Shibasaki, S.; Tanaka, A.; Ueda, M. Development of combinatorial bioengineering using yeast cell surface display—Order-made design of cell and protein for bio-monitoring. *Biosens. Bioelectron.* **2003**, *19*, 123–130.
83. Tamaru, Y.; Ohtsuka, M.; Kato, K.; Manabe, S.; Kuroda, K.; Sanada, M.; Ueda, M. Application of the arming system for the expression of the 380R antigen from red sea bream iridovirus (RSIV) on the surface of yeast cells: A first step for the development of an oral vaccine. *Biotechnol. Prog.* **2006**, *22*, 949–953.
84. Wasilenko, J.L.; Sarmento, L.; Spatz, S.; Pantin-Jackwood, M. Cell surface display of highly pathogenic avian influenza virus hemagglutinin on the surface of *Pichia pastoris* cells using α-agglutinin for production of oral vaccines. *Biotechnol. Prog.* **2010**, *26*, 542–547.
85. Lin, Y.; Shiraga, S.; Tsumuraya, T.; Matsumoto, T.; Kondo, A.; Fujii, I.; Ueda, M. Comparison of two forms of catalytic antibody displayed on yeast-cell surface. *J. Mol. Catal. B* **2004**, *28*, 241–246.
86. Lin, Y.; Tsumuraya, T.; Wakabayashi, T.; Shiraga, S.; Fujii, I.; Kondo, A.; Ueda, M. Display of a functional hetero-oligomeric catalytic antibody on the yeast cell surface. *Appl. Microbiol. Biotechnol.* **2003**, *62*, 226–232.
87. Okochi, N.; Kato-Murai, M.; Kadonosono, T.; Ueda, M. Design of a serine protease-like catalytic triad on an antibody light chain displayed on the yeast cell surface. *Appl. Microbiol. Biotechnol.* **2007**, *77*, 597–603.
88. Nakamura, Y.; Shibasaki, S.; Ueda, M.; Tanaka, A.; Fukuda, H.; Kondo, A. Development of novel whole-cell immunoadsorbents by yeast surface display of the IgG-binding domain. *Appl. Microbiol. Biotechnol.* **2001**, *57*, 500–505.
89. Matsui, K.; Kuroda, K.; Ueda, M. Creation of a novel peptide endowing yeasts with acid tolerance using yeast cell-surface engineering. *Appl. Microbiol. Biotechnol.* **2009**, *82*, 105–113.
90. Zou, W.; Ueda, M.; Tanaka, A. Screening of a molecule endowing *Saccharomyces cerevisiae* with *n*-nonane-tolerance from a combinatorial random protein library. *Appl. Microbiol. Biotechnol.* **2002**, *58*, 806–812.

91. Zou, W.; Ueda, M.; Yamanaka, H.; Tanaka, A. Construction of a combinatorial protein library displayed on yeast cell surface using DNA random priming method. *J. Biosci. Bioeng.* **2001**, *92*, 393–396.
92. Andreu, C.; del Olmo, M. Yeast arming by the Aga2p system: Effect of growth conditions in galactose on the efficiency of the display and influence of expressing leucine-containing peptides. *Appl. Microbiol. Biotechnol.* **2013**, doi:10.1007/s00253-013-5086-4.
93. Shiraga, S.; Ishiguro, M.; Fukami, H.; Nakao, M.; Ueda, M. Creation of *Rhizopus oryzae* lipase having a unique oxyanion hole by combinatorial mutagenesis in the lid domain. *Appl. Microbiol. Biotechnol.* **2005**, *68*, 779–785.
94. Kadonosono, T.; Kato-Murai, M.; Ueda, M. Alteration of substrate specificity of rat neurolysin from matrix metalloproteinase-2/9-type to -3-type specificity by comprehensive mutation. *Protein Eng. Des. Sel.* **2008**, *21*, 507–513.
95. Fushimi, T.; Miura, N.; Shintani, H.; Tsunoda, H.; Kuroda, K.; Ueda, M. Mutant firefly luciferases with improved specific activity and dATP discrimination constructed by yeast cell surface engineering. *Appl. Microbiol. Biotechnol.* **2013**, *97*, 4003–4011.
96. Van den Beucken, T.; Pieters, H.; Steukers, M.; van der Vaart, M.; Ladner, R.C.; Hoogenboom, H.R.; Hufton, S.E. Affinity maturation of Fab antibody fragments by fluorescent-activated cell sorting of yeast-displayed libraries. *FEBS Lett.* **2003**, *546*, 288–294.
97. Rajpal, A.; Beyaz, N.; Haber, L.; Cappuccilli, G.; Yee, H.; Bhatt, R.R.; Takeuchi, T.; Lerner, R.A.; Crea, R. A general method for greatly improving the affinity of antibodies by using combinatorial libraries. *Proc. Natl. Acad. Sci. USA* **2005**, *102*, 8466–8471.
98. Graff, C.P.; Chester, K.; Begent, R.; Wittrup, K.D. Directed evolution of an anti-carcinoembryonic antigen scFv with a 4-day monovalent dissociation half-time at 37 °C. *Protein Eng. Des. Sel.* **2004**, *17*, 293–304.
99. Razai, A.; Garcia-Rodriguez, C.; Lou, J.; Geren, I.N.; Forsyth, C.M.; Robles, Y.; Tsai, R.; Smith, T.J.; Smith, L.A.; Siegel, R.W.; *et al.* Molecular evolution of antibody affinity for sensitive detection of botulinum neurotoxin type A. *J. Mol. Biol.* **2005**, *351*, 158–169.
100. VanAntwerp, J.J.; Wittrup, K.D. Thermodynamic characterization of affinity maturation: The D1.3 antibody and a higher-affinity mutant. *J. Mol. Recognit.* **1998**, *11*, 10–13.
101. Wang, Z.; Kim, G.B.; Woo, J.H.; Liu, Y.Y.; Mathias, A.; Stavrou, S.; Neville, D.M., Jr. Improvement of a recombinant anti-monkey anti-CD3 diphtheria toxin based immunotoxin by yeast display affinity maturation of the scFv. *Bioconjug. Chem.* **2007**, *18*, 947–955.

9

Architecture of Amylose Supramolecules in Form of Inclusion Complexes by Phosphorylase-Catalyzed Enzymatic Polymerization

Jun-ichi Kadokawa [1,2,*]

[1] Graduate School of Science and Engineering, Kagoshima University, 1-21-40 Korimoto, Kagoshima 890-0065, Japan

[2] Research Center for Environmentally Friendly Materials Engineering, Muroran Institute of Technology, 27-1 Mizumoto-cho, Muroran, Hokkaido 050-8585, Japan

* Author to whom correspondence should be addressed; E-Mail: kadokawa@eng.kagoshima-u.ac.jp

---

Abstract: This paper reviews the architecture of amylose supramolecules in form of inclusion complexes with synthetic polymers by phosphorylase-catalyzed enzymatic polymerization. Amylose is known to be synthesized by enzymatic polymerization using α-D-glucose 1-phosphate as a monomer, by phosphorylase catalysis. When the phosphorylase-catalyzed enzymatic polymerization was conducted in the presence of various hydrophobic polymers, such as polyethers, polyesters, poly(ester-ether), and polycarbonates as a guest polymer, such inclusion supramolecules were formed by the hydrophobic interaction in the progress of polymerization. Because the representation of propagation in the polymerization is similar to the way that a vine of a plant grows, twining around a rod, this polymerization method for the formation of amylose-polymer inclusion complexes was proposed to be named "vine-twining polymerization". To yield an inclusion complex from a strongly hydrophobic polyester, the parallel enzymatic polymerization system was extensively developed. The author found that amylose selectively included one side of the guest polymer from a mixture of two resemblant guest polymers, as well as a specific range in molecular weights of the guest polymers poly(tetrahydrofuran) (PTHF) in the vine-twining polymerization. Selective inclusion behavior of amylose toward stereoisomers of chiral polyesters, poly(lactide)s, also appeared in the vine-twining polymerization.

Keywords: Amylose supramolecule; hydrophobic interaction; inclusion complex; vine-twining polymerization; selective inclusion

1. Introduction

Polysaccharides are naturally occurring carbohydrate polymers, where each monosaccharide residue is linked directly through a glycosidic linkage in the main-chain [1]. The glycosidic linkage is a type of covalent bond that joins a monosaccharide residue to another group, which is typically another saccharide residue. Natural polysaccharides are found in various sources such as plant, animal, seaweed, and microbial kingdoms, which have specific and very complicated structures owing, not only to a structural diversity of monosaccharide residues, but also to the differences in stereo- and regio-configurations of glycosidic linkages. The large diversity of polysaccharide structures contributes to serve as vital materials for a range of important *in vivo* functions in host organisms, e.g., providing an energy resource, acting as a structural material, and conferring specific biological properties, and a subtle change in the chemical structure has a profound effect on the properties and functions of the polysaccharides [2–4]. Therefore, the preparation of artificial polysaccharides has attracted increasing attention because of their potential applications as materials in the fields related to medicine, pharmaceutics, cosmetics, and food industries. Polysaccharides are theoretically produced by the repeated reactions for the formation of a glycosidic linkage, so-called glycosylation of a glycosyl acceptor with a glycosyl donor [5–8]. To develop a superior method for the synthesis of polysaccharides by such repeated glycosylations, the *in vitro* approach by enzymatic catalysis i.e., enzymatic polymerization, has been significantly investigated [9–15] as enzymes have remarkable catalytic advantages compared with other types of catalysts in terms of the stereo- and regioselectivities. The enzymatic polymerization, therefore, is a very powerful tool for the stereo- and regioselective construction of polysaccharides under mild conditions, where monomers can be employed in their unprotected forms, leading to the direct formation of the unprotected saccharide chains in aqueous media.

Amylose is a natural glucose polymer connected through α-(1→4)-glycosidic linkages (Figure 1) [1]. This is one component of starch and acts as an energy resource in nature with the other component of starch, that is, amylopectin, which has a branched structure composed of α-(1→4)-glucans with a small portion of α-(1→6)-glycosidic linkages [16]. Amylose has recently been recognized as a candidate as a high-performance polymeric material because it acts as a host molecule and forms polysaccharide supramolecules by inclusion complexation with various guest molecules of relatively low molecular weight (inclusion complexes) owing to the helical conformation (Figure 2a) [17]. The driving force for inclusion of guest molecules in the cavity is mainly host-guest hydrophobic interaction as the inside of the amylose helix has a hydrophobic nature due to the presence of hydrophilic hydroxy groups in the glucose residues on outer part of the helix. Therefore, hydrophobicity is generally in demand as the property of guest molecules to be included by amylose. Development of methods for the architecture of amylose supramolecules with polymeric guest molecules is a significant research topic to provide new self-assembled polymeric materials with regularly controlled nanostructures, which have potential to exhibit new high performance functions. However, only limited studies have been reported regarding the direct construction of inclusion complexes composed of amylose and polymeric molecules (Figure 2b) [18–26] as the driving force for the inclusion complexation of guest molecules into the cavity of amylose is the weak hydrophobic interaction as mentioned earlier, the amylose cavity does not have a sufficient ability to include the long chains of polymeric guests. The author has considered for the architecture of such amylose supramolecules, i.e., amylose-polymer inclusion complexes in the phosphorylase-catalyzed enzymatic polymerization field [14,15,27–30] as a structurally controlled amylose is efficiently synthesized by an enzymatic polymerization through phosphorylase catalysis [31–37]. Following the recent review article on the series of these studies [30], in this article, the author would like to deal with the comprehensive results and discussion of this approach, including the further progress of the investigation to precisely architect such amylose supramolecules in form of inclusion complexes between amylose and synthetic polymers by the phosphorylase-catalyzed enzymatic polymerization [38]. Specifically, the present review article is described on the basis of the viewpoint that the precision architecture of the regularly controlled polysaccharide supramolecules has been achieved by means of the enzymatic synthesis of structurally defined polysaccharides according to Section 2.

Figure 1. Structure of amylase.

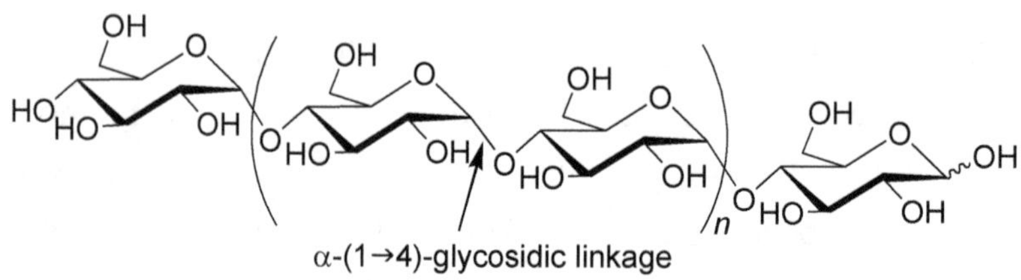

Figure 2. Amylose forms inclusion complex with relatively low molecular weight hydrophobic molecule (**a**); but, mostly, does not form it with polymeric molecule (**b**).

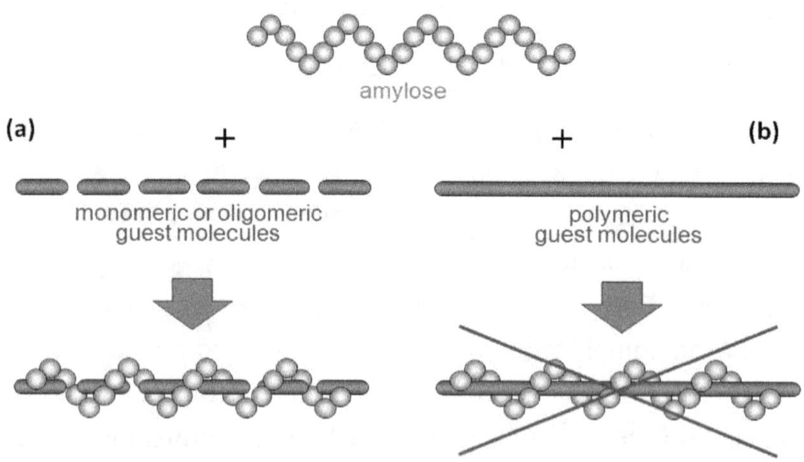

2. Characteristic Features of Phosphorylase-Catalyzed Enzymatic Polymerization to Produce Amylose

Phosphorylase catalyzes the reversible phosphorolysis of α-(1→4)-glucans at the nonreducing end, such as glycogen and starch, in the presence of inorganic phosphate to produce α-D-glucose 1-phosphate (G-1-P) [31]. By means of the reversibility of the enzymatic reaction, α-(1→4)-glycosidic linkage can be constructed by the phosphorylase-catalyzed glycosylation using G-1-P as a glycosyl donor. As a glycosyl acceptor, maltooligosaccharides with degrees of polymerization DPs higher than the smallest one recognized by phosphorylase are used. The smallest glycosyl acceptor for the phosphorylase-catalyzed glycosylation is typically known to be maltotetraose (G_4), whereas that for phosphorolysis is typically maltopentaose (G_5). In the glycosylation, a glucose unit is transferred from G-1-P to a nonreducing end of the glycosyl acceptor to form α-(1→4)-glycosidic linkage. When the excess molar ratio of G-1-P to the glycosyl acceptor is present in the reaction system, the successive glycosylations, i.e., the enzymatic polymerization of G-1-P as a monomer, occurs to produce the α-(1→4)-glucan chain, that is, amylose (Figure 3) [32–37]. The polymerization is initiated from a nonreducing end of the glycosyl acceptor, and thus, it is often called a "primer." Because the phosphorylase-catalyzed enzymatic polymerization proceeds analogously to a living polymerization, the polydispersity of the amylose produced is narrow (M_w/M_n < 1.2) and its molecular weight can be controlled by the G-1-P/primer feed molar ratios. Phosphorylase is the only enzyme that can produce amylose with the desired average molecular weight [39].

Figure 3. Phosphorylase-catalyzed enzymatic polymerization of G-1-P to form amylase.

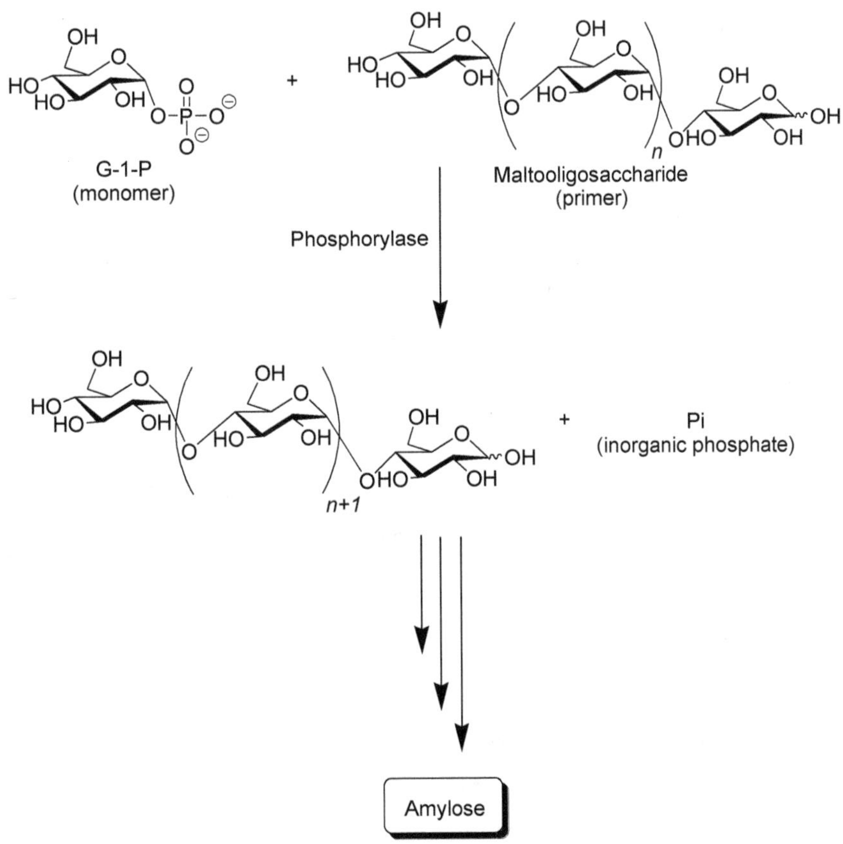

By means of the phosphorylase-catalyzed enzymatic polymerization for direct synthesis of amylose, the author has investigated developing an efficient method for the architecture of inclusion complexes of amylose with synthetic polymers. The representation of propagation in the polymerization system mirrors the way that the vine of a plant grows, twining around a rod. Accordingly, the author has proposed that this polymerization method for the architecture of amylose-polymer supramolecular inclusion complexes be named "vine-twining polymerization" (Figure 4) [14,15,27–30].

Figure 4. Image of "vine-twining polymerization".

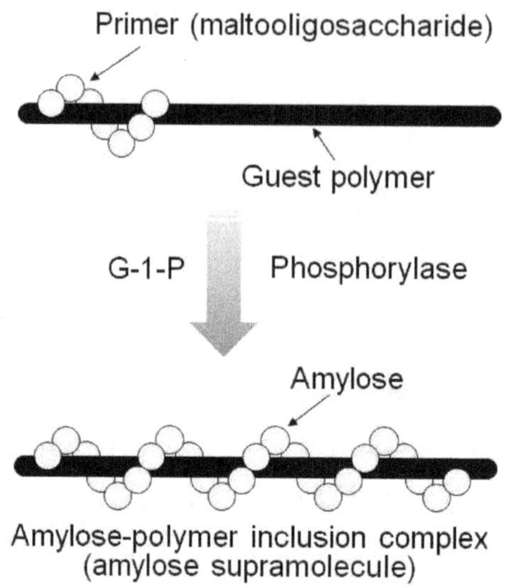

3. Architecture of Amylose-Poly(tetrahydrofuran) Inclusion Complex by Vine-Twining Polymerization

A first example of vine-twining polymerization was reported in the system using poly(tetrahydrofuran) (PTHF) as a hydrophobic guest polyether (Figure 5) [40]. When the phosphorylase-catalyzed enzymatic polymerization of G-1-P from maltoheptaose (G_7) as a primer was conducted in the presence of hydroxy-terminated telechelic PTHF, with M_n of 4000 in sodium citrate buffer, the product was gradually precipitated during the progress of the polymerization, which was isolated by filtration and characterized by ^{1}H NMR and powder X-ray diffraction (XRD) measurements.

Figure 5. Architecture of inclusion complexes by vine-twining polymerization using hydrophobic guest polyethers.

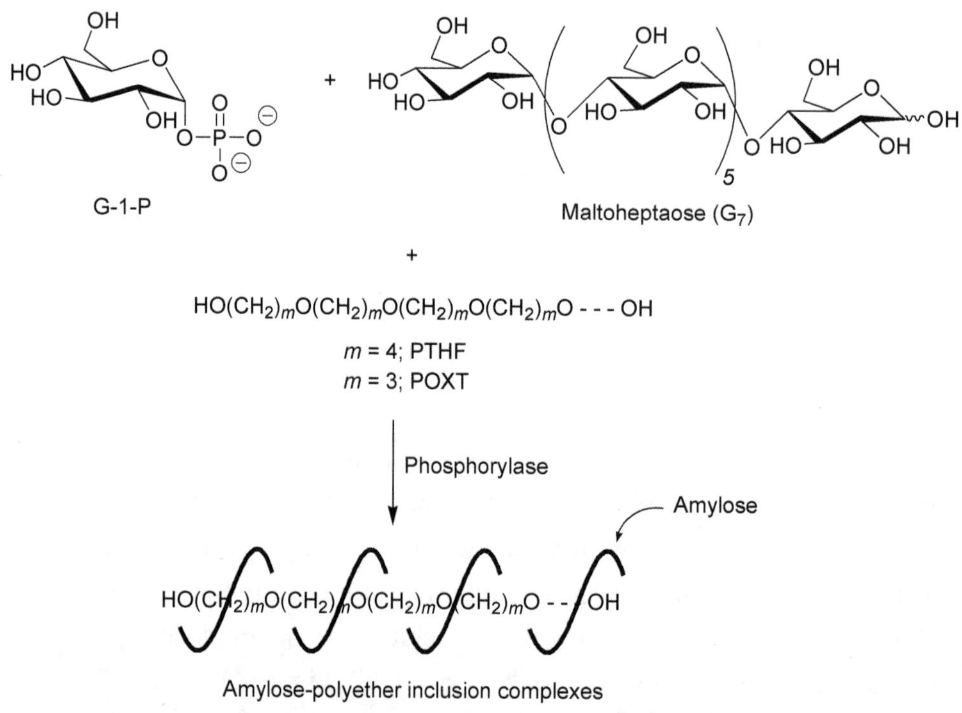

Figure 6. XRD profiles of amylose (**a**); amylose-PTHF inclusion complex (**b**); amylose-P(GA-*co*-CL) inclusion complex (**c**); the product obtained by vine-twining polymerization using P(GA-*b*-CL) (**d**); and amylose-PLLA inclusion complex (**e**).

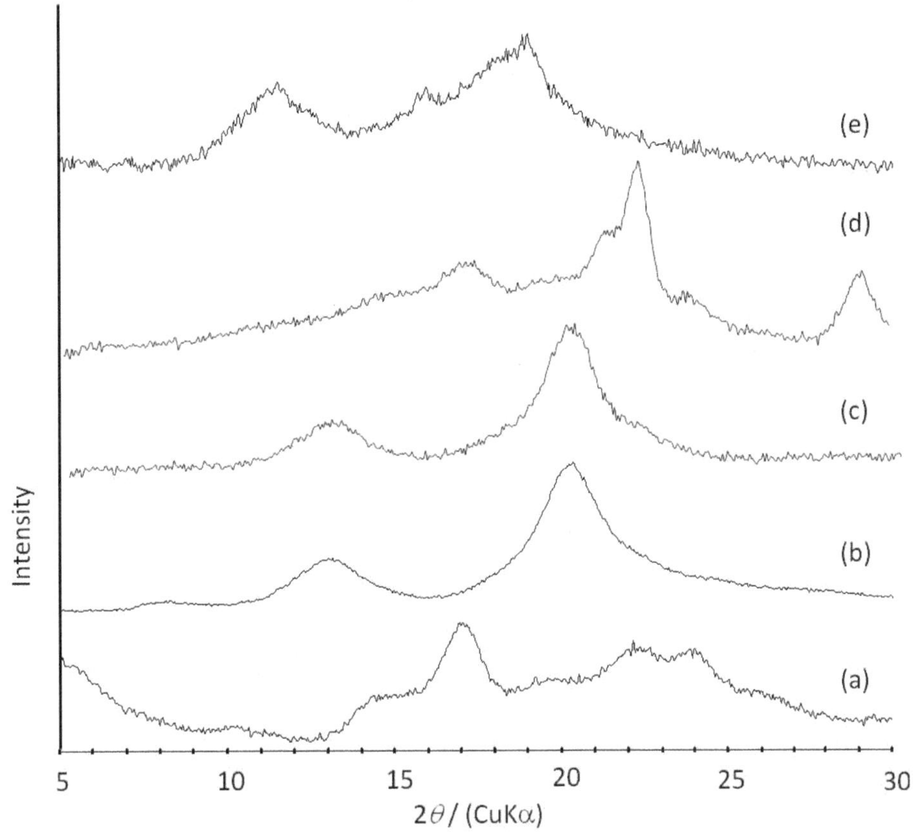

In the ^{1}H NMR spectrum of the product in DMSO-d_6, signals not only due to amylose but also due to PTHF were detected. Moreover, the methylene peak of PTHF was broadened and shifted to a higher magnetic field compared with that of a sole PTHF, suggesting that each methylene group in PTHF is immobile and interacts with the protons inside the amylose cavity by complexation. In addition, the NMR pattern assignable to amylose and PTHF was also observed by the measurement in NaOD/D_2O solvent. A sole PTHF was not dissolved with NaOD/D_2O, and thus, no peak due to PTHF appeared in the ^{1}H NMR spectrum of a mixture of PTHF with NaOD/D_2O. These NMR results indicated that PTHF in the product was solubilized in alkaline solution, probably by suppressing the formation of crystalline aggregates because of its inclusion complexation in the cavity of amylose. The XRD profile of the product showed two strong peaks at 2θ = 12.4 and 19.8° (Figure 6b), which was completely different from that of a sole amylose (Figure 6a), but similar to that of inclusion complexes of amylose with monomeric guest molecules reported in a previous study [41]. The above NMR and XRD results strongly supported that the amylose-PTHF inclusion complex was obtained by the vine-twining polymerization system.

The formation of an inclusion complex was not observed by mixing amylose and PTHF in a buffer solvent, strongly suggesting its formation during the progress of the enzymatic polymerization in the above system. To additionally study the relation between the formation of an inclusion complex and the enzymatic polymerization process, the following experiment was conducted. When PTHF was added to the reaction solution immediately after the enzymatic polymerization of G-1-P had initiated, the identical inclusion complex to the aforementioned system was produced. However, the contents of PTHF in the products decreased as the time between the initiation of the enzymatic polymerization and the addition of PTHF to the solution increased. These results revealed that amylose did not sufficiently include PTHF in the cavity after the polymerization produced amylose with relatively higher molecular weight. Accordingly, it was considered that the inclusion complex had only been formed simultaneously with the progress of the enzymatic polymerization according to the "vine-twining" process.

The effect of molecular weights of PTHFs on the formation of inclusion complexes in the vine-twining polymerization was investigated by using those with various molecular weights (1000, 2000, 10,000, and 14,000) [42]. When PTHF with M_n of 1000 or 2000 was employed as the guest polymer, inclusion complexes were obtained in the vine-twining polymerization. In contrast, the use of PTHFs with higher M_ns such as 10,000 and 14,000, in the vine-twining polymerization, did not induce the formation of inclusion complexes with amylose. The PTHFs with higher M_ns were not sufficiently dispersed in buffer of the polymerization solvent, resulting in difficulty of the inclusion by amylose. To yield inclusion complexes from these PTHFs, the diethyl ether/buffer two-phase system was attempted for the vine-twining polymerization. The higher molecular weight PTHFs were first dissolved in diethyl ether, and then, buffer solvent was added to the ether solution (diethyl ether:buffer = 1:5 (v/v)). Then, the phosphorylase-catalyzed enzymatic polymerization of G-1-P was carried out with vigorously stirring the two-phase mixture. Consequently, the XRD profiles of the products from the higher molecular weight PTHFs obtained by the two-phase system indicated the formation of the inclusion complexes.

4. Architecture of Inclusion Complexes by Vine-Twining Polymerization Using Other Polyethers as Guest Polymers

To investigate the effect of alkyl chain lengths in polyethers on the formation of inclusion complexes in the vine-twining polymerization, the phosphorylase-catalyzed enzymatic polymerization of G-1-P was performed in the presence of polyethers with different alkyl chain lengths from PTHF (the number of methylenes = 4), that is, poly(oxetane) (POXT, the number of methylenes = 3) and poly(ethylene glycol) (PEG, the number of methylenes = 2) [42]. The ^{1}H NMR spectrum of the product from POXT in DMSO-d_6 showed the signals due to both amylose and POXT and its XRD pattern was same as that of the aforementioned amylose-PTHF inclusion complex. These analytical data indicated that the inclusion complex was formed by the vine-twining polymerization using POXT as the guest polyether (Figure 5). In the ^{1}H NMR spectrum of the product from PEG, on the other hand, only the signals due to amylose were detected. Moreover, the XRD profile of the product from PEG showed the pattern for amylose, but did not exhibit that for an inclusion complex. These data indicated that amylose was produced by the enzymatic polymerization in this system, but that did not induce inclusion complexation with PEG. The above results were reasonably explained by the different hydrophobicities in the polyethers. The hydrophilic nature of PEG caused much less hydrophobic interaction with the cavity of amylase, resulting in no complexation by amylose, whereas the hydrophobic polyethers such as PTHF and POXT interacted with the cavity of amylose, leading to the formation of the inclusion complexes by the vine-twining polymerization. The above results suggested that the hydrophobicity of guest polymers strongly affected whether inclusion complexation takes place in the vine-twining polymerization system.

5. Architecture of Inclusion Complexes by Vine-Twining Polymerization Using Carbonyl-Containing Hydrophobic Polymers as Guest Polymers

Based on significance in the hydrophobicity of guest polymers on the formation of inclusion complexes, well-known hydrophobic polyesters, that is, hydroxy-terminated telechelic poly(ε-caprolactone) (PCL) and poly(δ-valerolactone) (PVL) were employed as the guest polymer in the vine-twining polymerization (Figure 7) [43,44]. The phosphorylase-catalyzed enzymatic polymerization of G-1-P from G_7 was conducted in the presence of PCL or PVL in sodium citrate buffer and the precipitated products were characterized by ^{1}H NMR and XRD measurements. The ^{1}H NMR spectra of the products from PCL with M_n of 1000 and PVL with M_n of 2000 showed signals not only due to amylose but also due to the polyesters. The XRD patterns of the products were completely different from those of a sole amylose and were similar as those of the aforementioned amylose-PTHF inclusion complex. Moreover, the IR spectrum of the original PVL exhibited a strong absorption at 1728 cm^{-1} (Figure 8a), corresponding to a carbonyl group of the crystalline PVL, and which of PVL in the product shifted to the region at 1736 cm^{-1} (Figure 8b) assignable to the non-crystalline PVL. This result suggested that the crystalline PVL did not present in the product because of inclusion of a PVL chain in the cavity of amylose, suppressing the formation of crystalline aggregates among PVL chains. All the above analytical results supported the structures of the inclusion complexes of amylose with PCL and PVL.

Figure 7. Architecture of inclusion complexes by vine-twining polymerization using carbonyl-containing hydrophobic polymers.

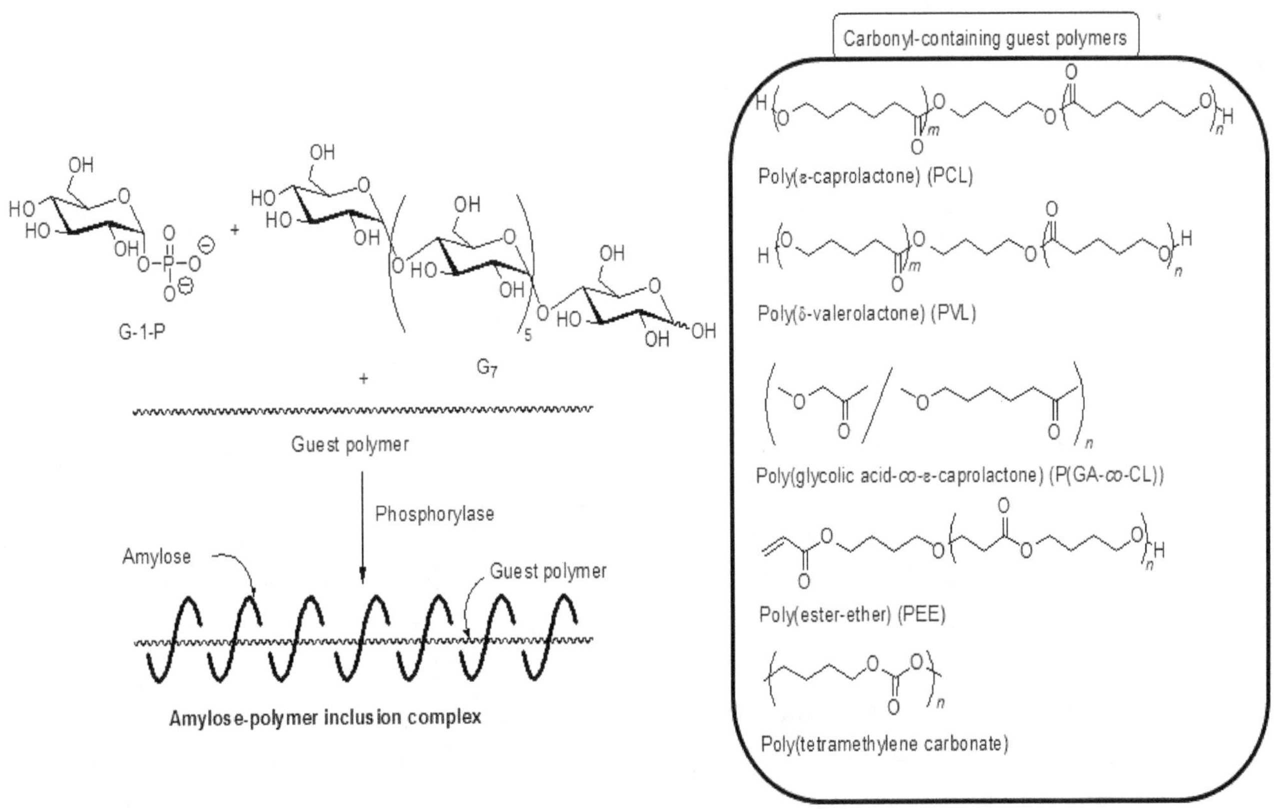

Figure 8. IR spectra of PVL (**a**); and amylose-PVL inclusion complex (**b**).

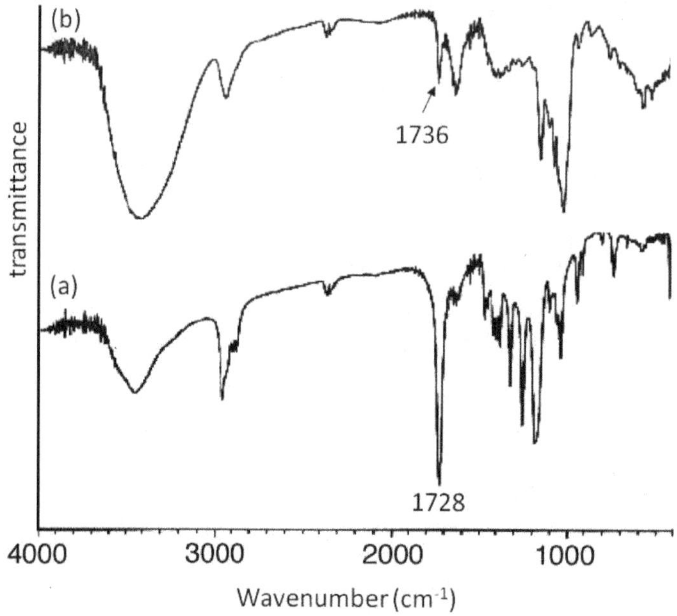

Although PCL with a higher M_n (2000) was employed as the guest polyester in the vine-twining polymerization, it was not sufficiently dispersed in sodium citrate buffer of the polymerization solvent, and accordingly, the inclusion complex was not produced. To yield the inclusion complex from such a PCL, the vine-twining polymerization was attempted in a mixed solvent of acetone and sodium citrate buffer (1:5 (v/v)) as PCL with the M_n was dispersed in the mixed solvent system. The resulting product was characterized by the ^{1}H NMR and XRD measurements to be the inclusion complex. As previously mentioned, the inclusion complex was formed from PVL with M_n of 2000 by the vine-twining polymerization in sodium citrate buffer, suggesting that PVL was more favorable as the guest polymer to form inclusion complex, compared with PCL by the vine-twining polymerization.

It was also found that amylose-poly(glycolic acid-*co*-ε-caprolactone) (P(GA-*co*-CL)) inclusion complexes were obtained by the vine-twining polymerization, using biodegradable P(GA-*co*-CL)s (Figure 7) [38]. As poly(glycolic acid) (PGA) shows high crystallinity and low dispersibility in aqueous buffer, resulting in the difficulty of the inclusion of PGA in the cavity of amylose, the copolyester, P(GA-*co*-CL) composed of the two biodegradable chains, was used. Thus, the vine-twining polymerization was carried out by the phosphorylase-catalyzed polymerization in the presence of P(GA-*co*-CL)s with various unit ratios in sodium acetate buffer. The ^{1}H NMR spectra of the products showed the signals due to amylose and P(GA-*co*-CL) and their XRD patterns showed the typical diffraction peaks, assignable to inclusion complexes (Figure 6c). These results indicated that the products were the inclusion complexes composed of amylose and P(GA-*co*-CL). The ratios of GA unit to CL unit in P(GA-*co*-CL)s did not affect inclusion complexation by amylose. On the other hand, an inclusion complex was not formed in the vine-twining polymerization using poly(glycolic acid-*block*-ε-caprolactone) (P(GA-*b*-CL)), as confirmed by the XRD profile of the product (Figure 6d). As P(GA-*co*-CL) and P(GA-*b*-CL) exhibit amorphous and crystalline natures, respectively, it has been considered that the crystallinity of the guest copolyesters affected inclusion complexation by amylose. In addition, lipase-catalyzed hydrolysis of P(GA-*co*-CL), in the inclusion complex, was partly inhibited, likely because amylose surrounded P(GA-*co*-CL) to prevent the approach of lipase.

As the other guest polyester, a hydrophobic poly(ester-ether) (–$CH_2CH_2C(=O)OCH_2CH_2CH_2CH_2O$–), which was composed of alternating ester and ether bonds, was employed for the vine-twining polymerization (Figure 7) [44]. The structure of the product was evaluated by the ^{1}H NMR and XRD measurements to be the inclusion complex. When a hydrophilic poly(ester-ether) (–$CH_2CH_2C(=O)OCH_2CH_2O$–), which had a shorter methylene length, was employed as the guest polymer, no inclusion complex was produced. This result also supported that the hydrophobicity of the guest polymers affected the formation of inclusion complexes by the vine-twining polymerization.

In addition, to no production of an inclusion complex from such a hydrophilic polymer, the formation of inclusion complexes, had not been achieved from polymers with strong hydrophobicity due to their aggregation in aqueous buffer solvent in the vine-twining polymerization, as an example using PCL with the higher M_n [44]. As the other example, the vine-twining polymerization using poly(oxepane), which was a more hydrophobic polyether owing to longer alkyl chains compared with PTHF and POXT, did not induce inclusion complexation by amylose [42].

To obtain the inclusion complex from a strongly hydrophobic polyester, the parallel enzymatic polymerization system, as an extensive approach of the vine-twining polymerization, was attempted [45]. In this system, two enzymatic polymerizations, which were the phosphorylase-catalyzed enzymatic polymerization of G-1-P from G_7, producing amylose and the lipase-catalyzed polycondensation of a dicarboxylic acid and a diol, giving an aliphatic polyester of the guest polymer [46,47], were simultaneously performed (Figure 9). As the monomers for the polycondensation, sebacic acid and 1,8-octanediol were used, which were converted to the strongly hydrophobic polyester by the lipase-catalyzed polycondensation under aqueous conditions. The product by the parallel enzymatic polymerization system was evaluated by the ^{1}H NMR and XRD measurements, which indicated the formation of the inclusion complex of amylose with the polyester. To confirm the fact that the inclusion complex was obtained only by the parallel enzymatic polymerization system, the following two experiments were conducted. When the phosphorylase-catalyzed enzymatic polymerization was carried out in the presence of such a strongly hydrophobic polyester, according to the vine-twining polymerization manner, amylose was produced, but it did not induce inclusion complexation with the polyester. As the other experiment, the lipase-catalyzed polycondensation of sebacic acid and 1,8-octanediol was carried out in the presence of amylose. Consequently, amylose included some monomers, but did not include the polyester although the enzymatic polycondensation progressed. These results concluded that the inclusion complex of amylose with such a strongly hydrophobic polyester was obtained only by the parallel enzymatic polymerization system.

Hydrophobic aliphatic polycarbonates were also used as the guest polymer in the vine-twining polymerization to prepare the corresponding inclusion complexes with amylose [48]. When the vine-twining polymerization was conducted, using poly(tetramethylene carbonate) as the guest polycarbonate in an acetone/aqueous buffer mixed solvent, the precipitated product was evaluated by ^{1}H NMR, XRD, and IR measurements to be the amylose-poly(tetramethylene carbonate) inclusion complex (Figure 7). The effect of the methylene chain lengths in the polycarbonates on inclusion complexation in the vine-twining polymerization was further examined by using poly(tetramethylene carbonate), poly(octamethylene carbonate), poly(decamethylene carbonate), and poly(dodecamethylene carbonate) as a guest polycarbonate. Compared with the polycarbonate with the shorter methylene chain length, such as the above mentioned poly(tetramethylene carbonate), only lesser amounts of the polycarbonates, having longer methylene chain lengths, were included in the cavity of amylose. Such strongly hydrophobic polycarbonates were not sufficiently dispersed in the acetone/aqueous buffer mixed solvent, resulting in difficulty in the inclusion complexation by amylose.

Figure 9. Architecture of inclusion complexes composed of amylose and strongly hydrophobic polyester in parallel enzymatic polymerization system.

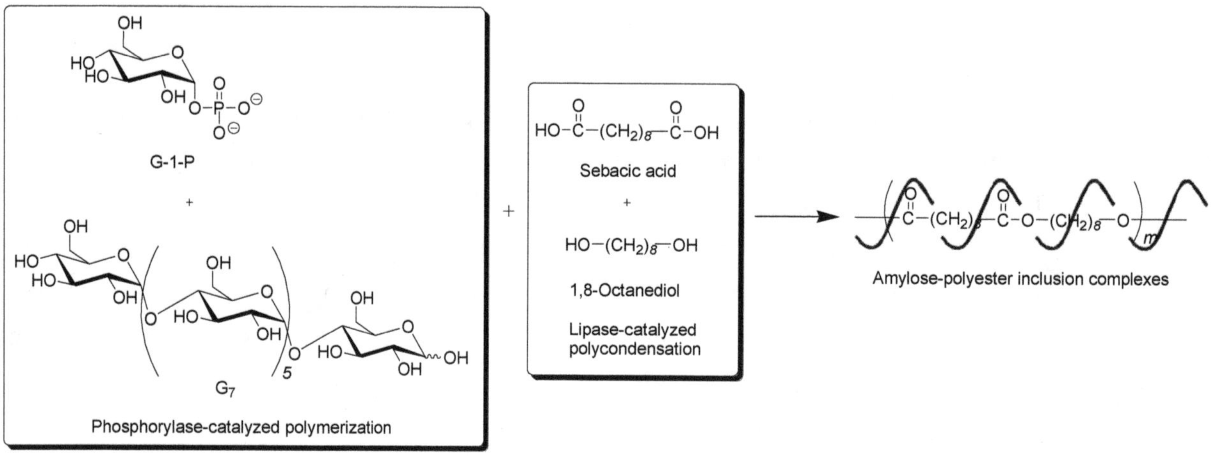

6. Selective Inclusion Complexation by Amylose in Vine-Twining Polymerization

On the basis of the aforementioned results, during the course of studies on the vine-twining polymerization, the author has considered that moderate hydrophobicity of guest polymers is in demand in deciding whether amylose includes them or not. Taking more precise information into account, amylose exhibits different inclusion behaviors depending on subtle changes in the structures of hydrophobic guest polymers. Such behavior of amylose was applied in the investigation to realize the selective inclusion complexation toward two resemblant guest polymers (Figure 10) [49,50]. For example, the selective inclusion of amylose was achieved by the vine-twining polymerization in the presence of a mixture of POXT (M_n = 1800) and PTHF (M_n = 1600) (unit molar ratio = 0.90:1.00) in sodium acetate buffer [49]. In the ^{1}H NMR spectrum of the product (DMSO-d_6), the signals due to PTHF and amylose were prominently detected, but the signals due to POXT, mostly, were not observed (POXT/PTHF = 0.02:1.00). This NMR result suggested that amylose almost selectively included PTHF from the mixture of the two resemblant polyethers in the vine-twining polymerization. The difference in the inclusion behavior of amylose toward the two polyethers was probably owing to the slight difference in their hydrophobicities. An attempt in the other selective inclusion system by amylose was made by the vine-twining polymerization in the presence of a mixture of two resemblant polyesters, that is, PVL (M_n = 830) and PCL (M_n = 930) (unit molar ratio = 1.00:0.92) [50]. In the ^{1}H NMR spectrum of the product (DMSO-d_6), the signals due to PVL and amylose were detected, whereas no signals due to PCL appeared, indicating that amylose selectively included PVL from the mixture of the two resemblant polyesters in the vine-twining polymerization.

Figure 10. Amylose selectively includes one of two resemblant polyethers or polyesters.

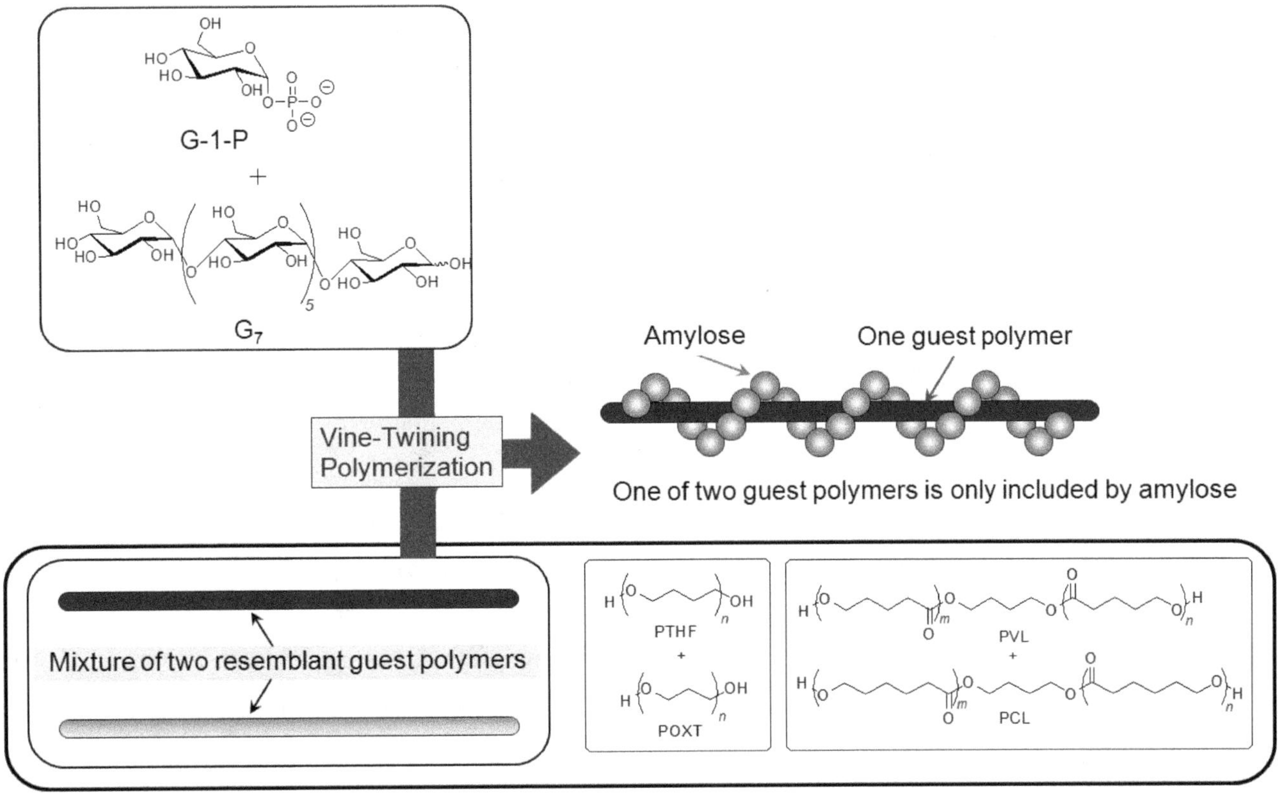

The selective inclusion complexation toward a specific range in molecular weights of synthetic guest polymers by amylose in the vine-twining polymerization was also investigated [51]. As synthetic polymers have molecular weight distribution, they are reasonably considered as mixtures of analogous molecules with different numbers of the repeating units. Moreover, the number of repeating units contributes to the exhibition of different properties of the polymers with the same repeating unit. For example, PTHF with considerably low molecular weight is soluble in water, whereas that with higher molecular weight is insoluble in water, suggesting that the molecular weight of PTHF affects its hydrophobicity. Therefore, the vine-twining polymerization in the presence of three PTHFs with different M_ns and (M_w/M_n)s (PTHF/1000; M_n and M_w/M_n = 1350 and 2.86, PTHF/3000; M_n and M_w/M_n = 3040 and 3.13, and PTHF/6000; M_n and M_w/M_n = 6330 and 2.45) was investigated. When PTHF/3000 with M_w/M_n = 3.13 was used as the guest polymer for the vine-twining polymerization, the M_w/M_n of PTHF included by amylose became narrower (M_w/M_n = 1.46) although its M_n (=3590) was comparable to that of the employed one (M_n = 3040). These results indicated that amylose selectively included a specific range in molecular weights of PTHF/3000 in the vine-twining polymerization. When the vine-twining polymerization using PTHF/1000, and PTHF/6000 as the guest polymer, was conducted, the M_ns and (M_w/M_n)s of PTHFs included by amylose were similar as those using PTHF/3000 as aforementioned (the included PTHF/1000; M_n and M_w/M_n = 3120 and 1.41, the included PTHF/6000; M_n and M_w/M_n = 3700 and 1.74). Thus, it was concluded that amylose selectively included a specific range in molecular weights of PTHFs probably by the specific hydrophobic interaction.

Figure 11. Stereoselective inclusion complexation by amylose in vine-twining polymerization using poly(L-lactide) (PLLA).

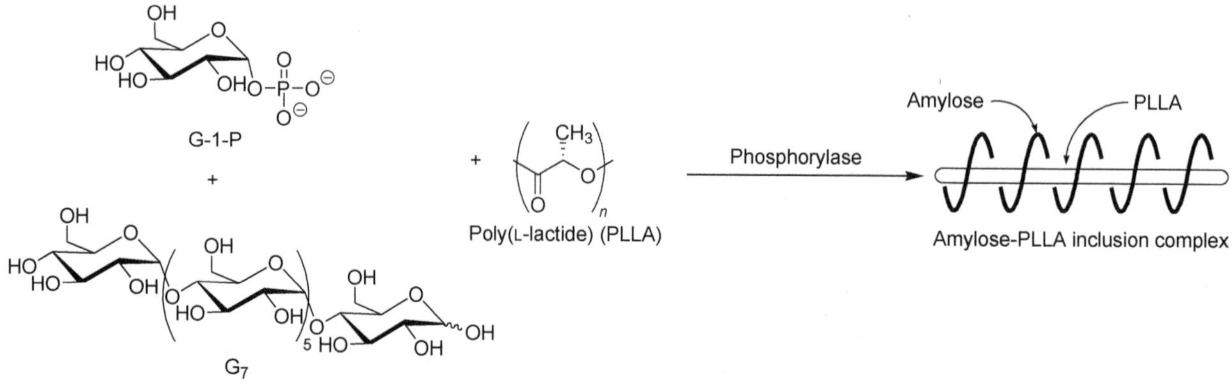

The selective inclusion complexation by amylose was also achieved in the vine-twining polymerization using chiral polyesters, i.e., poly(lactide)s (PLAs) as the guest polymer (Figure 11) [52]; there are three kinds of the stereoisomers, i.e., poly(L-lactide) (PLLA), poly(D-lactide) (PDLA), and racemic poly(DL-lactide) (PLDLA). When the vine-twining polymerization, using PLLA, was carried out, signals, not only due to amylose, but also due to PLLA, were observed in the ^{1}H NMR spectrum of the product in DMSO-d_6. The XRD pattern of the product showed two diffraction peaks at $2\theta = 11°–12°$ and 18°–19° (Figure 6e), which was completely different from that of a sole amylose (Figure 6a), supporting the inclusion complex structure of the product. Interestingly, the diffraction peaks of the product were detected at lower angles compared with those of amylase-PTHF inclusion complex (Figure 6b) [40]. This difference in the two XRD patterns reasonably suggested that the diameter of amylose helix in the product was larger by inclusion of the bulky PLLA in the cavity of amylose. These analytical results fully supported that the amylose-PLLA inclusion complex was obtained in the vine-twining polymerization using PLLA as the guest polyester. To investigate the effect of the chirality in PLAs on inclusion by amylose, the vine-twining polymerization was performed using PDLA and PDLLA. Consequently, the ^{1}H NMR spectra and the XRD patterns of the products indicated no formation of inclusion complexes. From the above results, it was concluded that amylose perfectly recognized the chirality in PLAs on inclusion complexation in the vine-twining polymerization.

7. Conclusions

In this article, the author reviewed the precision architecture of amylose supramolecules in the form of inclusion complexes by the phosphorylase-catalyzed enzymatic polymerization of G-1-P in the presence of hydrophobic synthetic polymers, such as polyethers, polyesters, a poly(ester-ether), and polycarbonates, according to the vine-twining polymerization manner. The corresponding amylose supramolecular inclusion complexes were formed in the process of the polymerization. The results in the vine-twining polymerization suggested that amylose exhibited different inclusion behaviors by the specific interactions with the guest polymers, depending on subtle changes in their structures. As of the production of the structurally defined polysaccharides by the enzymatic catalysis, the precision architecture of the regularly controlled amylose supramolecules, described herein, has been achieved. Moreover, the studies on the enzymatic architecture of the polysaccharide supramolecules have been based on the viewpoints that the greener and eco-friendly processes should be developed in the fields, not only of fundamental research, but also of practical application of the polymer and material chemistries. The polysaccharide supramolecules have increasingly been attracting a great deal of attention because of their potential for application as new functional materials in many research fields, such as medicine and pharmaceutics. Therefore, the present vine-twining polymerization method will be applied to the additional architecture of various amylose supramolecules with regularly controlled nanostructures, and accordingly contributes to providing new functional bio-based materials in the future [53,54].

Acknowledgments

The author is indebted to co-workers, whose names are found in references from his paper, for their enthusiastic collaborations. The author also gratefully thanks a Grant-in-Aid for Scientific Research from Ministry of Education, Culture, Sports, and Technology, Japan (Nos. 14550830, 17550118, 19550126, and 24350062) and Sekisui Foundation for financial supports.

References

1. Schuerch, C. Polysaccharides. In *Encyclopedia of Polymer Science and Engineering*, 2nd ed.; Mark, H.F., Bilkales, N., Overberger, C.G., Eds.; John Wiley & Sons: New York, NY, USA, 1986; Volume 13, pp. 87–162.
2. *Carbohydrates in Chemistry and Biology*; Ernst, B., Hart, G.W., Sinaÿ P., Eds.; Wiley-VCH: Weinheim, Germany, 2000.
3. *Glycoscience*, 2nd ed.; Fraser-Reid, B.O., Tatsuta, K., Thiem, J., Coté, G.L., Flitsch, S., Ito, Y., Kondo, H., Nishimura, S.-I., Yu, B., Eds.; Springer: Berlin, Germany, 2008.
4. *Essentials of Glycobiology*, 2nd ed.; Varki, A., Cummings, R.D., Esko, J.D., Freeze, H.H., Stanley, P., Bertozzi, C.R., Hart, G.W., Etzler, M.E., Eds.; Cold Spring Harbor Laboratory Press: New York, NY, USA, 2009.
5. Paulsen, H. Advances in selective chemical syntheses of complex oligosaccharides. *Angew. Chem. Int. Ed. Engl.* **1982**, *21*, 155–173.
6. Schmidt, R.R. New methods of the synthesis of glycosides and oligosaccharides—Are there alternative to the Koenigs-Knorr methods? *Angew. Chem. Int. Ed. Engl.* **1986**, *25*, 212–235.
7. Toshima, K.; Tatsuta, K. Recent progress in *O*-glycosylation methods and its application to natural products synthesis. *Chem. Rev.* **1993**, *93*, 1503–1531.
8. Mydock, L.K.; Demchenko, A.V. Mechanism of chemical *O*-glycosylation: From early studies to recent discoveries. *Org. Biomol. Chem.* **2010**, *8*, 497–510.
9. Kobayashi, S.; Uyama, H.; Kimura, S. Enzymatic polymerization. *Chem. Rev.* **2001**, *101*, 3793–3818.

10. Shoda, S.; Izumi, R.; Fujita, M. Green process in glycotechnology. *Bull. Chem. Soc. Jpn.* **2003**, *76*, 1–13.
11. Seibel, J.; Jördening, H.-J.; Buchholz, K. Glycosylation with activated sugars using glycosyltransferases and transglycosidases. *Biocatal. Biotranform.* **2006**, *24*, 311–342.
12. Kobayashi, S. New development of polysaccharide synthesis via enzymatic polymerization. *Proc. Jpn. Acad. Ser. B* **2007**, *83*, 215–247.
13. Kobayashi, S.; Makino, A. Enzymatic polymer synthesis: An opportunity for green polymer chemistry. *Chem. Rev.* **2009**, *109*, 5288–5353.
14. Kadokawa, J.; Kobayashi, S. Polymer synthesis by enzymatic catalysis. *Curr. Opin. Chem. Biol.* **2010**, *14*, 145–153.
15. Kadokawa, J. Precision polysaccharide synthesis catalyzed by enzymes. *Chem. Rev.* **2011**, *111*, 4308–4345.
16. Lenz, R.W. Biodegradable polymers. *Adv. Polym. Sci.* **1993**, *107*, 1–40.
17. Putseys, J.A.; Lamberts, L.; Delcour, J.A. Amylose-inclusion complexes: Formation, identity and physico-chemical properties. *J. Cereal Sci.* **2010**, *51*, 238–247.
18. Shogren, R.L.; Greene, R.V.; Wu, Y.V. Complexes of starch polysaccharides and poly(ethylene-*co*-acrylic acid)–structure and stability in solution. *J. Appl. Polym. Sci.* **1991**, *42*, 1701–1709.
19. Shogren, R.L. Complexes of starch with telechelic poly(ε-caprolactone) phosphate. *Carbohydr. Polym.* **1993**, *22*, 93–98.
20. Ikeda, M.; Furusho, Y.; Okoshi, K.; Tanahara, S.; Maeda, K.; Nishino, S.; Mori, T.; Yashima, E. A luminescent poly(phenylenevinylene)-amylose composite with supramolecular liquid crystallinity. *Angew. Chem. Int. Ed.* **2006**, *45*, 6491–6495.
21. Kida, T.; Minabe, T.; Okabe, S.; Akashi, M. Partially-methylated amyloses as effective hosts for inclusion complex formation with polymeric guests. *Chem. Commun.* **2007**, 1559–1561.
22. Kida, T.; Minabe, T.; Nakano, S.; Akashi, M. Fabrication of novel multilayered thin films based on inclusion complex formation between amylose derivatives and guest polymers. *Langmuir* **2008**, *24*, 9227–9229.
23. Frampton, M.J.; Claridge, T.D.W.; Latini, G.; Brovelli, S.; Cacialli, F.; Anderson, L. Amylose-wrapped luminescent conjugated polymers. *Chem. Commun.* **2008**, 2797–2799.
24. Kaneko, Y.; Kyutoku, T.; Shimomura, N.; Kadokawa, J. Formation of amylose-poly(tetrahydrofuran) inclusion complexes in ionic liquid media. *Chem. Lett.* **2011**, *40*, 31–33.
25. Rachmawati, R.; Woortman, A.J.J.; Loos, K. Facile preparation method for inclusion complexes between amylose and polytetrahydrofurans. *Biomacromolecules* **2013**, *14*, 575–583.
26. Rachmawati, R.; Woortman, A.J.J.; Loos, K. Tunable properties of inclusion complexes between amylose and polytetrahydrofuran. *Macromol. Biosci.* **2013**, *13*, 767–776.
27. Kaneko, Y.; Kadokawa, J. Vine-twining polymerization: A new preparation method for well-defined supramolecules composed of amylose and synthetic polymers. *Chem. Rec.* **2005**, *5*, 36–46.
28. Kaneko, Y.; Kadokawa, J. Synthesis of nanostructured bio-related materials by hybridization of synthetic polymers with polysaccharides or saccharide residues. *J. Biomater. Sci. Polym. Ed.* **2006**, *17*, 1269–1284.

29. Kaneko, Y.; Kadokawa, J. *Modern Trends in Macromolecular Chemistry*; Lee, J.N., Ed.; Nova Science Publishers, Inc.: Hauppauge, NY, USA, 2009; Chapter 8, pp. 199–217.
30. Kadokawa, J. Preparation and applications of amylose supramolecules by means of phosphorylase-catalyzed enzymatic polymerization. *Polymers* **2012**, *4*, 116–133.
31. Kitaoka, M.; Hayashi, K. Carbohydrate-processing phosphorolytic enzymes. *Trends Glycosci. Glycotechnol.* **2002**, *14*, 35–50.
32. Ziegast, G.; Pfannemüller, B. Phosphorolytic syntheses with di-, oligo- and multi-functional primers. *Carbohydr. Res.* **1987**, *160*, 185–204.
33. Gidley, M.J.; Bulpin, P.V. Aggregation of amylose in aqueous systems: The effect of chain length phase behavior and aggregation kinetics. *Macromolecules* **1989**, *22*, 341–346.
34. Niemann, C.; Sanger, W.; Pfannemüller, B.; Eigner, W.D.; Huber, A. Phospholytic synthesis of low-molecular-weight amyloses with modified terminal groups; Comparison of potato phosphorylase andmuscle phosphorylase b. In *ACS Symp. Ser. Volume 458, Biotechnology of amylodextrin oligosaccharides*; Friedman, R.B., Ed.; American Chemical Society: Washington, DC, USA, 1991; Chapter 13, pp. 189–204.
35. Fujii, K.; Takata, H.; Yanase, M.; Terada, Y.; Ohdan, K.; Takaha, T.; Okada, S.; Kuriki, T. Bioengineering and application of novel glucose polymers. *Biocatal. Biotransform.* **2003**, *21*, 167–172.
36. Yanase, M.; Takaha, T.; Kuriki, T. α-Glucan phosphorylase and its use in carbohydrate engineering. *J. Food Agric.* **2006**, *86*, 1631–1635.
37. Ohdan, K.; Fujii, K.; Yanase, M.; Takaha, T.; Kuriki, T. Enzymatic synthesis of amylose. *Biocatal. Biotransform.* **2006**, *24*, 77–81.
38. Nomura, S.; Kyutoku, T.; Shimomura, N.; Kaneko, Y.; Kadokawa, J. Preparation of inclusion complexes composed of amylose and biodegradable poly(glycolic acid-*co*-ε-caprolactone) by vine-twining polymerization and their lipase-catalyzed hydrolysis behavior. *Polym. J.* **2011**, *43*, 971–977.
39. Kitamura, S. Starch, Polymers, Natural and Synthetic. In *The Polymeric Materials Encyclopedia, Synthesis, Properties and Applications*; Salamone, C., Ed.; CRC Press: New York, NY, USA, 1996; Volume 10, pp. 7915–7922.
40. Kadokawa, J.; Kaneko, Y.; Tagaya, H.; Chiba, K. Synthesis of an amylose-polymer inclusion complex by enzymatic polymerization of glucose 1-phosphate catalyzed by phosphorylase enzyme in the presence of polyTHF: A new method for synthesis of polymer-polymer inclusion complexes. *Chem. Commun.* **2001**, *2001*, 449–450.
41. Seneviratne, H.D.; Biliaderis, C.G. Action of α-amylases on amylose-lipid complex superstructures. *J. Cereal Sci.* **1991**, *13*, 129–143.
42. Kadokawa, J.; Kaneko, Y.; Nagase, S.; Takahashi, T.; Tagaya, H. Vine-twining polymerization: Amylose twines around polyethers to form amylose-polyether inclusion complexes. *Chem. Eur. J.* **2002**, *8*, 3321–3326.
43. Kadokawa, J.; Kaneko, Y.; Nakaya, A.; Tagaya, H. Formation of an amylose-polyester inclusion complex by means of phosphorylase-catalyzed enzymatic polymerization of α-D-glucose 1-phosphate monomer in the presence of poly(ε-caprolactone). *Macromolecules* **2001**, *34*, 6536–6528.

44. Kadokawa, J.; Nakaya, A.; Kaneko, Y.; Tagaya, H. Preparation of inclusion complexes between amylose and ester-containing polymers by means of vine-twining polymerization. *Macromol. Chem. Phys.* **2003**, *204*, 1451–1457.
45. Kaneko, Y.; Saito, Y.; Nakaya, A.; Kadokawa, J.; Tagaya, H. Preparation of inclusion complexes composed of amylose and strongly hydrophobic polyesters in parallel enzymatic polymerization system. *Macromolecules* **2008**, *41*, 5665–5670.
46. Kobayashi, S.; Uyama, H.; Suda, S.; Namekawa, S. Dehydration polymerization in aqueous medium catalyzed by lipase. *Chem. Lett.* **1997**, 105–105.
47. Suda, S.; Uyama, H.; Kobayashi, S. Dehydration polycondensation in water for synthesis of polyesters by lipase catalyst. *Proc. Jpn. Acad. Ser. B* **1999**, *75*, 201–206.
48. Kaneko, Y.; Beppu, K.; Kadokawa, J. Preparation of amylose/polycarbonate inclusion complexes by means of vine-twining polymerization. *Macromol. Chem. Phys.* **2008**, *209*, 1037–1042.
49. Kaneko, Y.; Beppu, K.; Kadokawa, J. Amylose selectively includes one from a mixture of two resemblant polyethers in vine-twining polymerization. *Biomacromolecules* **2007**, *8*, 2983–2985.
50. Kaneko, Y.; Beppu, K.; Kyutoku, T.; Kadokawa, J. Selectivity and priority on inclusion of amylose toward guest polyethers and polyesters in vine-twining polymerization. *Polym. J.* **2009**, *41*, 279–286.
51. Kaneko, Y.; Beppu, K.; Kadokawa, J. Amylose selectively includes a specific range of molecular weights in poly(tetrahydrofuran)s in vine-twining polymerization. *Polym. J.* **2009**, *41*, 792–796.
52. Kaneko, Y.; Ueno, K.; Yui, T.; Nakahara, K.; Kadokawa, J. Amylose's recognition of chirality in polylactides on formation of inclusion complexes in vine-twining polymerization. *Macromol. Biosci.* **2011**, *11*, 1407–1415.
53. Kaneko, Y.; Fujisaki, K.; Kyutoku, T.; Furukawa, H.; Kadokawa, J. Preparation of enzymatically recyclable hydrogels through the formation of inclusion complexes of amylose in a vine-twining polymerization. *Chem. Asian J.* **2010**, *5*, 1627–1633.
54. Kadokawa, J.; Kaneko, Y. *Engineering of Polysaccharide Materials–by Phosphorylase-Catalyzed Enzymatic Chain-Elongation*; Pan Stanford Publishing Pte. Ltd.: Temasek Boulevard, Singapore, 2013.

10

Improvement of Biocatalysts for Industrial and Environmental Purposes by Saturation Mutagenesis

Francesca Valetti and Gianfranco Gilardi *

Department of Life Sciences and Systems Biology, University of Torino, via Accademia Albertina 13, Torino 10123, Italy

* Author to whom correspondence should be addressed; E-Mail: gianfranco.gilardi@unito.it

Abstract: Laboratory evolution techniques are becoming increasingly widespread among protein engineers for the development of novel and designed biocatalysts. The palette of different approaches ranges from complete randomized strategies to rational and structure-guided mutagenesis, with a wide variety of costs, impacts, drawbacks and relevance to biotechnology. A technique that convincingly compromises the extremes of fully randomized *vs.* rational mutagenesis, with a high benefit/cost ratio, is saturation mutagenesis. Here we will present and discuss this approach in its many facets, also tackling the issue of randomization, statistical evaluation of library completeness and throughput efficiency of screening methods. Successful recent applications covering different classes of enzymes will be presented referring to the literature and to research lines pursued in our group. The focus is put on saturation mutagenesis as a tool for designing novel biocatalysts specifically relevant to production of fine chemicals for improving bulk enzymes for industry and engineering technical enzymes involved in treatment of waste, detoxification and production of clean energy from renewable sources.

Keywords: biocatalysis; directed evolution; synthetic biology; protein engineering; industrial biotechnology; bioremediation; fine chemistry; saturation mutagenesis; screening methods

1. Introduction

Protein engineering allows exploration of mutational space under artificial evolutionary pressure and selection that could not be sampled by the natural environment of proteins. Advances in this field demonstrate how natural catalysts can be finely tuned to perform reactions that are new in terms of specificity [1–3], efficiency [4–6], stability of the enzyme, conditions [7,8] and chemistry of the reaction catalyzed [9–11]. Many of the biotechnological benefits of this "laboratory-driven evolution" have already been translated into practical applications, and many others can be foreseen to have a high impact in sustainable and innovative processes [12], environmental bioremediation, detoxification, and clean energy production [13–15].

New tools and strategies aiming at simplifying the experimental work required for a successful result in obtaining engineered enzymes are continuously being developed. The methods follow two main directions: the rational site-specific mutagenesis and the evolution-like random approach. Both are powerful but each suffer from different limitations in the performance of the outcome and in time necessary to achieve the results. The rational site-specific mutagenesis focuses on the mutation of one or more specific amino acids that are replaced with another residue. It needs to be supported by structural and functional data of the enzyme and it is frequently biased by the assumptions made by the researchers on the basis of previous knowledge. In this respect it might be less innovative and aim at less ambitious goals, although it remains a very precious strategy for testing hypothesis on the fine details and structural determinants of reaction mechanisms. Compared to the random evolution-like approach, it is less time consuming in the production of the mutants, but in the perspective of producing significantly improved biocatalysts for industrial applications, it often results in limited improvement of the desired property. Results are achieved through series of trial-and-error experiments that surely provide interesting data for theoretical speculation but that may require large amounts of time and resources.

On the other hand, laboratory evolution is based on the selection of random mutants with the desired features. It is not limited by the availability of the 3D structure of the enzyme and it mimics in the lab the evolution process that in nature has led to the selection of the best natural catalysts available: the enzymes. This approach establishes methods to introduce random genetic diversity in libraries of mutants (variants) that include various implementations of mutagenic PCR, oligonucleotide-assisted mutagenesis and *in vitro* recombination under mutagenic conditions, including DNA shuffling [16] and several specific techniques such as ITCHY [17], RACHITT [18] SHIPREC [19] and many others that have been extensively reviewed [20–25]. The time consuming process of obtaining the randomly mutated library and the requirement for a high-throughput screening procedure for selection of the desired properties among thousands of clones, is the severe drawback of a very powerful technique that otherwise has the advantage of providing entirely novel landscapes of mutants [26,27].

A specific type of laboratory-evolution method is the "targeted random mutagenesis" method, also called "saturation mutagenesis" that focuses on specific "hot spots" for mutational variability or on critical residues identified by structural comparison and modeling methods. It applies site-saturation mutagenesis (SSM), *i.e.*, the systematic replacement of one amino acid at a chosen site with all alternative encoded amino acids, to explore the performance of each possible variant in terms of structural or functional features of the resulting mutated enzyme. SSM may be applied at random positions but more often it is based on the assumption that most mutations are deleterious or neutral, and therefore the construction of mutant libraries by random methods is inefficient. Since the enzyme properties that are pursued are mainly codified in a small part of the enzyme corresponding to the active site or structural portions known to modulate protein stability, a rational choice of the sites to be targeted is usually preferred. This approach allows fine-tuning of the catalytic properties, particularly when performed as a refinement step after directed evolution. In fact, in fully random techniques a trade-off between the selected property and the overall enzyme performance might put an apparent threshold to the optimization of the target property [28]. Therefore, saturation mutagenesis is a precious tool for exploring and widening the landscape of the enzyme properties and applications. The advantages lie in a compromise solution combining the positive features of the rational mutagenesis and the random approach followed by laboratory selection, with minimum or negligible additive effect on the drawbacks. This is becoming clear in the last few years due to the increasing number of successful results obtained. A particular relevance is given in literature to positive results of this approach applied to enzymes used in fine chemical synthesis, industrial processing and bioremediation.

In order to improve the outputs and to obtain libraries with high abundance methods such as iterative Combinatorial Active Site Test (CAST) [29] and Iterative Saturation Mutagenesis (ISM) [30], all based on the same principle of SSM, were more recently implemented.

The methodology for Site Saturation Mutagenesis, Iterative Saturation Mutagenesis and other innovative methods will be presented highlighting advantages *vs.* site specific and random mutagenesis. Technical details and implications will be discussed, also tackling the issues of randomization and statistical evaluation of the library completeness and throughput efficiency of screening methods. Examples of successful applications covering different enzyme classes will be presented, focusing on cases that are relevant for the production of fine chemicals as well as bulk enzymes for industry, treatment of wastes, detoxification of pollutants and xenobiotics, and production of clean energy from renewable sources.

2. Experimental

Different methodologies pertaining saturation mutagenesis, leading to libraries of mutants relevant in terms of their size with minimal screening efforts, will be illustrated in the following paragraphs. The choice of alternative approaches bears crucial implications and must be carefully considered. Following the pattern of single site saturation mutagenesis and extending the strategies to various multiple combinations, a range of protocols have been proposed and tested. These are described and discussed here, together with the statistical analysis of library coverage and screening methods specifically for the saturation mutagenesis approaches.

2.1. Strategies for the Generation of Libraries of Mutants

2.1.1. Site Saturation Mutagenesis (SSM)

The SSM libraries are usually generated with protocols that follow the commercially available QuikChange™ kit commercialized by Stratagene [31] or using equivalent in house procedures [30]. Mutagenic and complementary primers that carry the desired mutation (Figure 1) are used in a PCR reaction to amplify the plasmid with high fidelity and thus inserting the desired mutations. The position chosen for mutagenesis can be randomized with the codon NNN (where N is any nucleotide), or with a codon NNK (where K is either a T or a G) that can produce codons for all the 20 amino acids and a stop codon. Compared to NNN degeneration, NNK has the advantage that it will produce 32 variants instead of 64, reducing screening effort and inserting one stop codon instead of three. The mutagenic primers are designed with the targeted position in the middle and at least 15 non-mutated bases before and after the point of mutation. The PCR product is digested with DpnI, a restriction enzyme that recognizes and cleaves the methylated template DNA, while the non-methylated newly synthesized and mutated DNA strands are not recognized nor digested. The mutated nicked plasmid is transformed in highly competent *E. coli* strain DH5α or XL1-Blue.

Figure 1. Scheme of site saturation mutagenesis approach following the QuikChange^{TM} kit.

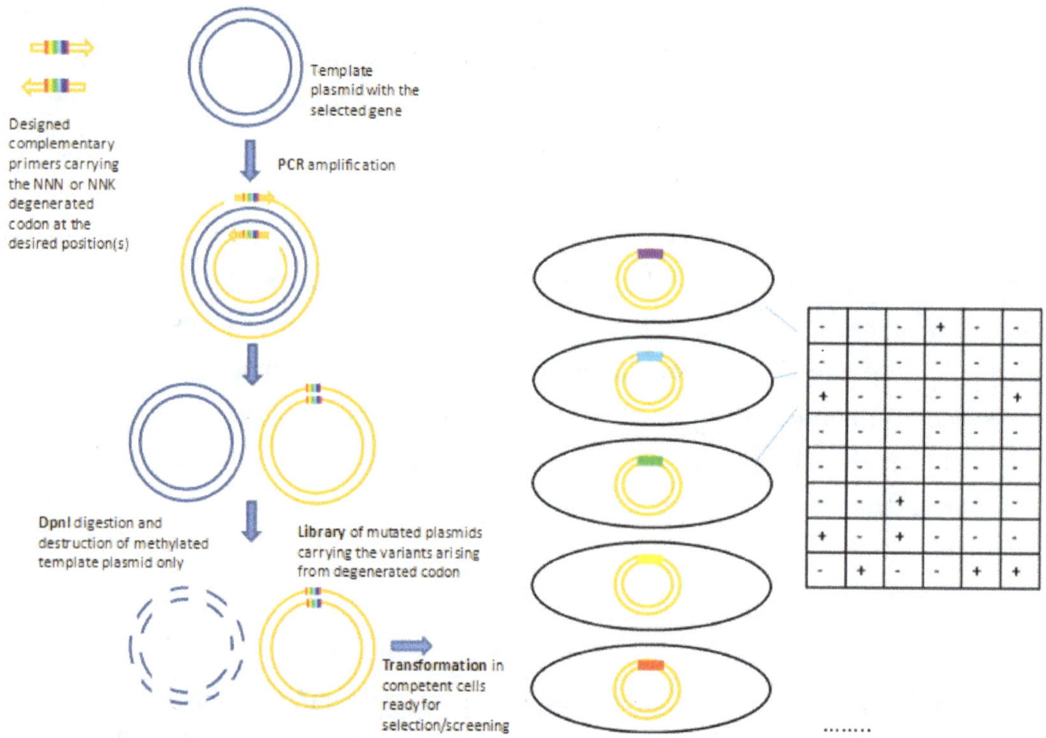

2.1.2. Iterative Saturation Mutagenesis (ISM)

The Iterative Saturation Mutagenesis (ISM) was proposed by Reetz and coworkers [30] and it combines, in an iterative manner, the SSM described above. While other strategies simply add mutations at rationally-chosen single sites by producing double or triple mutants that simply contain the positive mutation 1, 2, 3 *etc.*, in the ISM approach, a few sites in the protein sequence are identified as crucial by means of structural data or modeling, requiring a partially rational approach as in SSM, but saturation mutagenesis is then applied at the chosen sites in a combinatorial pattern. The site can be represented by a single amino acid or by a few neighboring amino acids, ideally not more than three, keeping in mind that an increase in the number of variants will then require screening of a large number of clones. These sites are then mutated according to the saturation mutagenesis approach. The novelty of this approach resides in the iterative feature given by selecting the best hit of the library obtained at each target site. For example, assume sites X, Y, Z have been selected for mutagenesis. These sites will lead to three libraries X, Y, Z, each giving as best variant X1, Y1, and Z1. Saturation mutagenesis is applied at the respective other sites: X1 will be subjected to SSM at site Y, providing library X1Y, and at site Z, providing library X1Z, as shown in the scheme of Figure 2.

Figure 2. Scheme of iterative saturation mutagenesis showing the branching process and highlighting the productive pathway (in green), non-productive mutants that stop the process simplifying the screening procedure (in red). Highlighted in yellow are mutants produced with moderate to low improvement that can be discarded or reconsidered for further processing in a second phase.

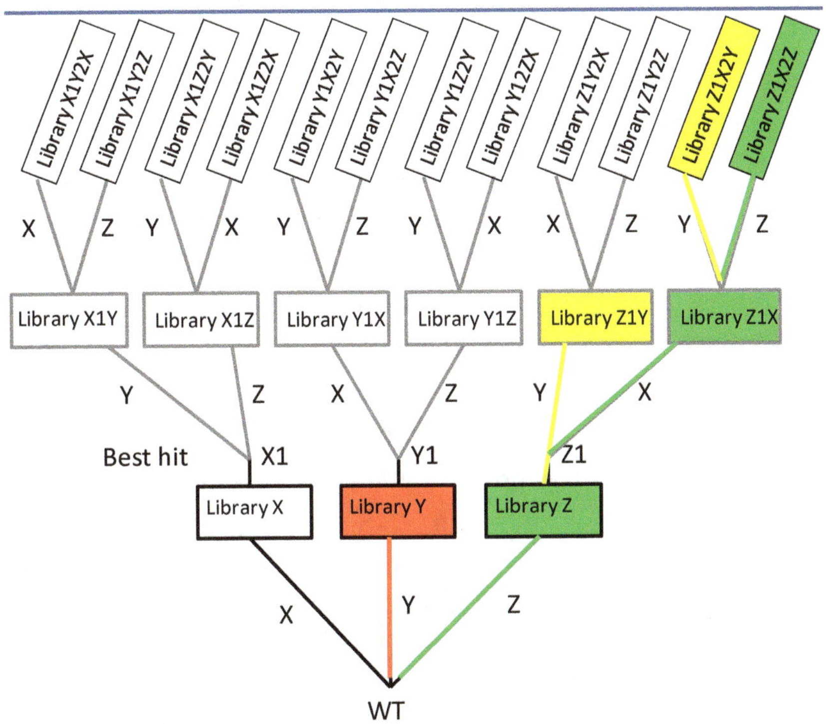

This branching process, iterated by applying SSM to a single site one or more times, can theoretically extend very quickly. For example, iterating each SSM at three sites results in 12 libraries as shown in Figure 2. In practice, non-productive branches will stop the process, as the example highlighted in red in Figure 2, while a pathway leading to synergistically improved mutants (*i.e.*, not resulting from the simple sum of single mutations) can be efficiently defined, reducing the library size. The productive pathway is highlighted in green, while yellow is shown as another branch producing variants with limited improvement. Each new cycle of ISM maximizes the probability of obtaining additive and/or cooperative effects of newly introduced mutations, which optimize the fitness landscape in a defined region of protein sequence space. This is not the case when the best-hit mutant of each library is simply added to a double or multiple mutant, where the effect can be non additive or even detrimental to the desired protein property. ISM has been demonstrated to achieve impressive results especially in enhancing enantio-selectivity [32] and thermostability of enzymes. Notably, the ISM strategy was also tested on libraries that initially did not contain improved variants, by applying the iterative cycle even to inferior mutants as templates. This was done within the systematical testing of features of 24 alternative pathways to improved variants of a biocatalyst (epoxide hydrolase from *Aspergillus niger*) and the performance evaluation of the ISM when reaching a local minimum [33]. The results showed that applying ISM resulted in successfully escaping from the local minimum.

2.1.3. Combinatorial Active-Site Saturation Test (CAST)

Strategies for evolving properties such as substrate recognition and catalysis specificity, including stereochemistry of reaction and selection of improved enzyme for resolution of racemic mixtures and precise enantio-selectivity, have been proposed over the last 5 years by selecting one or more amino acids in the active site pocket or in its close proximity. The best example of a systematic approach to this end is represented by the Combinatorial Active-site Saturation Test (CAST). In this approach, pairs of amino acid residues pointing towards the active site of an enzyme are chosen for complete randomization. The selection of residues is made on the basis of geometric assumptions that suggest choosing amino acid pairs along the sequence of loops, helices or β-sheets. For example, two residues pointing both to the catalytic pocket will for instance be n and n + 2 along the sequence in a β-sheet and n and n + 4 in a α-helix. The randomization of each pair generates a CAST library with 20^2, *i.e.*, 400 possible variants. The limited size of each CAST library allows an oversampling of 3,000 clones for statistically significant screening coverage, thus drastically reducing time and cost efforts. The results from each CAST library can then be combined pairwise by multiple mutations or by iterative strategies and re-randomized as explained above. The impressive results obtained with enzyme specificity and enantio-selectivity [32,34] highlight the suitability of the method to evolve new functions for biocatalysts. Lipases are a good example of the powerful application of CAST. The results achieved on lipases support the suitability of SSM-based methods for biocatalysts improvement, as lipases certainly constitute the core business in key industrial processes such as detergent additives, food processing and biomass pretreatment, bearing a significant impact on the global biocatalysts market that is expected to reach $7.6 billion by 2015 [35]. The availability of the CASTER software provides a very powerful tool for assigning the residue pairs for randomization on the basis of a crystal structure or a homology model. This makes the approach easy to test with several different enzymes in reproducible conditions. The main group working with this approach is that of Reetz [29], but recent applications from other research groups highlighted equally important results [36]. The limitation of the method, that requires as ideal starting point a substrate-bound crystal structure of the biocatalyst to be targeted, can be overcome in most cases by homology modeling, docking tools and in general by available bio-computing techniques.

2.1.4. B-Factor Iterative Test (B-FIT)

The focus of the B-factor iterative test (B-FIT) is the protein scaffold stability, more so than the detail of the catalytic pocket, and therefore it can guide the improvement of parameters such as thermostability that is known to not necessarily relate to the active site residues. The B-factor, or "temperature-factor", can be calculated from crystallographic data and indicates the static or dynamic mobility of an atom or groups of atoms. The B-FIT approach therefore relies on the principle of ISM combined with criteria for selecting the crucial sites that are based on the availability of B-factor values, *i.e.*, on information about the protein scaffold mobility. The hot spots are identified using software called the B-FITTER. This averages the B-factors available from X-ray crystallographic structures and it relies on the principle that high B-factors are signature of very flexible regions of the protein scaffold. Iterative mutagenesis at these flexible sites of the enzyme aims at increasing their rigidity and therefore improving the thermal stability of the enzymes to be used in industrial processes or bulk applications. The test cases that show the most impressive results to date are once more regarding the lipases, enzymes that need to be thermostable as they are typically added to detergents for mid to high temperature biological activity. The availability of the crystal structure is much more of a prerequisite here and therefore the bottleneck. Another limiting point could be the matching of an increased rigidity for thermostability [37,38] and also for stability against denaturating agents such as organic solvents [39], together with an adequate dynamic range necessary for the structural rearrangements that occur in many enzymes during catalysis. The interesting further proof that the approach is based on a measurable parameter directly correlated with flexibility and thermolability is the re-engineering of the thermostable lipase from *Pseudomonas aeruginosa.* This enzyme maintains catalytic features while dramatically decreasing its thermal stability, with Tm halved from 72 °C to 36 °C. This was achieved by reversing the approach illustrated above, by selecting and randomizing few chosen positions with a lower B-value according to the B-FITTER software to achieve destabilization of the original enzyme [40].

Another recent strategy able to select regions of potential protein flexibility and therefore hot spots to be subjected to saturation mutagenesis for tuning thermostability was named Coevolving-Site Saturation Mutagenesis (CSSM) [41]. The method relies on computational algorithm [42] and sequence alignment to select coevolving residues and/or pairs of co-evolutionary interactions that are then targeted with saturation mutagenesis to generate variants selected for improved thermostability.

2.1.5. Cassette Mutagenesis and Other Approaches for Multisite Saturation Mutagenesis

Cassette mutagenesis is one of the classical approaches for systematic mutagenesis at fixed positions [43] that can be chosen for multisite saturation mutagenesis. It is usually applied when a relatively short DNA sequence is to be mutated by synthetic oligonucleotide primers designed to introduce multiple mutations at targeted amino acids in the same stretch of primary sequence. The excision and re-introduction of the mutated cassette by molecular biology techniques, such as introduction of restriction sites and ligation in the original vector, makes it a time consuming procedure. Likewise, methods that follow the classical Kunkel mutagenesis approach using ssDNA also suffer from the same drawbacks. However, a recent novel approach named PFunkel [44] has been proposed that re-interprets the Kunkel methodology and that can be performed in one day in a single test tube (Figure 3). This was applied to create a library with site-saturation at four distal sites and it was tested on TEM-1 β-lactamase gene to produce a library of 18,081 designed variants: library sequencing attested that a 97% coverage of the expected variants were present in the library, and this was then screened for variants resistant to the ß-lactamase inhibitor tazobactam.

Another recent strategy to simultaneously introduce saturation mutagenesis at multiple sites (up to five codons) was proposed by Schwaneberg and co-workers [45]. The scheme of this approach, named OmniChange is reported below (Figure 4).

Figure 3. Schematic of PFunkel mutagenesis strategy (adapted from Firnberg *et al.* [44]).

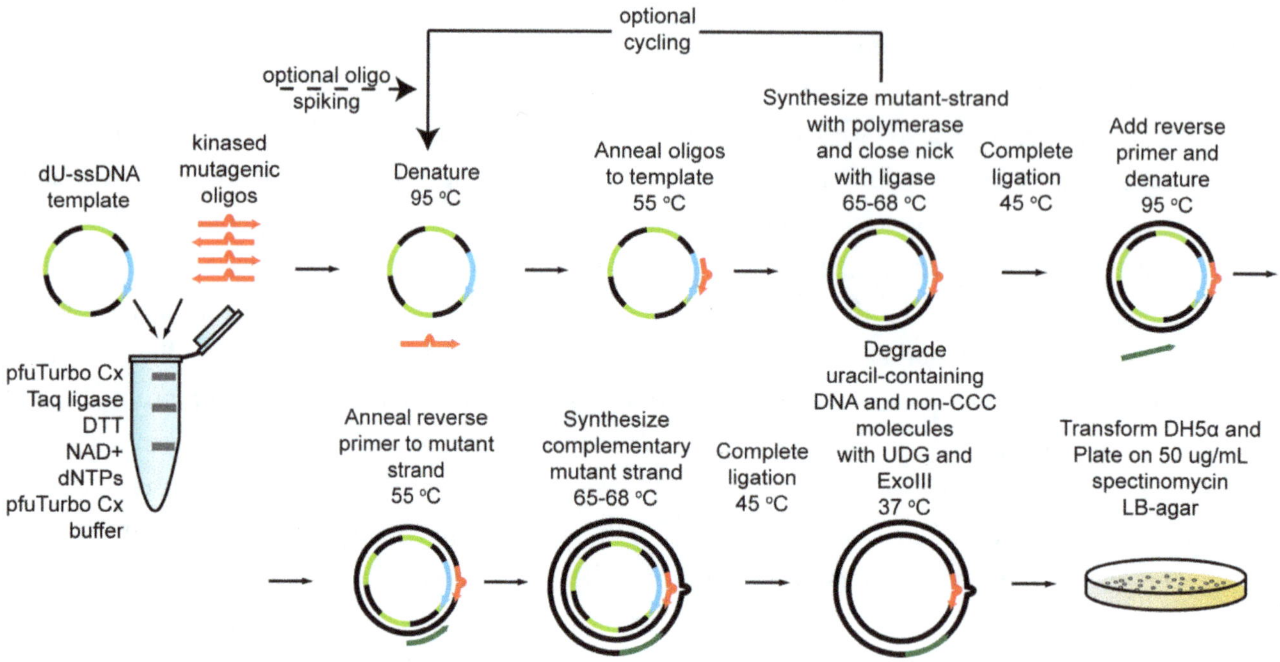

Figure 4. The 4-step strategy for the simultaneous saturation of 5 independent codons by OmniChange (adapted from Dennig *et al.* [45]).

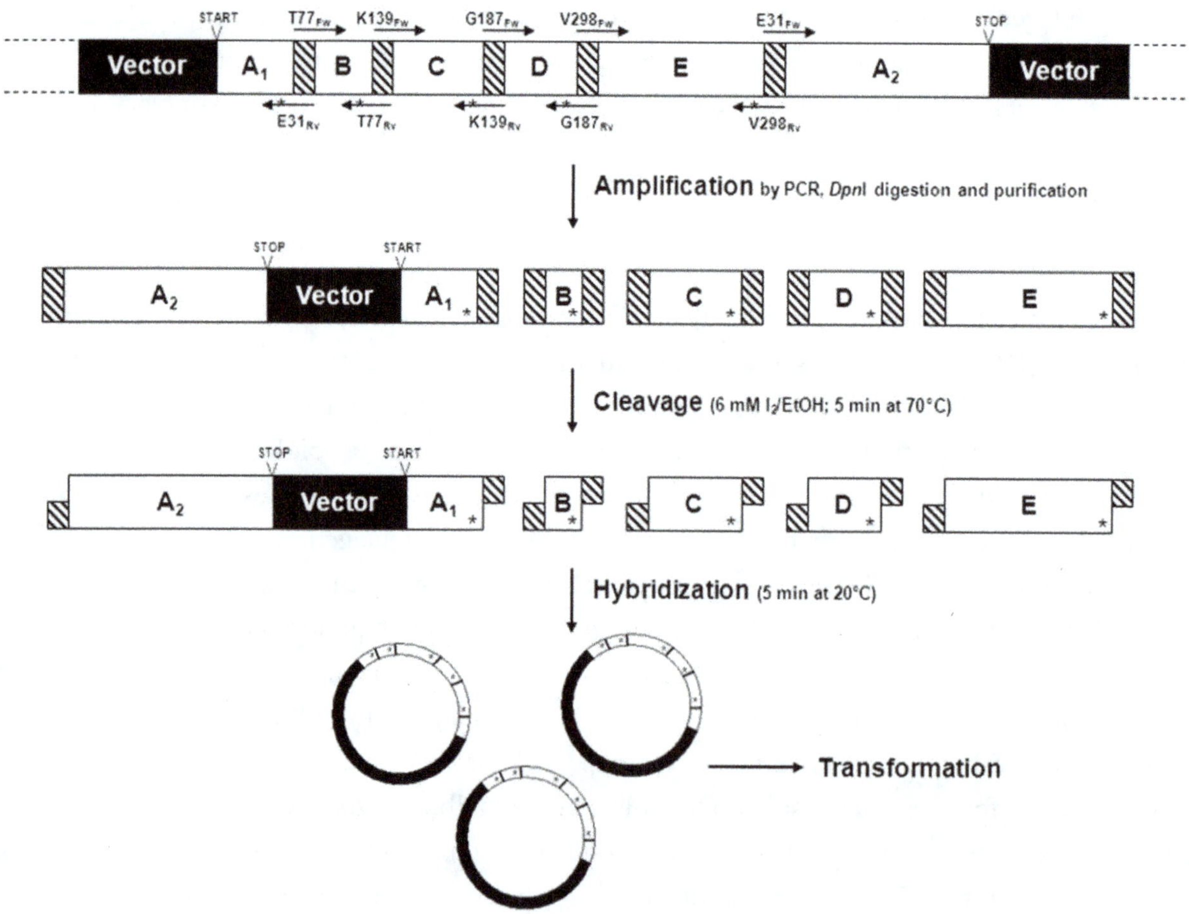

2.1.6. Reducing Amino Acid Alphabet

The query for strategies that can reduce library size without limiting functional variants has led to several attempts to restrict the amino acid alphabet. All "reducing amino acid alphabet" approaches aim at defining a small set of a few representative amino acids that can efficiently function as building blocks for all proteins. Saturation mutagenesis performed with a restricted alphabet at multiple sites has the advantage of generating smaller and potentially smarter libraries. The risk is to over-simplify the subset and exclude subtle and specific properties of some amino acids. The design of the subset chosen is therefore a very delicate step. The main efforts in this direction came from Hilvert [46] and Reetz [47] and co-workers who respectively proposed a reduced alphabet of 9 and 12 representative amino acids applied to the design of an enzyme able to function as chorismate mutase [46] and to the engineering of the active site of an epoxide hydrolase [47]. Although the function of these enzymes can efficiently be complied by this simplified catalyst, the stability of the protein was not entirely satisfactory, as an undesired enhanced flexibility was observed in the enzyme designed with the 9 amino acid reduced alphabet [46]. In other cases the reduced amino acid alphabet was specifically designed on the basis of sequence alignment and consensus variants and the strategy applied to the focused mutagenesis of a phenyl acetone monooxygenase [48]. The main advantage of this method is well highlighted by the rigorous comparison of library coverage when randomizing multiple positions with the alternative codon NNK for the 20 amino acids and with the codon NDT (D: adenine/guanine/thymine) encoding for the reduced 12 amino acid alphabet. The number of variants to be screened in the NDT library for 95% coverage is less than 500 for a two position randomized mutant and 5,000 for a three position mutant. In the case of NNK library for a two positions mutant a screening of 3,000 is required, while for a three positions mutant the screening of 10,000 variants only covers 25% [49]. For the purpose of reducing library redundancy, and consequently screening efforts, a more convincing strategy has recently been proposed by designing appropriate mutagenic primers that can cover the 20 amino acids with only 22 codons [50].

2.2. Statistical Robustness of the Method and Requirements for Library Screenings

A key point of all laboratory evolution techniques is library screening and variant selection, which is tightly intertwined with the statistical analysis of library coverage. Although SSM is a focused strategy among the wider landscape of directed evolution approaches, the importance of these two aspects is crucial and bears implications for judging SSM and evaluating its potential application. Therefore a brief coverage of the topic will be presented below with a focus on relevance to SSM strategy.

The saturation mutagenesis methods usually aim at the production of relatively small and high quality libraries, whose screening could cover all different variants with an established degree of confidence. It is therefore crucial to acknowledge the importance of statistics [51–54] for estimating the number of analyses to be performed and determining the sample size to be screened. In most methods, with the exception of recently proposed alphabet reducing [47] and redundancy reducing approaches [50], the distribution of encoded amino acids is impaired in frequency due to the genetic code redundancy. Thus a library constructed with NNN configuration will have leucine represented six times for every tryptophan. As a result, the sample size should always be calculated on the basis of nucleotide rather than amino acid diversity. The statistics of the process is described by the following equation [54].

$$L \approx -V \ln\left(-\frac{\ln P_c}{V}\right) \tag{1}$$

where V is the number of possible variants (64 for NNN degeneracy, 32 for NNK degeneracy of a single codon), L is the number of clones in the library, Pc is the probability of completeness of the library. Thus, the equation correlates the number of clones in the library with the probability that each clone is actually present in the library at least once. The same holds for the screening. As an example, the screening of 360 clones obtained by a NNK degenerated library at a single site, providing 32 different codon variants, ensures a probability of 99.96% that each variant has been tested at least once, while lowering the screened clones to 247 lowers the probability to 98.59%. The assumption is of course that the NNK or NNN degeneracy and the SSM protocol applied is not affected by biases and that the incorporation of each codon is equally possible. This is not always the case and controls of library completeness can be performed by sequencing the entire library mixture (Figure 5) and/or randomly selecting a few clones (either positive or negative) to demonstrate that a good variability of codons for different amino acids are actually present [55].

Figure 5. DNA sequencing of the three libraries produced for evaluation of the randomization efficiency on selected position in hydrogenase gene: the targeted position is properly randomized for NNK in library A and C (K either a T or a G), while only partial degeneration is present in library B.

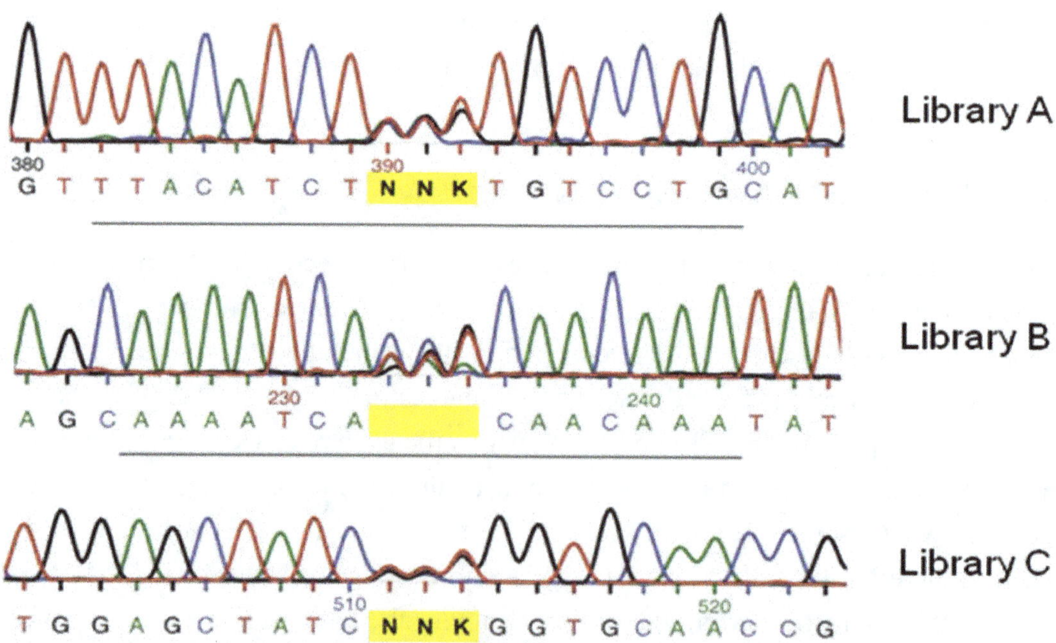

The higher the degeneracy of the library, the higher the number of clones to be screened in order to have a significant probability of coverage of all mutants. For example, to achieve 95% (the threshold for significance is usually set to this value) probability to cover all mutants in a 1,024-fold degenerated library, it has been estimated that about 3,000 clones should be screened. Most SSM experiments reported in the literature cover the mutated library between two to four times on a basis of nucleotide diversity (e.g., 64–128 clones are usually screened for a 32-fold degenerated library). Often incomplete screenings of large libraries can allow identification of variants with desirable features [56]. This strategy is, however, prone to the statistical uncertainty of missing clones with remarkable properties. To reduce the library size and overcome genetic code redundancy, mixtures of highly specific primers can be used instead of a degenerated primer. Therefore, 19 primers (one for each specific amino acid alternative to the WT amino acid) can be used to randomize each codon. This can also be applied when a bias in codon incorporation is present (Figure 3 library B) and a properly randomized library cannot be synthesized.

Recent novel techniques and designed primers were proposed to further reduce codon redundancy and to ensure equal probability of coding each amino acid, by limiting the code to 22 triplets covering the 20 standard amino acids [50].

The researcher in the laboratory designs the selection of desired variants by the application of an appropriate screening method. The general rule that "you get what you screen for" indicates that this step is a particularly crucial one and often represents the bottleneck for the success of directed evolution in developing improved or new biocatalysts. The selection method must be rapid, sensitive and allow for the clear identification of the desired properties, implying that the screening must not be marred by undesired selection criteria.

The fully randomized methods of shuffling or error prone PCR implies the production of very large libraries and therefore the requirement for equally powerful high-throughput screening techniques, such as phage display or other more recent molecular display methods [57,58]. These methods enable the screening of up to 10^{12} protein variants, but usually rely on costly equipment and are only suitable for very focused applications. On the contrary, *in vivo* selection of suitable enzymes by setting experimental parameters so that conditional cell survival is linked to the desired biocatalyst function usually is low cost and allows high-throughput performance. Unfortunately, instances have been reported in literature in which surviving cells bypassed the desired enzyme expression. Also by setting a high threshold there is the probability that low activity variants with potential interest are excluded.

The application of spectrophotometric [59] or fluorimetric [60] platform that can screen for the desired product formation or at least for substrates and co-substrates consumption by the biocatalyst of interest is a more versatile option that can be extended to very specific catalysis, such as stereo-specific production of chiral compounds [61], biodegradation of recalcitrant poly-aromatic hydrocarbons [62], for the synthesis of drug metabolites [63,64] and the turnover of novel chemical entities for drug synthesis, such as 1,2,5-Oxadiazole derivatives [65], for hydrogen evolution and uptake [55,66]. The superior specificity and versatility of such assays is reached at the cost of lowering the through-put efficiency, even for quick assays that can be performed on multi-well plates, directly on cell lysates or colonies (Figure 6) [55]. Compared to fully randomized methods, saturation mutagenesis, which provides small but high quality libraries, allows the application of such focused and function-specific screenings whilst maintaining statistically sound library coverage.

Figure 6. (**a**) Scheme of the principle of on-colonies activity test for a [FeFe] hydrogenase [55]; (**b**) Example of the screening results.

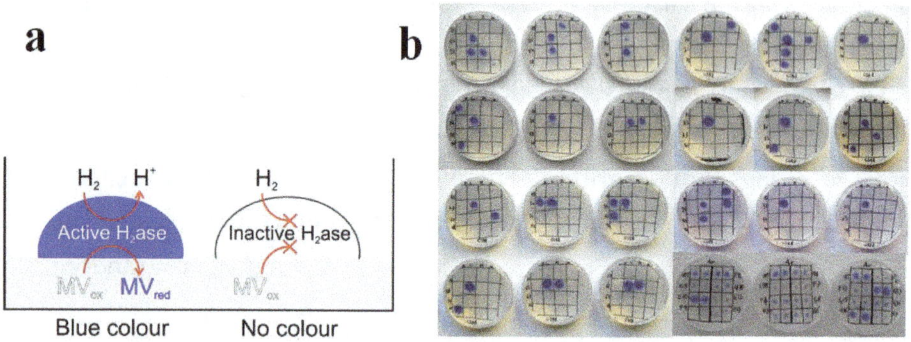

3. Recent Successful Applications

An increasing number of recent papers proposes the application of saturation mutagenesis to biocatalysts of applicative interest, for "greener" industrial processes [67,68] improved bulk enzymes [41,69,70], biotechnology [71,72] bioremediation [73], fine chemical synthesis [74–80], biofuels production [55,81,82] and biomass exploitation [82–85].

A selection of relevant successful examples published in the past 5 years is presented below. The report is divided in two sections: (1) enzyme classes with high impact on industrial processes and fine chemistry (*3.1*) and (2) enzyme classes with applications in environmental care and production of clean energy (*3.2*). Some classes, for example oxygenases, are relevant to both and therefore are listed twice.

3.1. Enzymes Relevant for Industry

3.1.1. Lipases

Lipases are considered as benchmark enzymes for biocatalysis: Lipolase®Ultra and LipoPrime® are the first examples of engineered lipases for commercial distribution in detergent industry. They are also exploited in other industrial large-scale processes and as dedicated catalysts for highly stereo-specific catalysis in fine chemistry. Saturation mutagenesis has played a key role in engineering several lipases both for thermal stability and enantio-selectivity, with at least 20 research papers published in the last 5 years. Among groups involved in lipases engineering, Reetz and co-workers achieved relevant results [29,32,34] by applying SSM, ISM and CAST for enhancing enantio-selectivity and B-FIT for tuning thermal stability properties. The SSM approach was applied to *Pseudomonas aeruginosa* lipase, a well-known catalyst applied to hydrolysis of carboxylic acid esters and transesterification of primary and secondary alcohols, with the aim of redesigning the substrate recognition pocket to enable catalysis on more bulky substrates, such as benzoic acid esters. Ser 82, the key residue for the stabilization of the oxyanion intermediate, was not addressed by the mutagenesis since it structurally belongs to a more distant portion of the enzyme, while the CAST strategy guided the selection of five pairs of residues pointing towards the active site and defining the recognition determinants of the hydrophobic portion of the ester. Five libraries were produced by simultaneous saturation mutagenesis at the two defined positions, that is library A to E: Met16/Leu17, Leu118/Ile 121, Leu131/Val 135, Leu159/Leu162, and Leu231/Val 232 (Figure 7).

Figure 7. Scheme of the structure of *P. aeruginosa* lipase active site pocket (PDB: 1EX9) with the targeted sites (library A: Met16/Leu17 in red; library B: Leu118/Ile 121 in orange; library C: Leu131/Val 135 in yellow; library D: Leu159/Leu162 in green; and library E: Leu231/Val 232 in cyan). Ser82, Asp229, and His251 (in violet) represent the catalytic triad. A substrate analogue (RC-(RP,SP)-1,2-dioctylcarbamoyl-glycero-3-O-octylphosphonate) covalently bound to Ser 82 is shown in blue.

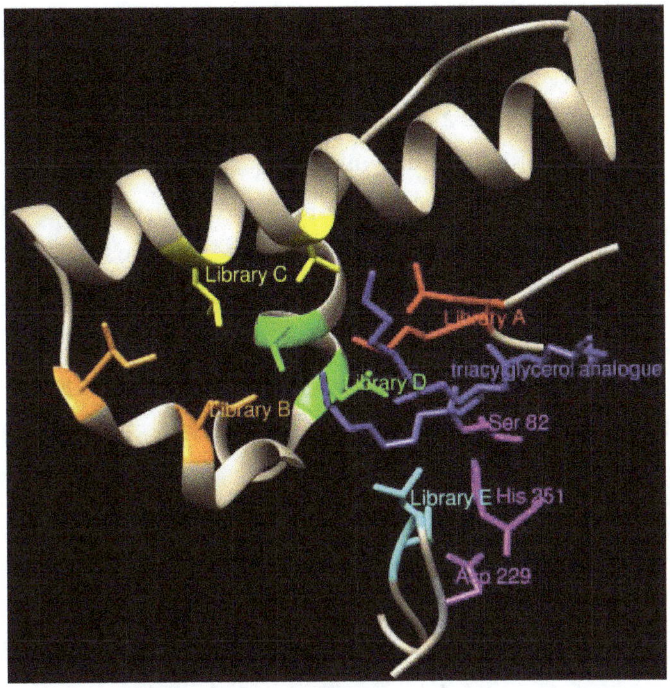

The five libraries of 3,000 variants each were then screened with a spectrophotometric method by testing 11 different substrates. The total reactions performed (165,000) allowed to select eight hits from libraries A and D, consistent with the focus on hot spots even within the restricted region analyzed. Although the success rate in this case was lower than for other SSM reported approaches, the few selected variants showed an impressive gain in function, for instance by binding adamantyl carboxylic acid esters that are not recognized by the WT, as well as showing a 100 fold increase in the rate of hydrolytic activity on substrates that are poorly recognized by the WT [29]. Further works on the same enzyme by ISM highlighted the enormous potentiality of iterative saturation versus other methods such as error prone PCR, shuffling and even the previous SSM, in particular for enhancing the stereospecificity of reactions. In fact, a more recent paper reports, on the same enzyme, the gain of function for the bulky 2-phenylalkanoic acid esters that are not recognized by the WT and the selection of variants with enantio-selectivity of E = 436, achieved with only small mutant libraries and thus a minimum of screening effort [34].

Also, the selection of a *Pseudomonas aeruginosa* lipase engineered variant with an enatioselectivity of E = 594 for the kinetic resolution of a chiral ester from an ISM library upon screening only 10,000 transformants is an unprecedented result [32], given that by directed evolution based on DNA shuffling, only a best variant value of E = 51 (ee > 95% at 24% conversion) could be obtained by screening about 50,000 transformants [86]. In the specific case of variant 1B2, characterized by a high E value of 594, this was produced by ISM starting from three libraries with simultaneous randomization at two near sites each, namely library A (Met16/Leu17), B (Leu159/Leu162), and C (Leu231/Val232). The selection of a best hit from library B with E = 8 (Leu162Asn) was followed by a second round of randomization on library A with DNT codon that simplifies amino acid alphabet by excluding Leu and therefore back-mutating to the original amino acid Leu17. This led to the highly optimized 1B2 variant (Met16Ala/Leu17Phe/Leu162Asn).

A very recent paper [70] reports the application of ISM and CAST to the engineering of *Candida antarctica* lipase B (CALB; Novozyme 435), a top industrial biocatalyst applied in kinetic resolution of racemic alcohols and amines, desymmetrization of diols and in other stereoselective synthesis of chiral intermediates for pharmaceuticals, polymer chemistry, and protection/deprotection technology. CAST guided selection of active site residues and ISM cycles with restricted alphabet using NDT degeneracy allowed for the isolation of two best mutants that were tested on several substrates for enhancement of activity and S- or R-stereospecificity. These two best hits, named RG401 and SG303 were tested on four representative chiral α-substituted carboxylic acid esters. Specificity constants k_{cat}/K_M from 13 to 270 fold higher than WT were achieved for SG303 with E (S) up to 64. The other mutant, RG401, acquired an enantio-specificity with E (R) up to 68 although the specificity constants were only slightly higher or of the same order of magnitude of WT. On *Candida antarctica* lipase A (CALA), Bäckvall and co-workers [36] applied the CAST strategy to enhance the performance of the catalyst by building two reduced libraries based on the NDT degeneracy: library FI (Phe149 and Ile150) with side chains directed toward the R-methyl group of the substrate and library FG (Phe233 and Gly237) with side chains defining the acyl-recognizing pocket of the active site. The reduced library size allowed a high coverage (>95%) by screening only 600 variants per library and allowed to select variants with E values of 45–276 (WT E value is only up to 20) and up to 30 fold increased activity for seven different esters used for the preparation of enantiomerically pure 2-arylpropionic acids, important building blocks for the synthesis of non-steroidal anti-inflammatory drugs such as Naproxen, Ibuprofen, and Flurbiprofen. The same group recently reported a further enhancement where CALA variants with high activity and E value of 100 towards an ester of ibuprofen were obtained. This substrate had failed to be recognized efficiently and with high stereospecificity by variants selected previously [80].

The robustness of the saturation mutagenesis methods, in particular with the B-FIT strategy, for thermal stabilization and destabilization of lipases for catalysis at desired optimal temperature, has already been discussed (Section 2.1.4) [37,41] and the same approach has proven to be suitable for stabilization towards other denaturing agents such as organic solvents [39].

3.1.2. Esterases and Other Hydrolases

Esterases are also extensively used in biocatalysis: saturation mutagenesis strategies have been applied to some enzymes of this class, in particular for the esterase from *Pseudomonas fluorescens*. Enhancement of enantio-selectivity [87] of this enzyme was pursued by the use of simultaneous saturation mutagenesis at four hot spots, with restricted alphabets chosen on the basis of more frequently represented amino acids in structurally equivalent positions on the basis of 1,750 known sequences. This approach granted variants with improved rates (up to 240-fold) and enantioselectivities (up to E(true) = 80) towards 3-phenylbutyric acid esters with the advantage of a relatively limited effort for screening these "small but smart" libraries. As for thermal stabilization, the same enzyme was targeted at three sites, selected by B-FIT strategies, granting an enhanced stability of almost 10 °C higher than the starting catalyst [88].

Other class 3 hydrolytic enzymes that were targeted by saturation mutagenesis for improved catalysts development include epoxide hydrolases, already mentioned as test cases for the development of focused restricted alphabet libraries [46]. A limonene epoxide hydrolase from *Rhodococcus erythropolis*, performing a rare one-step mechanism, was also targeted by ISM to select variants with high stereoselectivity on substrates different from the natural limonene epoxide. Active site binding pocket residues were selected and the codons randomized with a reduced amino acid alphabet strategy. Variants obtained from 5,000 screened hits can catalyze the desymmetrization of cyclopentene-oxide with stereoselective production of (*R,R*)- or (*S,S*)-enantiomers, the desymmetrization of other meso-epoxides and kinetic resolution of racemic substrates [89].

Because of its potential usefulness in β-lactam antibiotics synthesis, α-amino acid ester hydrolases were also chosen for improvement by saturation mutagenesis. A study was performed on 13 residues not directly involved in substrate recognition (based on the crystal structure of a protein-cefprozil complex) that were individually randomized in the enzyme from *Xanthomonas rubrillineans.* Mutants were selected with improved synthetic activity of *p*-hydroxylcephalosporins with a 23%, 17% and 64% increase in product yield for cefadroxil, cefprozil and cefatrizine, respectively [90].

Another biocatalyst relevant for bulk applications and belonging to the hydrolase class is represented by phytase, commercialized as an additive to poultry and swine feeding preparation in order to enhance digestibility of phytate and increase phosphorus assimilation. The challenge for enzyme engineering here is to enhance the stability of the catalyst not only to temperature but also to gastric degradation and to very low pH environment of the digestive tract so that the enzyme can still be active during the feeding process. Industry interest in this biocatalyst and in mutagenesis approaches aiming at improving its performance is testified by a paper dating back to 2004 [91] published on a research carried out by the company Diversa Corporation, San Diego, CA, USA. The dhlA phytase encoding gene from *Rhodococcus* was chosen to apply saturation mutagenesis with NNK codon systematically to all 431 positions of the protein sequence and screening was performed on at least 150 clones for each individually produced library. By isolating the best single mutants for enhanced low pH stability after heat treatment of the variants, therefore combining a selection for two desired properties, the authors selected 14 single mutants with improved properties and performed a combinatorial strategy and a second screening to isolate synergic and additive effects of multiple mutations. Variant Phy9X, with eight combined mutations, led to a novel biocatalyst with the ability to reversibly renature upon heat treatment and also function at process temperatures of 65 °C, with specific activity at the same level of WT but extending to below pH 2.5 and a 3.5 fold enhanced stability to gastric degradation.

3.1.3. Oxygenases and Other Redox Enzymes

Among redox enzymes, oxygenases have been key examples of the possible improvements brought by protein engineering to the efficiency of enzymes, and particularly of biocatalysts: the focus on cytochromes P450 and Baeyer-Villiger monooxygenases has always been maintained when proposing rational, semi-rational and randomized techniques of laboratory evolution with the seeding work of the groups of Arnold and Reetz, respectively and of many other groups that proposed directed evolution of these versatile biocatalysts. More recently, in particular for P450s, an increasing number of papers proposed saturation mutagenesis alone, or in combination with random techniques, to refine particular applications supported by this class of enzymes in fine chemical synthesis. This also extends to other non-heme iron oxygenases used for enantioselective synthesis of pharmaceutical compounds and chiral sulfoxides [74].

Starting from P450s, saturation mutagenesis seems to be the preferred method to enable enhancement of both regio- and stereo-selectivity for the C-H hydroxylation reactions that are of interest in fine chemistry. Steroid hydroxylation by cytochromes P450 in controlled positions leading to enantiomerically pure products is one of the most targeted goals of industry. The results achieved with saturation mutagenesis in the last few years benefit from the knowledge in terms of key spots relevant to improving enzyme performance acquired through directed evolution. Further specific improvements have been made possible by saturation mutagenesis. A very recent work by Glieder and co-workers [75] addressed the two active site residues 216 and 483 by saturation mutagenesis to generate all 400 possible combinations of amino acids. A double mutant of WT CYP2D6 resulted in a high regio-selectivity for hydroxylation at the 2β-position, instead of the 6β-position, suggesting that the mutation F483G could be preferential to the reported F483I for regio-selectivity in the well-known protein hot spot F483. Moreover, a previously obtained mutant F87A of P450 BM3, was further targeted by ISM for selective hydroxylation of testosterone in either of the two possible products 2β- and 15β-alcohols [92]. The CAST approach was applied to choose appropriate sites surrounding the binding pocket. The 20 residues selected as possible candidates for ISM were grouped into nine sites of neighboring amino acids, as this is known to maximize the cooperativity more than the additive effects and it is obviously useful to reduce the library size. Site A (Arg47, Thr49, Tyr51), and site B (Val78, Ala82) were targeted (Figure 8) first with NDC codon degeneracy at the three spots of site A with the need to screen only 430 transformants for a 95% coverage.

The two-residues at site B were randomized using NNK codon degeneracy. From this first screening, highly 2β-selective mutants (97%) were obtained from library A while 15β-selective variants, also reaching 91% regio-selectivity on testosterone, were found mainly in library B. The best variant from library B was then subjected to randomization at site A with some variants reaching 96% regio-selectivity on testosterone (R47Y/T49F/V78L/A82M/F87A) while a variant from library B only selected on testosterone (V78V/A82N/F87A) was able to reach a 100% regio-selectivity on other steroidal substrates such as progesterone. Moreover, some mutated variants displayed increased coupling of product formation with NADPH consumption. This ISM approach was also characterized by a limited amount of screening, the step that is normally considered the bottleneck of directed evolution.

Figure 8. Structure of P450 BM3 heme domain (PDB: 2HPD) showing the target sites A (Arg47, Thr49, Tyr51) in green and B (Val78, Ala82) in blue. Heme is shown in red, the Fe coordinating Cys 400 is in magenta.

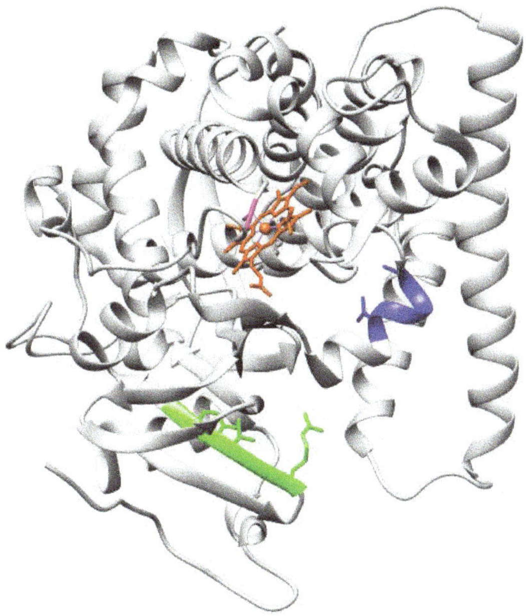

A refinement of previously evolved mutants of P450 BM3 was also proposed in 2008 for production of indigo and indirubin by indole hydroxylation [93]. Starting from a variant A74G/F87V/L188Q obtained by random methods and directed evolution, and by applying saturation mutagenesis as a refinement of catalyst properties, granted two variants with increased catalytic efficiency up to six times that of the starting variant, with improved regio-selectivity for 3-hydroxyindole, leading to 93% indigo production *vs.* the initial 72%. One of the variants also showed increased coupling efficiency with NADH. The overall result nicely supports the importance of synergy of random and saturation mutagenesis approaches for optimized catalysts production.

Recently, another study has been published [94] on P450sca-2 from *Streptomyces carbophilus* to be employed in the synthesis of the cholesterol-lowering drug pravastatin. Here the saturation mutagenesis was applied to enhance electron transfer efficiency in a hybrid P450sca-2/Pdx/Pdr functional system by targeting residues at the interface between the electron transfer moiety putidaredoxin (Pdx) and the catalytic P450sca-2. Three rounds of ISM granted a variant with a 10 fold improved catalytic performance.

The other important enzymes belonging to the oxygenase class and successfully targeted for improvement by saturation mutagenesis [95,96] are represented by the Baeyer-Villiger monoxygenases (BVMO), able to perform specific reactions on racemic mixture of various ketones to obtain enantiopure lactones, conversion of prochiral ketones in chiral lactones and oxidation of organic sulfides. Although novel Baeyer-Villiger monoxygenases with tuned substrate specificity can be found in diverse microbial populations [97–99], there is the need to evolve BVMOs with specific performance in biosynthesis. This can be done with random or SSM laboratory techniques.

A thermostable phenylacetone monoxygenase (PAMO) belonging to the BVMO group was successfully engineered by saturation mutagenesis to perform catalysis on 2-aryl, 2-alkylcyclohexanones and a bicyclic ketone that are not recognized as substrates by the WT enzyme [96]. Given that a CAST approach previously applied to positions 441–444 belonging to a loop next to the binding pocket, were only partially successful [48], only positions 440 and 437 were targeted instead, where the first amino acid is located in the second sphere, and therefore not in direct contact with the substrate (Figure 9). Pro440 was identified to play a key role, since several mutants generated at this position granted an enhanced percentage of conversions and improved enantio-selectivity for substrates not recognized by the WT. Since in this case the library was apparently not covering the entire range of variants at position 440, the missing variants Pro440Tyr and Pro440Trp were produced by site-specific mutagenesis, with the aim of exploring the entire range of amino acid properties at this position for the enhancement of the biocatalyst performance. Further work on the same enzyme [95] targeted positions 93 and 94, located in site distal from the binding pocket chosen on the basis of the crystal structure, with a simultaneous saturation mutagenesis using a NDT codon to reduce degeneracy. A double mutant Gln93Asn/Pro94Asp was selected for its acquired activity on an otherwise inert 2-substituted cyclohexanone derivatives and it was found to be able to catalyze the conversion to the corresponding lactones with high enantio-selectivity. These results have been rationalized by a rearrangement of the H-bonds and salt-bridge networks in the protein, much alike an induced allosteric effect.

Figure 9. Scheme of the active site of PAMO (PDB: 1W4X) with targeted residues Pro440, Pro437, Gln93 and Pro94 (in black). FAD is shown in orange; Arg 337, involved in catalysis, is shown in blue.

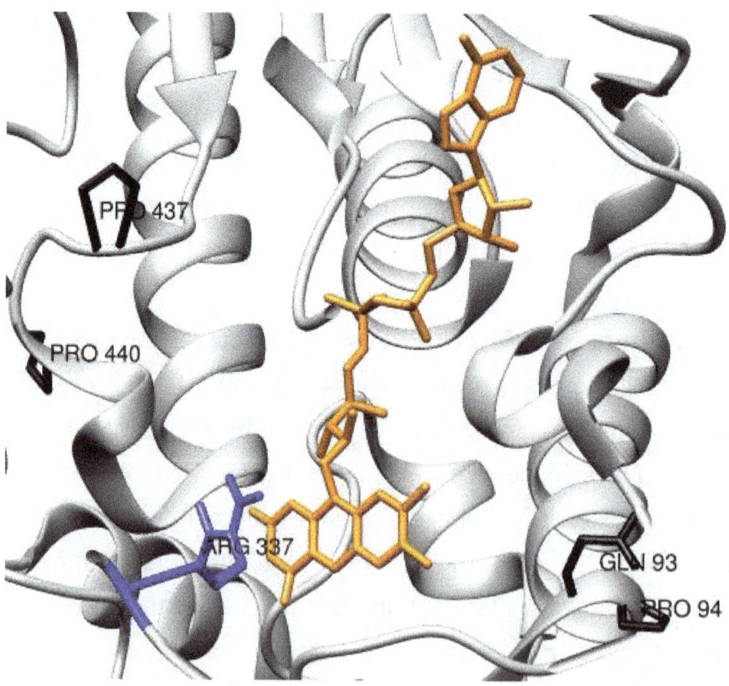

In order to enhance the performance of biocatalysts for fine chemistry, for example, for the synthesis of chiral sulfoxides and asymmetric ketone reduction, other redox enzymes such as nitrobenzene dioxygenase [77], alcohol dehydrogenase [78] and carbonyl reductase [79] were also recently optimized by saturation mutagenesis.

An interesting example of active site saturation mutagenesis recently published, targeted an unusual non-heme iron dioxygenase, belonging to the class of α-ketoglutarate dependent dioxygenase [74].

This enzyme is involved in the biosynthesis of carbapenem-3-carboxylic acid, the core building block of the all carbapenems, including Meropenem and Imipenem. This is a relatively new class of β-lactam antibiotics of great importance as therapeutic agents given the increasing bacterial resistance to an older class of antibiotics. In order to dissect and better understand the molecular determinants of the biocatalyst that promote the epimerization and desaturation crucial for the biosynthesis of the core of cabapenem, SSM was applied to six active sites and four second sphere residues of the dioxygenase, generating point as well as double mutant libraries. The importance of Tyr 67 for catalyst engineering was highlighted together with the advantage of promoting a two step reaction mechanism, including epimerization and desaturation, with release and rebound of the intermediate to ensure complete desaturation and avoid the frequent aborted cycles that are observed in the native enzyme due to a difficult rotation of the intermediate required in the catalytic pocket in the full reaction.

Other redox enzymes optimized by SSM include dehydrogenases as the previously cited alcohol dehydrogenase from *Thermoethanolicus brockii* [78] and the meso-diaminopimelate dehydrogenase from *Symbiobacterium thermophilum* [100] successfully exploited for the synthesis of D-phenylalanine, thanks to a 35-fold increase in specific activity of the variant compared to the WT.

3.2. Enzymes Relevant to Environmental and Clean Energy Approaches

The use of enzymes in environmental applications include biocatalysts able to detoxify pesticides such as atrazine, chlorinated polyaromatic hydrocarbons, DDT, toxic compounds in industrial wastes such as phenols, organic solvents, aniline, drugs, explosives and chemicals resulting from military operations, among which trinitrotoluene (TNT) and G-series organophosphorus toxins contained in nerve agents like Sarin and Cyclosarin. These are usually xenobiotics particularly recalcitrant to degradation by bacteria and fungi, given that their natural enzymes, though powerful catalysts for bioremediation [101], have not evolved under the selective pressure of such compounds, as these organisms were not massively exposed to these compounds until very recently. In this respect, protein engineering by laboratory-driven evolution is of unique importance for what it can deliver. Several important results have been achieved in this respect by random directed evolution approaches both on P450 enzymes acting on pollutants [1,4,62] and on hydrolytic enzymes, for example on paraoxonases (PON) for detoxification of organophosphorus toxins [102], but an increasing number of works have recently tackled the same problem by applying SSM methods.

SSM relevance for improvement of lipase applications as an industrial catalyst has already been discussed in Section 3.1. Lipases are also relevant for clean energy issues in the transesterification of triacylglyerol with methanol for biodiesel production [81,103]. These have many advantages over traditional base or acid catalyzed approaches, but natural lipases often lack the required stability and efficiency in the high methanol concentrations used for biodiesel synthesis, limiting their practical use. Directed evolution techniques were very recently applied to the lipase from *Proteus mirabilis* to enhance methanol tolerance and allow its industrial application as a biocatalyst. The dieselzyme variant 4, evolved by randomized methods (error prone PCR) and site-directed mutagenesis to combine beneficial mutations, shows a 30-fold increase in the half-inactivation time to temperature (50 °C) and a 50-fold longer half-inactivation time in 50% aqueous methanol [81]. Although saturation mutagenesis was not the chosen technique for this approach, the authors foresee the application of CAST and structure guided ISM for further refinement of the obtained catalyst.

Enhancement of performance of enzymes such as cellulases and ligninases, present in nature in a restricted number of organisms, is of high relevance to the production of clean and sustainable energy from renewable sources. These enzymes offer precious tools for waste and poor-value biomass recycling, acting both on recovery of resources for energy production and on management of wastes [104].

The frontier of environmental care and clean energy production is the setup of hybrid systems based on biocatalysts, often interfaced with semiconductor materials [105,106] with the ability to mimic nature in efficient solar energy harvesting and energy storage in transportable fuels of low impact to the delicate equilibrium of our planet. In this respect, photosystems, light activated proteins, CO_2 fixing enzymes and biocatalysts able to produce fuels such as biohydrogen, bioethanol and biodiesel, are the ideal target of engineering approaches. Many clean-energy production related enzymes (in particular photosystems and hydrogenases) are generally difficult to purify, manipulate and engineer, and therefore the laboratory evolution approaches are still at their first steps of development, but it is foreseen that increasing interest will be devoted to engineering, particularly with SSM methods applied to hydrogenases, nitrogenases, formate-dehydrogenase.

Here a choice of examples, grouped as in Section *3.1* by enzyme classes or subclasses, focus on the three aspects: detoxification, biomass degradation and clean energy production.

3.2.1. Oxygenases and Other Oxidoreductases for Bioremediation

Oxygenases and more in general redox enzymes represent a class of biocatalysts spanning from P450s to non-heme iron mono- and di-oxygenases and flavoenzymes widely used for the oxidation of toxic compounds. The addition of one or two hydroxyls to a poorly reactive C-H bond, for example in aromatic and aliphatic hydrocarbons, is usually crucial for the initiation of the detoxification and clearance process. The increasing amount of pollutants with halogenated substitutions in aromatic rings, for example in pesticides, and the presence of compounds recalcitrant to biodegradation, poses difficult challenges to protein engineers. SSM techniques are often the selected method to test and modify redox enzymes to recognize a broader substrate range and to attack xenobiotics with a sustainable approach, recovering carbon sources for safe microorganism growth. The catabolic pathways that enable many microorganisms to degrade large classes of aromatic pollutants, often relay on non-heme iron dioxygenases and monoxygenases. These include di-iron oxo-bridged monoxygenases such as methane-monoxygenase, phenol hydroxylase, toluene 4-monoxygenase and toluene-o-xylene monoxygenase. The last two enzymes have been target of early applications of SSM [107,108], as well as refinement of previous successful directed evolution approaches [109]. Further work, more focused on developing enzyme catalysts for bioremediation, has been developed on dioxygenases containing a single iron atom such as ring-cleaving dioxygenases acting on polychlorinated biphenyls [110], aniline [111,112], dinitrotoluene [113] and chlorinated catechols [114]. The engineering of the extradiol dioxygenase (DoxG) that displays a low activity in 3,4-dihydroxybiphenyls ring cleavage was achieved by a combination of error-prone PCR, SSM at hot spots and DNA shuffling applied in sequence. Four residues located within 14 Å of the enzyme active site iron, highlighted by error prone PCR to be relevant for enzyme activity on the screening substrate, were targeted by saturation mutagenesis applied in pairs, grouping Ile-154 and Leu242, Leu-190 and Ser-191. The two resulting libraries were screened with coverage of 99.9% of the possible diversity resulting in variants with 2–10-fold increases in 3,4-dihydroxybiphenyl cleavage rates. After DNA shuffling, a further improvement generated a variant with a k_{cat}/K_M towards 3,4-dihydroxybiphenyl increased by 770 fold when compared to WT, confirming the feasibility and advantage of a coupled random and saturation mutagenesis approach in biocatalysts activity enhancement. SSM was also applied to an aniline dioxygenase isolated from *Acinetobacter* sp. strain YAA [112]. Substrate-binding pocket residues were selected and the V205A mutation that is possibly responsible for enlarging the binding pocket, was highlighted to lead the oxidation of 2-isopropylaniline, a substrate not recognized by the WT enzyme. The same mutants also shift the substrate specificity from 2,4-dimethylaniline, a good substrate for WT, to 2-isopropylaniline. Another variant, I248L, improved activity towards aniline and 2,4-dimethylaniline by approximately 2 fold. Both residues I248 and V205 were not previously reported to influence substrate recognition, therefore the finding also granted basic information on the enzyme active site determinants for substrate specificity. A further refinement by random mutagenesis on mutant V205A generated variant 3-R21, with improvement in activity towards the carcinogenic 2,4-dimethylaniline of 3.5 fold and retaining WT activity levels towards the natural substrate aniline. Therefore it can be concluded that the laboratory evolution of this biocatalyst generated a powerful tool to detoxify highly hazardous compounds. Another pollutant that has received attention in view of bioremediation strategies is 2,4,6-Trinitrotoluene (TNT), the most common explosive found in past and present war sites, and the intermediates of its synthesis 2,6-dinitrotoluene (2,6-DNT) and 2,4-dinitrotoluene (2,4-DNT) found as soil and water contaminants at TNT production facilities. 2,4-DNT dioxygenase of *Burkholderia* sp. strain DNT can catalyze the oxidation of 2,4-DNT to form 4-methyl-5-nitrocatechol and nitrite, but it has poor activity on other DNTs and nitrotoluens. By applying saturation mutagenesis at position I204 of the catalytic subunit and selecting for nitro-catechol producing mutants (signature of activity on the screened substrate), variants I204L and I204Y were identified [113]. These showed unprecedented activity on 2,3-DNT and 2,5-DNT and 2 to 8 fold improved activity towards 2,4-DNT, 2,6-DNT, 2NT and 4NT. The activity reported on 2,5-DNT, never observed for an enzyme, confirms that new biocatalysts unexplored by natural evolution can be generated by laboratory-driven evolution.

A gain of function on unnatural substrates and an inversion of specificity were also achieved by site directed and site-saturation mutagenesis on a catechol 1,2 dioxygenase from *Acinetobacter radioresistens* S13 [114]. Catechols are the converging metabolites of several aromatic degrading pathways, although natural enzymes usually cannot efficiently oxidize highly chlorinated or variously substituted catechols originated from chloroaromatic, biphenil and nitroaromatic compounds. The advantage of catechol dioxygenases is that these enzymes do not require any supply of reducing equivalent to perform the dihydroxylation and ring-cleavage of substrates, and therefore have a simpler architecture, higher stability and no need for expensive cofactors such as NAD(P)H to perform catalysis. Encapsulated and immobilized forms are also available [115], making them ideal biocatalysts. Mutagenesis on the active site was performed on residues L69 and A72 with a combined site-specific and SSM approach. This led to a series of variants with improved activity on the rarely recognized substrate 4,5-dichlorocatechol (by 2 fold in variant A72S), inversion of specificity for 4-chlorocatechol instead of catechol (variants L69A and L69A-A72G) and gain-of-function for recognition and catalysis on 4-*tert*-butyl catechol, a contaminant of cosmetics and foodstuff banned by EU since it can give sensitizations in patch testing at low concentrations (1%). The effect of active site re-shaping of the chosen mutational sites is shown in Figure 10, together with an example of the SSM obtained variants. An influence on the oxygen binding properties of mutants *vs.* WT was recently highlighted [116] and further work is ongoing in our labs for SSM at other catalytic pocket sites and for production of multiple site variants.

Figure 10. (**a**) Structure of active site of catechol 1,2 dioxygenase highlighting the residue that define the active site pocket (PDB file from crystallographic structure in [117]); (**b**) The effect of reshaping by mutagenesis and SSM on model of substrate/pocket interaction; (**c**) The list of identified and characterized mutants for SSM on position 72 are reported in the table (related to studies published in [114]).

a

b

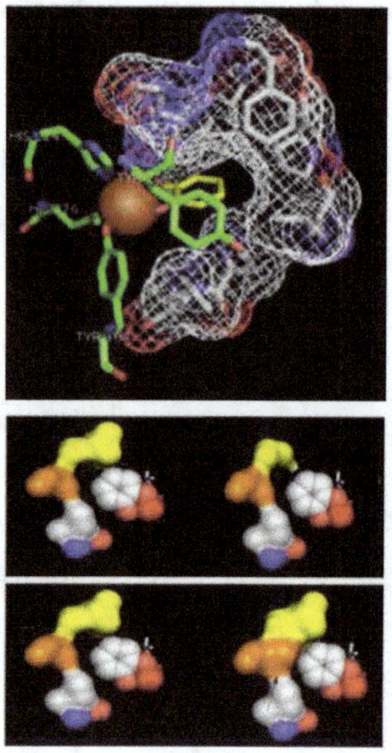

c

VARIANT	IDENTIFIED MUTATION	CODON
SSM 5	A72K	AAG
SSM 9	A72N	AAT
SSM 14	A72T	ACG
SSM 15	A72S	AGT
SSM 17	A72T	ACG
SSM 23		
SSM 24		
SSM 26		
SSM 27	A72S	AGT
SSM 34		
SSM 36	A72G	GGT
SSM 37	A72T	ACT
SSM 45	A72D	GAT
SSM 47		
SSM 50	A72N	AAT
SSM 57		
SSM 62	A72P	CCG
SSM 63		
SSM 64	A72S	AGT
SSM 68	A72N	AAT

Cytochrome P450 enzymes were also targeted by directed evolution for enhancing the degradation of recalcitrant aromatic and aliphatic pollutants. A recent work by Arnold and co-workers [118] compares combinatorial SSM strategies to the results obtained by random directed evolution. Although in this case it seems that the notable achievements obtained by error prone PCR and several round of random mutagenesis cannot be fully matched by SSM, the paper reports an improved activity on propane and ethane hydroxylation. In this case, nonetheless, a simpler approach by two rounds of error-prone PCR and back-crossing with parental DNA devised in our group on the same P450 BM3, led to variants that are active on highly recalcitrant polyaromatic hydrocarbon (PAH) pollutants, more relevant for environmental concerns, such as chrysene and pyrene [62].

A detoxification activity specifically improved by SSM on lactaldehyde oxidoreductase [72] is of relevance for detoxification of furfural, a toxic compound that originates from pre-treatment of cellulosic material. In this perspective the optimized catalyst obtained by SSM, a L7F mutant with a 10-fold higher activity than WT, is crucial both for lowering a toxic compound in an environment and for direct application in cell factory systems to enable cells to improve growth on treated lignocellulosic material. In the cited paper the variant obtained by SSM was also tested for performance in *E. coli* cells and showed a 2-fold higher rate of furfural metabolism during fermentation.

3.2.2. Cellulases, Haloalkane Dehalogenase and Other Hydrolases for Waste Degradation

Hydrolytic enzymes such as cellulases, endoglucanases, xylosidases and β-glycosidases, are increasingly being applied in lignocellulosic waste pre-treatment in combination or in alternative to steam-explosion and chemical treatments for enhanced saccharification of the biomass and lowered environmental impact. The SSM approach to enhance applicability of this class of biocatalysts has been focusing on improvement of thermal stability by the same research group both for an endoglucanase [84] and more recently on a β-glycosidase [85]. In the first case the endoglucanase CelA from *Clostridium thermocellum* was chosen for SSM at protein surface position Ser329. All the variants with improved thermal stability (approximately 5-fold increase in half-life of inactivation) and maintaining hydrolytic activity at WT levels, showed the presence of the S329G mutation. This finding suggested a systematic analysis of other possible substitutions to Gly of surface Ser residues, in line with reported works that Ser to Gly mutations on protein surface may improve thermostability. Thr and His surface residues were also selected on the basis that His and Thr, along with Ser, are generally substituted by Gly on the surface of proteins with enhanced thermal stability compared to their thermo-sensitive homologous. Few residues were also tested for substitution to Pro. A final variant S329G/S269P/H194G, generated by a combination of SSM and site-directed mutagenesis resulted in a 10-fold increase in half-life of inactivation at 86 °C.

A more recent paper from the same group [85] reports a consensus-based semi rational approach that benefits from the results of the previous work to enhance thermal stability of a β-glycosidase BglY from *Thermus thermophilus*.

An SSM approach applied to β-D-Xylosidase/α-L-arabinofuranosidase from *Selenomonas ruminantium* to residue W145 was instead focused on modulating the inhibitory effect of glucose and xylose on this enzyme for application to the saccharification of lignocellulosic waste biomass for biofuels production and as microbial substrate for other biotechnological processes [119]. While the β-D-Xylosidase/α-L-arabinofuranosidase can promote the hydrolytic cleavage of 1,4-β-D-xylooligosaccharides to D-xylose, the high affinity for the product D-xylose as well as for D-glucose hinders its excellent performance as a catalyst. Three variants isolated by screening the SSM library, W145F, W145L, and W145Y, showed decreased inhibition by the monosaccharides and increased catalytic activities up to 70% greater than that of the WT enzyme.

Another hydrolase applied to a different perspective of waste recycling is represented by a haloalkane dehalogenase DhaA from *Rhodococcus rhodochrous*. This enzyme is able to convert 1,2,3-trichloropropane (TCP) into (*R*)- or (*S*)-2,3-dichloropropan-1-ol, which can be converted into optically active epichlorohydrins, industrially important building blocks for the synthesis of fine chemicals. Enatioselectivity of the WT DhaA was further improved [120] by a pair-wise SSM approach applied to 16 active-site residues not directly involved in the catalytic reaction. A further refinement was then applied to the best R- and S-enantioselective variants by site directed mutagenesis including residues that are not part of the active site. A multi-site mutagenesis protocol with restricted codon usage allowed to finalize two variants, r5-90R and r5-97S with 13 and 17 mutations, that generate (*R*)-epichlorohydrin with 90% ee and (*S*)-epichlorohydrin with 97% ee, respectively.

3.2.3. Hydrogenases and Other Enzymes Relevant to Clean Energy Production

The energy issue has driven a great interest towards hydrogenases as powerful and efficient catalysts for hydrogen production in cell factory systems and in biohybrid fuel cells or solar harvesting devices as catalysts instead of platinum or other expensive rare-metal based materials [105]. Among the three classes of reported hydrogenases, [FeFe] hydrogenases are in this respect the most efficient catalysts due to their high turnover numbers, reaching turnover frequencies up to 10^4 s^{-1} [121] with a bias toward hydrogen production, but with relatively low overpotential needed as a driving force for catalysis in either direction. Interestingly, perspectives are also discussed in literature on [NiFe] hydrogenases for application in molecular hydrogen conversion for biofuel cells and in $NAD(P)^+$ cofactor regeneration.

Some limiting features, such as oxygen sensitivity, and the interest to further investigate the complex mechanism of the catalytic center, are increasingly pointing towards the application of saturation mutagenesis techniques to refine hydrogenases for desired applications. Although until now not many papers have been published on this topic [55,122] and in general on mutagenesis and laboratory evolution of all classes of hydrogenases [66,123–128], a very recent review from a leading group in the field of [FeFe] hydrogenase foresees imminent development in this direction [129]. In our group we applied saturation mutagenesis to a key residue in the active site of [FeFe] hydrogenase from *C. acetobutylicum* (CaHydA) recombinantly expressed in *E. coli.* This residue, namely cysteine 298, is involved in proton delivery to the active site; therefore it is crucial for substrate supply and product release, since the protons are converted reversibly to molecular hydrogen. In this case, accounting for proton pathways and local delivery engineering, means not only a matter of pH stability and fine regulation, but also of controlling the substrate concentration. We are also pursuing the same SSM strategy on other active site positions. The results of the focused approach on the conserved residue Cys 298, the final amino acid of a proton transfer chain to the active H-cluster [124] and believed to relay proton to the dithiolate bridging group that funnel them to the distal Fe, are reported in a recent publication [55]. Upon saturation mutagenesis with the NNK codon, a colorimetric screening performed on colonies allowed to reach 99.8% coverage of the library. Clones containing an active enzyme (with a detection threshold of 14% of original WT activity) were identified resulting in selection of only WT revertants or Cys-to-Asp mutants. The C298D variant shows a retained activity of 50%, which is interesting since the Cys residue is fully conserved in evolution, and therefore novel mutational spaces were explored, attesting that Asp can functionally replace Cys in proton relay and is structurally compatible (Figure 11).

Figure 11. Model of CaHydA structure illustrating C298 (left) and replacement at 298 position with aspartic acid (right) (adapted from Morra *et al.* [55]).

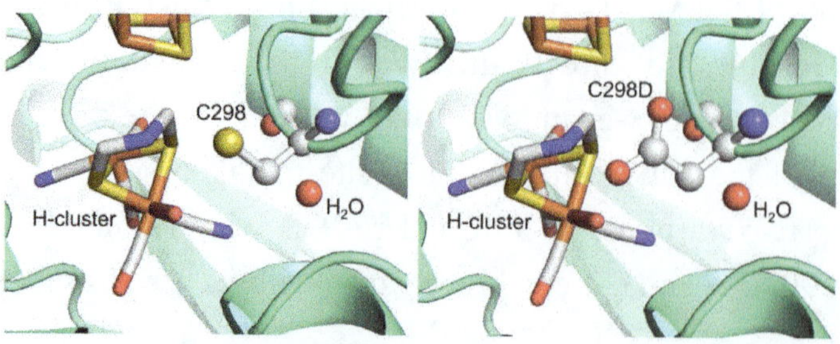

The frequency of WT revertants and Asp mutants matched the expected value on the basis of encoding codon frequency. To confirm the library completeness, selected clones were sequenced, showing a good and balanced codon randomization [55].

The SSM approach reported on [NiFe]-hydrogenase [122], performed in combination with directed evolution techniques such as error-prone PCR and shuffling is, as a matter of fact, the first random protein engineering of a hydrogenase. This work targeted the large subunit (HycE) of *Escherichia coli* hydrogenase 3. Hydrogenase 3 is responsible for synthesizing hydrogen from $2H^+$ and $2e^-$ within the supramolecular complex of formate hydrogen lyase (FHL), that also contains a formate dehydrogenase-H for forming $2H^+$, $2e^-$, and CO_2 from HCOOH: the overall FHL catalyses therefore hydrogen production from formate. A C-terminal truncated variant of HycE, generated by this combined random and SSM approach, showed increased hydrogen production by 30-fold.

Formate processing enzymes other than the cited subunit of *E. coli* hydrogenase are also relevant for the energy and sustainable process issues in their reversible activity of CO_2 conversion. The possibility of CO_2 sequestering and conversion of formate to methanol and methane is an intriguing perspective for research and applications [130]. Also the formate/CO_2 conversion is coupled to reduction of NAD^+ to NADH. Formate dehydrogenase from the yeast *Candida boidinii* catalyses the reaction with a selectivity for NAD^+ only, while $NADP^+$ is not recognized as a productive cofactor for the redox reaction and only gives minimal activity. SSM applied to two specific residues, Asp195 and Tyr196, of the dinucleotide-binding region, allowed an improvement in catalytic efficiency with $NADP^+$ of the order of 10^7 [131]. The selected variant Asp195Gln/Tyr196His is relevant for cofactor recycling systems with specificity for NADPH, preferred in enzymes such as cytochrome P450 monooxygenases that are largely applied in industry. The recovery of reduced cofactor is basically the natural strategy for storing solar energy in photosynthetic and chemical energy in chemosynthetic organisms, and therefore the control of biocatalysts performance in this reaction is a step forward in the direction of exploiting and mimicking nature in a sustainable manner.

4. Conclusions

The huge number of successful applications of SSM methods to enzymes reported in the last years underlines the feasibility of a semi-random approach to enzyme engineering. The results in activity, specificity and stability enhancement obtained are in several respects more cost-effective and less time-consuming than their counterparts, purely based on random approach and directed evolution. A factor of about ten, comparing enhancement of 20–50 fold by directed evolution and up to 700 for SSM, put SSM far ahead of fully randomized methods achievement-wise. In addition, the number of screened variants required for sound library coverage is generally 2–3 orders of magnitude smaller, allowing for application of more specific screening methods, able to precisely select the desired feature. Generating small and smart libraries is certainly a common and important goal also for the random directed evolution approach. The positive and negative results of both strategies in this direction can give important inputs and shared benefits. The drawback of SSM, *i.e.*, the required knowledge of structural data, is becoming less relevant given the increasingly available 3D models that can be calculated by homology with existing structural data and/or *ab-initio* modeling methods. These indirect structural data might not provide details of mechanisms and functions, but they are very indicative for intelligent planning of experimental approaches in SSM. Therefore, at least for technical enzymes, the SSM methods can be foreseen playing a major role in enzyme evolution. In combination with site-directed and random approaches, the methods have the potential to make a difference in exploring novel landscapes for biocatalysts most ambitious refinement, enhancement and application. The challenges remain in the development of biocatalysts performing entirely novel activities. In this respect, the importance of information in the details of mechanism of natural and successfully engineered enzymes is crucial. The role played by rational site-directed mutagenesis in elucidating mechanism and substrate specificity has been of paramount importance. A very recent review focused on an important class of enzymes foresees a similarly important role for SSM [132].

The next generation of engineered biocatalysts can certainly reach unprecedented performances [133] and this can be achieved due to the choices available to scientists to select among different strategies, whose advantages and limitations have to be carefully balanced. The versatility of SSM and the various modifications of this general approach, together with the chance to combine with other strategies, equip protein engineers with an already powerful toolbox. How to apply the tools is not simple to rationalize or give rules for, but this is certainly the undefined area that must remain open, in which scientists can propose original experimental design and improved methods.

References

1. Peters, M.W.; Meinhold, P.; Glieder, A.; Arnold, F.H. Regio- and enantioselective alkane hydroxylation with engineered cytochromes P450 BM-3. *J. Am. Chem. Soc.* **2003**, *125*, 13442–13450.
2. Bocola, M.; Otte, N.; Jaeger, K.E.; Reetz, M.T.; Thiel, W. Learning from directed evolution: Theoretical investigations into cooperative mutations in lipase enantioselectivity. *Chembiochem* **2004**, *5*, 214–223.
3. Bartsch, S.; Kourist, R.; Bornscheuer, U.T. Complete inversion of enantioselectivity towards acetylated tertiary alcohols by a double mutant of a *Bacillus subtilis* esterase. *Angew. Chem. Int. Ed.* **2008**, *47*, 1508–1511.
4. Glieder, A.; Farinas, E.T.; Arnold, F.H. Laboratory evolution of a soluble, self-sufficient, highly active alkane hydroxylase. *Nat. Biotechnol.* **2002**, *20*, 1135–1139.
5. Schmidt, D.M.Z.; Mundorff, E.C.; Dojka, M.; Bermudez, E.; Ness, J.E.; Govindarajan, S.; Babbitt, P.C.; Minshull, J.; Gerlt, J.A. Evolutionary potential of (β/α)(8)-barrels: Functional promiscuity produced by single substitutions in the enolase superfamily. *Biochemistry* **2003**, *42*, 8387–8393.
6. Bosma, T.; Danborsky, J.; Stucki, G.; Janssen, D.B. Biodegradation of 1,2,3-trichloropropane through directed evolution and heterologous expression of a haloalkane dehalogenase gene. *Appl. Environ. Microbiol.* **2002**, *68*, 3582–3587.

7. Reetz, M.T.; Soni, P.; Acevedo, J.P.; Sanchis, J. Creation of an amino acid network of structurally coupled residues in the directed evolution of a thermostable enzyme. *Angew. Chem. Int. Ed. Engl.* **2009**, *48*, 8268–8272.
8. Zumarraga, M.; Bulter, T.; Shleev, S.; Polaina, J.; Martinez-Arias, A.; Plou, F.J.; Ballesteros, A.; Alcalde, M. *In vitro* evolution of a fungal laccase in high concentrations of organic cosolvents. *Chem. Biol.* **2007**, *14*, 1052–1064.
9. Siegel, J.B.; Zanghellini, A.; Lovick, H.M.; Kiss, G.; Lambert, A.R.; St. Clair, J.L.; Gallaher, J.L.; Hilvert, D.; Gelb, M.H.; Stoddard, B.L.; *et al.* Computational design of an enzyme catalyst for a stereoselective bimolecular Diels-Alder reaction. *Science* **2010**, *329*, 309–313.
10. Khersonsky, O.; Kiss, G.; Röthlisberger, D.; Dym, O.; Albeck, S.; Houk, K.N.; Baker, D.; Tawfik, D.S. Bridging the gaps in design methodologies by evolutionary optimization of the stability and proficiency of designed Kemp eliminase KE59. *Proc. Natl. Acad. Sci. USA* **2012**, *109*, 10358–10363.
11. Merski, M.; Shoichet, B.K. Engineering a model protein cavity to catalyze the Kemp elimination. *Proc. Natl. Acad. Sci. USA* **2012**, *109*, 16179–16183.
12. Savile, C.K.; Janey, J.M.; Mundorff, E.C.; Moore, J.C.; Tam, S.; Jarvis, W.R.; Colbeck, J.C.; Krebber, A.; Fleitz, F.J.; Brands, J.; *et al.* Biocatalytic asymmetric synthesis of chiral amines from ketones applied to sitagliptin manufacture. *Science* **2010**, *329*, 305–309.
13. Janssen, D.B. Evolving haloalkane dehalogenases. *Curr. Opin. Chem. Biol.* **2004**, *8*, 150–159.
14. Pavlova, M.; Klvana, M.; Prokop, Z.; Chaloupkova, R.; Banas, P.; Otyepka, M.; Wade, R.C.; Tsuda, M.; Nagata, Y.; Damborsky, J. Redesigning dehalogenase access tunnels as a strategy for degrading an anthropogenic substrate. *Nat. Chem. Biol.* **2009**, *5*, 727–733.
15. Fasan, R.; Meharenna, Y.T.; Snow, C.D.; Poulos, T.L.; Arnold, F.H. Evolutionary history of a specialized p450 propane monooxygenase. *J. Mol. Biol.* **2008**, *383*, 1069–1080.
16. Stemmer, W.P. Rapid evolution of a protein *in vitro* by DNA shuffling. *Nature* **1994**, *370*, 389–391.
17. Ostermeier, M.; Shim, J.H.; Benkovic, S.J. A combinatorial approach to hybrid enzymes independent of DNA homology. *Nat. Biotechnol.* **1999**, *17*, 1205–1209.
18. Pelletier, J.N. A RACHITT for our toolbox. *Nat. Biotechnol.* **2001**, *19*, 314–315.
19. Sieber, V.; Martinez, C.A.; Arnold, F.H. Libraries of hybrid proteins from distantly related sequences. *Nat. Biotechnol.* **2001**, *19*, 456–460.
20. Wang, M.; Si, T.; Zhao, H. Biocatalyst development by directed evolution. *Bioresour. Technol.* **2012**, *115*, 117–125.
21. Bornscheuer, U.T.; Huisman, G.W.; Kazlauskas, R.J.; Lutz, S.; Moore, J.C.; Robins, K. Engineering the third wave of biocatalysis. *Nature* **2012**, *485*, 185–194.
22. Yuan, L.; Kurek, I.; English, J.; Keenan, R. Laboratory-directed protein evolution. *Microbiol. Mol. Biol. Rev.* **2005**, *69*, 373–392.
23. Bloom, J.D.; Meyer, M.M.; Meinhold, P.; Otey, C.R.; MacMillan, D.; Arnold, F.H. Evolving strategies for enzyme engineering. *Curr. Opin. Struct. Biol.* **2005**, *15*, 447–452.
24. Valetti, F.; Gilardi G. Directed evolution of enzymes for product chemistry. *Nat. Prod. Rep.* **2004**, *21*, 490–511.
25. Farinas, E.T.; Bulter, T.; Arnold, F.H. Directed enzyme evolution. *Curr. Opin. Biotechnol.* **2001**, *12*, 545–551.

26. Romero, P.A.; Arnold, F.H. Exploring protein fitness landscapes by directed evolution. *Nat. Rev. Mol. Cell. Biol.* **2009**, *10*, 866–876.
27. Turner, N.J. Directed evolution drives the next generation of biocatalysts. *Nat. Chem. Biol.* **2009**, *5*, 567–573.
28. Guo, F.; Xu, H.; Xu, H.; Yu, H. Compensation of the enantioselectivity-activity trade-off in the directed evolution of an esterase from *Rhodobacter sphaeroides* by site-directed saturation mutagenesis. *Appl. Microbiol. Biotechnol.* **2013**, *97*, 3355–3362.
29. Reetz, M.T.; Bocola, M.; Carballeira, J.D.; Zha, D.; Vogel, A. Expanding the range of substrate acceptance of enzymes: Combinatorial active-site saturation test. *Angew. Chem. Int. Ed. Engl.* **2005**, *44*, 4192–4196.
30. Reetz, M.T.; Carballeira, J.D. Iterative saturation mutagenesis (ISM) for rapid directed evolution of functional enzymes. *Nat. Protoc.* **2007**, *2*, 891–903.
31. Loke, P.; Sim, T.S. A comparison of three site-directed mutagenesis kits. *Z. Naturforsch. C* **2001**, *56*, 810–813.
32. Reetz, M.T.; Prasad, S.; Carballeira, J.D.; Gumulya, Y.; Bocola, M. Iterative saturation mutagenesis accelerates laboratory evolution of enzyme stereoselectivity: Rigorous comparison with traditional methods. *J. Am. Chem. Soc.* **2010**, *132*, 9144–9152.
33. Gumulya, Y.; Sanchis, J.; Reetz, M.T. Many pathways in laboratory evolution can lead to improved enzymes: How to escape from local minima. *Chembiochem* **2012**, *13*, 1060–1066.
34. Prasad, S.; Bocola, M.; Reetz, M.T. Revisiting the lipase from *Pseudomonas aeruginosa*: Directed evolution of substrate acceptance and enantioselectivity using iterative saturation mutagenesis. *Chemphyschem* **2011**, *12*, 1550–1557.
35. http://www.reportlinker.com/p0747897-summary/World-Enzymes-Industry.html
36. Engström, K.; Nyhlén, J.; Sandström, A.G.; Bäckvall, J.E. Directed evolution of an enantioselective lipase with broad substrate scope for hydrolysis of alpha-substituted esters. *J. Am. Chem. Soc.* **2010**, *132*, 7038–7042.
37. Wen, S.; Tan, T.; Zhao, H. Improving the thermostability of lipase Lip2 from *Yarrowia lipolytica. J. Biotechnol.* **2012**, *164*, 248–253.
38. Gumulya, Y.; Reetz, M.T. Enhancing the thermal robustness of an enzyme by directed evolution: Least favorable starting points and inferior mutants can map superior evolutionary pathways. *Chembiochem* **2011**, *12*, 2502–2510.
39. Reetz, M.T.; Soni, P.; Fernández, L.; Gumulya, Y.; Carballeira, J.D. Increasing the stability of an enzyme toward hostile organic solvents by directed evolution based on iterative saturation mutagenesis using the B-FIT method. *Chem. Commun. (Camb.)* **2010**, *46*, 8657–8658.
40. Reetz, M.T.; Soni, P.; Fernández, L. Knowledge-guided laboratory evolution of protein thermolability. *Biotechnol. Bioeng.* **2009**, *102*, 1712–1717.
41. Wang, C.; Huang, R.; He, B.; Du, Q. Improving the thermostability of alpha-amylase by combinatorial coevolving-site saturation mutagenesis. *BMC Bioinformatics* **2012**, *13*, e263.
42. Gouveia-Oliveira, R.; Pedersen, A.G. Finding coevolving amino acid residues using row and column weighting of mutual information and multi-dimensional amino acid representation. *Algorithms Mol. Biol.* **2007**, *2*, 12.

43. Derbyshire, K.M.; Salvo, J.J.; Grindley, N.D. A simple and efficient procedure for saturation mutagenesis using mixed oligodeoxynucleotides. *Gene* **1986**, *46*, 145–152.
44. Firnberg, E.; Ostermeier, M. PFunkel: Efficient, expansive, user-defined mutagenesis. *PLoS One* **2012**, *7*, e52031.
45. Dennig, A.; Shivange, A.V.; Marienhagen, J.; Schwaneberg, U. OmniChange: The sequence independent method for simultaneous site-saturation of five codons. *PLoS One* **2011**, *6*, e26222.
46. Walter, K.U.; Vamvaca, K.; Hilvert, D. An active enzyme constructed from a 9-amino acid alphabet. *J. Biol. Chem.* **2005**, *280*, 37742–37746.
47. Reetz, M.T.; Kahakeaw, D.; Sanchis, J. Shedding light on the efficacy of laboratory evolution based on iterative saturation mutagenesis. *Mol. Biosyst.* **2009**, *5*, 115–122.
48. Reetz, M.T.; Kahakeaw, D.; Lohmer, R. Addressing the numbers problem in directed evolution. *Chembiochem* **2008**, *9*, 1797–1804.
49. Reetz, M.T.; Wu, S. Greatly reduced amino acid alphabets in directed evolution: Making the right choice for saturation mutagenesis at homologous enzyme positions. *Chem. Commun. (Camb.)* **2008**, 5499–5501. DOI: 10.1039/B813388C.
50. Kille, S.; Acevedo-Rocha, C.G.; Parra, L.P.; Zhang, Z.G.; Opperman, D.J.; Reetz, M.T.; Acevedo, J.P. Reducing codon redundancy and screening effort of combinatorial protein libraries created by saturation mutagenesis. *ACS Synth. Biol.* **2013**, *2*, 83–92.
51. Bosley, A.D.; Ostermeier, M. Mathematical expressions useful in the construction, description and evaluation of protein libraries. *Biomol. Eng.* **2005**, *22*, 57–61.
52. Mena, M.A.; Daugherty, P.S. Automated design of degenerate codon libraries. *Protein Eng. Des. Sel.* **2005**, *18*, 559–561.
53. Firth, A.E.; Patrick, W.M. Statistics of protein library construction. *Bioinformatics* **2005**, *21*, 3314–3315.
54. Patrick, W.M.; Firth, A.E.; Blackburn, J.M. User-friendly algorithms for estimating completeness and diversity in randomized protein-encoding libraries. *Protein Eng.* **2003**, *16*, 451–457.
55. Morra, S.; Giraudo, A.; di Nardo, G.; King, P.W.; Gilardi, G.; Valetti, F. Site saturation mutagenesis demonstrates a central role for cysteine 298 as proton donor to the catalytic site in CaHydA [FeFe]-hydrogenase. *PLoS One* **2012**, *7*, e48400.
56. Koga, Y.; Kato, K.; Nakano, H.; Yamane, T. Inverting enantioselectivity of *Burkholderia cepacia* KWI-56lipase by combinatorial mutation and high-throughput screening using single-molecule PCR and *in vitro* expression. *J. Mol. Biol.* **2003**, *331*, 585–592.
57. Levin, A.M.; Weiss, G.A. Optimizing the affinity and specificity of proteins with molecular display. *Mol. Biosyst.* **2006**, *2*, 49–57.
58. Granieri, L.; Baret, J.C.; Griffiths, A.D.; Merten, C.A. High-throughput screening of enzymes by retroviral display using droplet-based microfluidics. *Chem. Biol.* **2010**, *17*, 229–235.
59. Tsotsou, G.E.; Cass, A.E.G.; Gilardi, G. High throughput assay for cytochrome P450 BM3 for screening libraries of substrates and combinatorial mutants. *Biosens. Bioelectron.* **2002**, *17*, 119–131.
60. Despotovic, D.; Vojcic, L.; Prodanovic, R.; Martinez, R.; Maurer, K.H.; Schwaneberg, U. Fluorescent assay for directed evolution of perhydrolases. *J. Biomol. Screen.* **2012**, *17*, 796–805.

61. Sass, S.; Kadow, M.; Geitner, K.; Thompson, M.L.; Talmann, L.; Bottcher, D.; Schmidt, M.; Bornscheuer, U.T. A high-throughput assay method to quantify Baeyer-Villiger monooxygenase activity. *Tetrahedron* **2012**, *68*, 7575–7580.
62. Sideri, A.; Goyal, A.; di Nardo, G.; Tsotsou, G.E.; Gilardi, G. Hydroxylation of non-substituted polycyclic aromatic hydrocarbons by cytochrome P450 BM3 engineered by directed evolution. *J. Inorg. Biochem.* **2013**, *120*, 1–7.
63. Tsotsou, G.E.; Sideri, A.; Goyal, A.; di Nardo, G.; Gilardi, G. Identification of mutant Asp251Gly/Gln307His of cytochrome P450 BM3 for the generation of metabolites of diclofenac, ibuprofen and tolbutamide. *Chemistry* **2012**, *18*, 3582–3588.
64. Di Nardo, G.; Gilardi, G. Optimization of the Bacterial Cytochrome P450 BM3 System for the production of human drug metabolites. *Int. J. Mol. Sci.* **2012**, *13*, 15901–15924.
65. Tsotsou, G.E.; di Nardo, G.; Sadeghi, S.J.; Fruttero, R.; Lazzarato, L.; Bertinaria, M.; Gilardi, G. A rapid screening for cytochrome P450 catalysis on new chemical entities: Cytochrome P450 BM3 and 1,2,5-oxadiazole derivatives. *J. Biomol. Screen.* **2013**, *18*, 211–218.
66. Stapleton, J.A.; Swartz, J.R. A cell-free microtiter plate screen for improved [FeFe] hydrogenases. *PLoS One* **2010**, *5*, e10554.
67. Chuah, J.A.; Tomizawa, S.; Yamada, M.; Tsuge, T.; Doi, Y.; Sudesh, K.; Numata, K. Characterization of site-specific mutations in a short-chain-length/medium-chain-length polyhydroxyalkanoate synthase: *In vivo* and *in vitro* studies of enzymatic activity and substrate specificity. *Appl. Environ. Microbiol.* **2013**, *79*, 3813–3821.
68. Jakoblinnert, A.; van den Wittenboer, A.; Shivange, A.V.; Bocola, M.; Heffele, L.; Ansorge-Schumacher, M.; Schwaneberg, U. Design of an activity and stability improved carbonyl reductase from *Candida parapsilosis*. *J. Biotechnol.* **2013**, *165*, 52–62.
69. Vojcic, L.; Despotovic, D.; Maurer, K.H.; Zacharias, M.; Bocola, M.; Martinez, R.; Schwaneberg, U. Reengineering of subtilisin Carlsberg for oxidative resistance. *Biol. Chem.* **2013**, *394*, 79–87.
70. Wu, Q.; Soni, P.; Reetz, M.T. Laboratory evolution of enantiocomplementary *Candida antarctica* lipase B mutants with broad substrate scope. *J. Am. Chem. Soc.* **2013**, *135*, 1872–81.
71. Nallaseth, F.S.; Anderson, S. A screen for over-secretion of proteins by yeast based on a dual component cellular phosphatase and immuno-chromogenic stain for exported bacterial alkaline phosphatase reporter. *Microb. Cell Fact.* **2013**, *12*, e36.
72. Zheng, H.; Wang, X.; Yomano, L.P.; Geddes, R.D.; Shanmugam, K.T.; Ingram, L.O. Improving *Escherichia coli* FucO for furfural tolerance by saturation mutagenesis of individual amino acid positions. *Appl. Environ. Microbiol.* **2013**, *79*, 3202–3208.
73. Zhou, H.; Qu, Y.; Kong, C.; Shen, E.; Wang, J.; Zhang, X.; Ma, Q.; Zhou, J. The key role of a non-active-site residue Met148 on the catalytic efficiency of meta-cleavage product hydrolase BphD. *Appl. Microbiol. Biotechnol.* **2013**, doi:10.1007/s00253-013-4814-0.
74. Phelan, R.M.; Townsend, C.A. Mechanistic insights into the bifunctional non-heme iron oxygenase carbapenem synthase by active site saturation mutagenesis. *J. Am. Chem. Soc.* **2013**, *135*, 7496–7502.

75. Geier, M.; Braun, A.; Fladischer, P.; Stepniak, P.; Rudroff, F.; Hametner, C.; Mihovilovic, M.D.; Glieder, A. Double site saturation mutagenesis of the human cytochrome P450 2D6 results in regioselective steroid hydroxylation. *FEBS J.* **2013**, *280*, 3094–3108.
76. Molloy, E.M.; Field, D.; O'Connor, P.M.; Cotter, P.D.; Hill, C.; Ross, R.P. Saturation mutagenesis of lysine 12 leads to the identification of derivatives of nisin A with enhanced antimicrobial activity. *PLoS One* **2013**, *8*, e58530.
77. Shainsky, J.; Bernath-Levin, K.; Isaschar-Ovdat, S.; Glaser, F.; Fishman, A. Protein engineering of nitrobenzene dioxygenase for enantioselective synthesis of chiral sulfoxides. *Protein Eng. Des. Sel.* **2013**, *26*, 335–345.
78. Agudo, R.; Roiban, G.D.; Reetz, M.T. Induced axial chirality in biocatalytic asymmetric ketone reduction. *J. Am. Chem. Soc.* **2013**, *135*, 1665–1668.
79. Jakoblinnert, A.; Wachtmeister, J.; Schukur, L.; Shivange, A.V.; Bocola, M.; Ansorge-Schumacher, M.B.; Schwaneberg, U. Reengineered carbonyl reductase for reducing methyl-substituted cyclohexanones. *Protein Eng. Des. Sel.* **2013**, *26*, 291–298.
80. Sandström, A.G.; Wikmark, Y.; Engström, K.; Nyhlén, J.; Bäckvall, J.E. Combinatorial reshaping of the *Candida antarctica* lipase A substrate pocket for enantioselectivity using an extremely condensed library. *Proc. Natl. Acad. Sci. USA* **2012**, *109*, 78–83.
81. Korman, T.P.; Sahachartsiri, B.; Charbonneau, D.M.; Huang, G.L.; Beauregard, M.; Bowie J.U. Dieselzymes: Development of a stable and methanol tolerant lipase for biodiesel production by directed evolution. *Biotechnol. Biofuels* **2013**, *6*, e70.
82. Anbar, M.; Bayer, E.A. Approaches for improving thermostability characteristics in cellulases. *Methods Enzymol.* **2012**, *510*, 261–271.
83. Sygmund, C.; Santner, P.; Krondorfer, I.; Peterbauer, C.K.; Alcalde, M.; Nyanhongo, G.S.; Guebitz, G.M.; Ludwig, R. Semi-rational engineering of cellobiose dehydrogenase for improved hydrogen peroxide production. *Microb. Cell Fact.* **2013**, *12*, e38.
84. Yi, Z.L.; Pei, X.Q.; Wu, Z.L. Introduction of glycine and proline residues onto protein surface increases the thermostability of endoglucanase CelA from *Clostridium thermocellum. Bioresour. Technol.* **2011**, *102*, 3636–3638.
85. Yi, Z.L.; Zhang, S.B.; Pei, X.Q.; Wu, Z.L. Design of mutants for enhanced thermostability of β-glycosidase BglY from *Thermus thermophilus. Bioresour. Technol.* **2013**, *129*, 629–633.
86. Reetz, M.T.; Wilensek, S.; Zha, D.; Jaeger, K.E. Directed evolution of an enantioselective enzyme through combinatorial multiple-cassette mutagenesis. *Angew. Chem. Int. Ed. Engl.* **2001**, *40*, 3589–3591.
87. Jochens, H.; Bornscheuer, U.T. Natural diversity to guide focused directed evolution. *Chembiochem* **2010**, *11*, 1861–1866.
88. Jochens, H.; Aerts, D.; Bornscheuer, U.T. Thermostabilization of an esterase by alignment-guided focussed directed evolution. *Protein Eng. Des. Sel.* **2010**, *23*, 903–909.
89. Zheng, H.; Reetz, M.T. Manipulating the stereoselectivity of limonene epoxide hydrolase by directed evolution based on iterative saturation mutagenesis. *J. Am. Chem. Soc.* **2010**, *132*, 15744–15751.

90. Ye, L.J.; Wang, L.; Pan, Y.; Cao, Y. Changing the specificity of α-amino acid ester hydrolase toward para-hydroxyl cephalosporins synthesis by site-directed saturation mutagenesis. *Biotechnol. Lett.* **2012**, *34*, 1719–1724.
91. Garrett, J.B.; Kretz, K.A.; O'Donoghue, E.; Kerovuo, J.; Kim, W.; Barton, N.R.; Hazlewood, G.P.; Short, J.M.; Robertson, D.E.; Gray, K.A. Enhancing the thermal tolerance and gastric performance of a microbial phytase for use as a phosphate-mobilizing monogastric-feed supplement. *Appl. Environ. Microbiol.* **2004**, *70*, 3041–3046.
92. Kille, S.; Zilly, F.E.; Acevedo, J.P.; Reetz, M.T. Regio- and stereoselectivity of P450-catalysed hydroxylation of steroids controlled by laboratory evolution. *Nat. Chem.* **2011**, *3*, 738–743.
93. Li, H.M.; Mei, L.H.; Urlacher, V.B.; Schmid, R.D. Cytochrome P450 BM-3 evolved by random and saturation mutagenesis as an effective indole-hydroxylating catalyst. *Appl. Biochem. Biotechnol.* **2008**, *144*, 27–36.
94. Ba, L.; Li, P.; Zhang, H.; Duan, Y.; Lin, Z. Semi-rational engineering of cytochrome P450sca-2 in a hybrid system for enhanced catalytic activity: Insights into the important role of electron transfer. *Biotechnol. Bioeng.* **2013**, doi:10.1002/bit.24960.
95. Wu, S.; Acevedo, J.P.; Reetz, M.T. Induced allostery in the directed evolution of an enantioselective Baeyer-Villiger monooxygenase. *Proc. Natl. Acad. Sci. USA* **2010** 107, 2775–2780.
96. Reetz, M.T.; Wu, S. Laboratory evolution of robust and enantioselective Baeyer-Villiger monooxygenases for asymmetric catalysis. *J. Am. Chem. Soc.* **2009**, *131*, 15424–15432.
97. Willetts, A.; Joint, I.; Gilbert, J.A.; Trimble, W.; Mühling, M. Isolation and initial characterization of a novel type of Baeyer-Villiger monooxygenase activity from a marine microorganism. *Microb. Biotechnol.* **2012**, *5*, 549–559.
98. Minerdi, D.; Zgrablic, I.; Sadeghi, S.J.; Gilardi, G. Identification of a novel Baeyer-Villiger monooxygenase from *Acinetobacter radioresistens*: Close relationship to the *Mycobacterium tuberculosis* prodrug activator EtaA. *Microb. Biotechnol.* **2012**, *5*, 700–716.
99. Mascotti, M.L.; Juri Ayub, M.; Dudek, H.; Sanz, M.K.; Fraaije, M.W. Cloning, overexpression and biocatalytic exploration of a novel Baeyer-Villiger monooxygenase from *Aspergillus fumigatus* Af293. *AMB Express* **2013**, *3*, e33.
100. Gao, X.; Huang, F.; Feng, J.; Chen, X.; Zhang, H.; Wang, Z.; Wu, Q.; Zhu, D. Engineering the meso-diaminopimelate dehydrogenase from *Symbiobacterium thermophilum* by site-saturation mutagenesis for D-phenylalanine synthesis. *Appl. Environ. Microbiol.* **2013**, *79*, 5078–5081.
101. Paul, D.; Pandey, G.; Pandey, J.; Jain, R.K. Accessing microbial diversity for bioremediation and environmental restoration. *Trends Biotechnol.* **2005**, *23*, 135–142.
102. Goldsmith, M.; Ashani, Y.; Simo, Y.; Ben-David, M.; Leader, H.; Silman, I.; Sussman, J.L.; Tawfik, D.S. Evolved stereoselective hydrolases for broad-spectrum G-type nerve agent detoxification. *Chem. Biol.* **2012**, *19*, 456–466.
103. Du, W.; Li, W.; Sun, T.; Chen, X.; Liu, D. Perspectives for biotechnological production of biodiesel and impacts. *Appl. Microbiol. Biotechnol.* **2008**, *79*, 331–337.
104. Parawira, W. Enzyme research and applications in biotechnological intensification of biogas production. *Crit. Rev. Biotechnol.* **2012**, *32*, 172–186.
105. King, P.W. Designing interfaces of hydrogenase-nanomaterial hybrids for efficient solar conversion. *Biochim. Biophys. Acta* **2013**, *1827*, 949–957.

106. Morra, S.; Valetti, F.; Sadeghi, S.J.; King, P.W.; Meyer, T.; Gilardi, G. Direct electrochemistry of an [FeFe]-hydrogenase on a TiO_2 electrode. *Chem. Commun. (Camb.)* **2011**, *47*, 10566–10568.
107. Vardar, G.; Wood, T.K. Protein engineering of toluene-o-xylene monooxygenase from *Pseudomonas stutzeri* OX1 for synthesizing 4-methylresorcinol, methylhydroquinone, and pyrogallol. *Appl. Environ. Microbiol.* **2004**, *70*, 3253–3562.
108. Tao, Y.; Fishman, A.; Bentley, W.E.; Wood, T.K. Altering toluene 4-monooxygenase by active-site engineering for the synthesis of 3-methoxycatechol, methoxyhydroquinone, and methylhydroquinone. *J. Bacteriol.* **2004**, *186*, 4705–4713.
109. Canada, K.A.; Iwashita, S.; Shim, H.; Wood, T.K. Directed evolution of toluene ortho-monooxygenase for enhanced 1-naphthol synthesis and chlorinated ethene degradation. *J. Bacteriol.* **2002**, *184*, 344–349.
110. Fortin, P.D.; MacPherson, I.; Neau, D.B.; Bolin, J.T.; Eltis, L.D. Directed evolution of a ring-cleaving dioxygenase for polychlorinated biphenyl degradation. *J. Biol. Chem.* **2005**, *280*, 42307–42314.
111. Ang, E.L.; Obbard, J.P.; Zhao, H. Directed evolution of aniline dioxygenase for enhanced bioremediation of aromatic amines. *Appl. Microbiol. Biotechnol.* **2009**, *81*, 1063–1070.
112. Ang, E.L.; Obbard, J.P.; Zhao, H. Probing the molecular determinants of aniline dioxygenase substrate specificity by saturation mutagenesis. *FEBS J.* **2007**, *274*, 928–939.
113. Leungsakul, T.; Keenan, B.G.; Yin, H.; Smets, B.F.; Wood, T.K. Saturation mutagenesis of 2,4-DNT dioxygenase of *Burkholderia* sp. strain DNT for enhanced dinitrotoluene degradation. *Biotechnol. Bioeng.* **2005**, *92*, 416–426.
114. Caglio, R.; Valetti, F.; Caposio, P.; Gribaudo, G.; Pessione, E.; Giunta, C. Fine-tuning of catalytic properties of catechol 1,2-dioxygenase by active site tailoring. *Chembiochem* **2009**, *10*, 1015–1024.
115. Di Nardo, G.; Roggero, C.; Campolongo, S.; Valetti, F.; Trotta, F.; Gilardi, G. Catalytic properties of catechol 1,2-dioxygenase from Acinetobacter radioresistens S13 immobilized on nanosponges. *Dalton Trans.* **2009**, *7*, 6507–6512.
116. Caglio, R.; Pessione, E.; Valetti, F.; Giunta, C.; Ghibaudi, E. An EPR, thermostability and pH-dependence study of wild-type and mutant forms of catechol 1,2-dioxygenase from *Acinetobacter radioresistens* S13. *Biometals* **2013**, *26*, 75–84.
117. Micalella, C.; Martignon, S.; Bruno, S.; Pioselli, B.; Caglio, R.; Valetti, F.; Pessione, E.; Giunta, C.; Rizzi, M. X-ray crystallography, mass spectrometry and single crystal microspectrophotometry: A multidisciplinary characterization of catechol 1,2 dioxygenase. *Biochim. Biophys. Acta* **2011**, *1814*, 817–823.
118. Chen, M.M.; Snow, C.D.; Vizcarra, C.L.; Mayo, S.L.; Arnold, F.H. Comparison of random mutagenesis and semi-rational designed libraries for improved cytochrome P450 BM3-catalyzed hydroxylation of small alkanes. *Protein Eng. Des. Sel.* **2012**, *25*, 171–178.
119. Jordan, D.B.; Wagschal, K.; Fan, Z.; Yuan, L.; Braker, J.D.; Heng, C. Engineering lower inhibitor affinities in β-D-xylosidase of *Selenomonas ruminantium* by site-directed mutagenesis of Trp145. *J. Ind. Microbiol. Biotechnol.* **2011**, *38*, 1821–1835.

120. Van Leeuwen, J.G.; Wijma, H.J.; Floor, R.J.; van der Laan, J.M.; Janssen, D.B. Directed evolution strategies for enantiocomplementary haloalkane dehalogenases: From chemical waste to enantiopure building blocks. *Chembiochem* **2012**, *13*, 137–148.
121. Frey, M. Hydrogenases: Hydrogen-activating enzymes. *Chembiochem* **2002**, *3*, 153–160.
122. Maeda, T.; Sanchez-Torres, V.; Wood, T.K. Protein engineering of hydrogenase 3 to enhance hydrogen production. *Appl. Microbiol. Biotechnol.* **2008**, *79*, 77–86.
123. Buhrke, T.; Lenz, O.; Krauss, N.; Friedrich, B.J. Oxygen tolerance of the H_2-sensing [NiFe] hydrogenase from *Ralstonia eutropha* H16 is based on limited access of oxygen to the active site. *Biol. Chem.* **2005**, *280*, 23791–23796.
124. Cornish, A.J.; Gartner, K.; Yang, H.; Peters, J.W.; Hegg, E.L. Mechanism of proton transfer in [FeFe]-hydrogenase from *Clostridium pasteurianum*. *J. Biol. Chem.* **2011**, *286*, 38341–38347.
125. Knorzer, P.; Silakov, A.; Foster, C.E.; Armstrong, F.A.; Lubitz, W.; Happe, T. Importance of the protein framework for catalytic activity of [FeFe]-hydrogenases. *J. Biol. Chem.* **2012**, *286*, 38341–38347.
126. Lautier, T.; Ezanno, P.; Baffert, C.; Fourmond, V.; Cournac, L.; Fontecilla-Camps, J.C.; Soucaille, P.; Bertrand, P.; Meynial-Salles, I.; Léger, C. The quest for a functional substrate access tunnel in FeFe hydrogenase. *Faraday Discuss* **2011**, *148*, 385–407.
127. Stapleton, J.A.; Swartz, J.R. Development of an *in vitro* compartmentalization screen for high-throughput directed evolution of [FeFe] hydrogenases. *PLoS One* **2010**, *5*, e15275.
128. Bingham, A.S.; Smith, P.R.; Swartz, J.R. Evolution of an [FeFe] hydrogenase with decreased oxygen sensitivity. *Int. J. Hydrogen Energy* **2012**, *37*, 2965–2976.
129. Winkler, M.; Esselborn, J.; Happe, T. Molecular basis of [FeFe]-hydrogenase function: An insight into the complex interplay between protein and catalytic cofactor. *Biochim. Biophys. Acta* **2013**, *1827*, 974–985.
130. Reda, T.; Plugge, C.M.; Abram, N.J.; Hirst, J. Reversible interconversion of carbon dioxide and formate by an electroactive enzyme. *Proc. Natl. Acad. Sci. USA* **2008**, *105*, 10654–10658.
131. Andreadeli, A.; Platis, D.; Tishkov, V.; Popov, V.; Labrou, N.E. Structure-guided alteration of coenzyme specificity of formate dehydrogenase by saturation mutagenesis to enable efficient utilization of $NADP^+$. *FEBS J.* **2008**, *275*, 3859–3869.
132. Andrews, F.H.; McLeish, M.J. Using site-saturation mutagenesis to explore mechanism and substrate specificity in thiamin diphosphate-dependent enzymes. *FEBS J.* **2013**, doi:10.1111/febs.12459.
133. Goldsmith, M.; Tawfik, D.S. Directed enzyme evolution: Beyond the low-hanging fruit. *Curr. Opin. Struct. Biol.* **2012**, *22*, 406–412.

11

Pyranose Dehydrogenase from *Agaricus campestris* and *Agaricus xanthoderma*: Characterization and Applications in Carbohydrate Conversions

Petra Staudigl †, Iris Krondorfer †, Dietmar Haltrich and Clemens K. Peterbauer *

Food Biotechnology Laboratory, BOKU–University of Natural Resources and Life Sciences Vienna, Muthgasse 11, Vienna 1190, Austria; E-Mails: petra.wuehrer@boku.ac.at (P.S.); iris.krondorfer@boku.ac.at (I.K.); dietmar.haltrich@boku.ac.at (D.H.)

† These authors contributed equally to this work.

* Author to whom correspondence should be addressed; E-Mail: clemens.peterbauer@boku.ac.at

---

Abstract: Pyranose dehydrogenase (PDH) is a flavin-dependent sugar oxidoreductase that is limited to a rather small group of litter-degrading basidiomycetes. The enzyme is unable to utilize oxygen as an electron acceptor, using substituted benzoquinones and (organo) metal ions instead. PDH displays a broad substrate specificity and intriguing variations in regioselectivity, depending on substrate, enzyme source and reaction conditions. In contrast to the related enzyme pyranose 2-oxidase (POx), PDHs from several sources are capable of oxidizing α- or β-1→4-linked di- and oligosaccharides, including lactose. PDH from *A. xanthoderma* is able to perform C-1 and C-2 oxidation, producing, in addition to lactobionic acid, 2-dehydrolactose, an intermediate for the production of lactulose, whereas PDH from *A. campestris* oxidizes lactose nearly exclusively at the C-1 position. In this work, we present the isolation of PDH-encoding genes from *A. campestris* (Ac) and *A. xanthoderma* (Ax) and a comparison of other so far isolated PDH-sequences. Secretory overexpression of both enzymes in *Pichia pastoris* was successful when using their native signal sequences with yields of 371 $U \cdot L^{-1}$ for AxPDH and 35 $U \cdot L^{-1}$ for AcPDH. The pure enzymes were characterized biochemically and tested for applications in carbohydrate conversion reactions of industrial relevance.

Keywords: pyranose dehydrogenase; heterologous expression; agaricus; lactose conversion; lactobionic acid

1. Introduction

Pyranose dehydrogenase (PDH, EC 1.1.99.29) is a monomeric extracellular glycoprotein of around 75 kDa, carrying a covalently bound FAD cofactor (8α-N3-histidyl-FAD) [1]. It is a member of the glucose-methanol-choline (GMC) family together with other sugar oxidoreductases like the catalytically related enzymes glucose oxidase, cellobiose dehydrogenase and pyranose-2 oxidase. PDH was first described in 1997 when it was isolated from the edible basidiomycete fungus *Agaricus bisporus* [2]. Later, the enzyme from other members of the family of *Agaricaceae* like *Macrelepiota rhacodes* [3], *A. xanthoderma* [4] and *A. meleagris* [5,6] was investigated. Recently, the crystal structure of *A. meleagris* PDH (AmPDH) was resolved and revealed a two-domain structure consisting of the ADP-binding Rossman domain and a sugar-binding domain [1].

The biological function of PDH is still not fully clear. As the enzyme is limited to litter-decomposing fungi of the family *Agaricaceae* and is not able to utilize molecular oxygen as electron acceptor, the reduction of quinones and radicals formed during lignin degradation were proposed as its natural role [3]. Other possible functions like a participation in Fenton's reaction or the defense against antimicrobial (quinone) substances produced by plants were reported [7].

As the production of the enzyme in basidiomycete fungi is quite laborious and time-consuming [2,4,5], approaches for heterologous expression in *Aspergillus nidulans* and *A. niger* [8], *E. coli* and *P. pastoris* [9] were tested. Attempts to solubly express PDH in *E. coli* did not succeed due to the formation of inactive inclusion bodies whereas an expression in *P. pastoris* yielded high levels of recombinant protein with properties equal to the wild-type [9]. Therefore, the methylotrophic yeast was the expression host of choice for the heterologous production of *A. campestris* and *A. xanthoderma* PDH in this study.

The oxidation products of PDH depend on the source of the enzyme, the substrate and the reaction conditions. The enzyme is able to oxidize free, non-phosphorylated sugars in pyranose form, heteroglycosides, disaccharides and glucooligosaccharides at the C-1, C-2, C-3 and also at C-1,2, C-2,3 and C-3,4 atom [2,3,10–14]. The oxidation products of D-glucose and D-galactose, 2-keto-D-glucose and 2-keto-D-galactose, represent industrially relevant intermediates for the production of the high-value sugars D-fructose and D-tagatose [15,16]. An easily available disaccharide lactose, can be oxidized by PDH to the corresponding C-1, C-2 or C-2,3 product. Depending on the source of the enzyme, 2-keto-lactose and lactobionic acid, the hydrolysis product of lactobionolactone, are formed in different ratios. Volc and coworkers screened various *Agaricus sp.* for their PDH oxidation products and observed that *A. campestris* PDH almost exclusively oxidizes lactose at the C-1 position, yielding lactobionic acid, whereas 2-keto-lactose was the main product of *A. xanthoderma* PDH [12]. Lactobionic acid has numerous applications in the pharmaceutical-, cosmetic- and food-industry, such as in organ preservation solutions and macrolide antibiotics, skin care cosmetics or as an acidulant or flavor enhancer in food [17]. The C-2 oxidation product 2-keto-lactose can be used for the production of lactulose, a prebiotic carbohydrate administered against obstipation and hepatic encephalopathy which has beneficial effects on the gastrointestinal microbiota [18]. Here, we describe for the first time the heterologous expression of PDH genes from *Agaricus campestris* and *Agaricus xanthoderma* in the methylotrophic yeast *Pichia pastoris* and present a detailed characterization of both enzymes. Furthermore, we performed comprehensive studies on the conversion of lactose and present a novel alternative for the production of lactobionic acid and 2-dehydrolactose, a key intermediate for the isomerization to lactulose.

2. Results and Discussion

2.1. Expression of Acpdh and Axpdh in P. pastoris

To obtain the PDH-encoding gene from *A. campestris,* different oligonucleotide primers were designed based on conserved regions from already known sequences (accession numbers are given in paragraph 3.8.). PCRs were performed using different forward primers, the anchor primer and first-strand-cDNA as template. Resulting fragments were sequenced and used to design sequence specific primers for identification of the 5'-flanking region by primer walking using the DNA Walking SpeedUp Premix Kit. The nucleotide sequence of the AcPDH cDNA contains an ORF of 1,788 bp encoding a polypeptide of 595 amino acids. Two primers based on the cDNA sequence and containing restriction sites for ligation into the pPICZb vector were designed, and used to re-amplify the cDNA and construct the expression vector under control of the methanol-inducible AOX promoter.

The previously unknown signal sequence and the 5'-flanking region of AxPDH were analogously identified using the DNA Walking SpeedUp Premix Kit and three specific reverse primers (AxTSP1-3). The purified fragments were sequenced, and based on these results, the full-length cDNA could be amplified. The AxPDH encoding cDNA contains an ORF of 1,803 bp encoding a polypeptide of 600 amino acids. The cDNA fragment was re-amplified with two primers containing restrictions sites for ligation into the pPICZb vector. The plasmids were transformed into *E. coli* NEB5α for proliferation. Isolated plasmids were linearized with *SacI* and transformed into the expression host *P. pastoris* and cultivated in 96-well deep well plates and screened for PDH activity. To confirm the results from the first round of screening, a rescreening experiment with multiple parallel determinations was performed. The clones with the highest activity were selected for further studies.

2.2. Multiple Sequence Alignment

To compare the amino acid sequences of the PDHs isolated and characterized so far in our group (*A. meleagris* PDH1 [5,6,9], *A. bisporus* (our unpublished information) *A. campestris* (this work) and *A. xanthoderma* PDH [4]), a multiple sequence alignment was constructed using the MUSCLE algorithm (Figure S1). The PDHs from different sources show a sequence identity between 74% and 78%. From the crystal structure of AmPDH1 [1], His 512 and His 556 were identified as the catalytic pair of major importance for sugar substrate oxidation. These two amino acids and His 103, where the FAD cofactor is covalently bound, are highly conserved among PDHs (Figure S1, highlighted in red). Docking experiments with several electron donors in different oxidation poses revealed that the principal sugar interaction partners are the two catalytic histidines but also Gln 392 and Tyr 510 (Figure S1, highlighted in green). The fact that these amino acids are conserved in all four PDHs supports these findings.

All PDHs are glycoproteins with different degrees of glycosylation. Putative N- and O-glycosylation sites predicted by NetNGlyc 1.0 server and NetOGlyc 3.1 server [19] are highlighted in red and green, respectively.

2.3. Heterologous Protein Production

The cultivation of *P. pastoris* cells expressing the *acpdh*- and *axpdh*-encoding gene was carried out in a 7-L stirred and aerated bioreactor and lasted 187 and 161 h, respectively (Figure 1). The initial glycerol batch phase produced 113 g·L^{-1} and 75 g·L^{-1} wet biomass in 27 h and 25 h (its end was indicated by an increase in dissolved oxygen concentration). The following feed with 50% glycerol was maintained for 17 h and 19 h and resulted in a final biomass of 156 g·L^{-1} and 155 g·L^{-1}, respectively. A methanol feed was initiated for induction, and at the end of this phase, the biomass reached a level of 260 g·L^{-1} and 293 g·L^{-1}. The activity of the extracellular enzyme fraction finally reached 35 U·L^{-1} for AcPDH and 371 U·L^{-1} for AxPDH, while the level of extracellular protein increased to 345 mg·L^{-1} and 350 mg·L^{-1}.

Figure 1. Large scale production of pyranose dehydrogenases (PDHs) in *P. pastoris*. Black circles, wet biomass; grey triangles, extracellular protein concentration; black squares, volumetric activity.

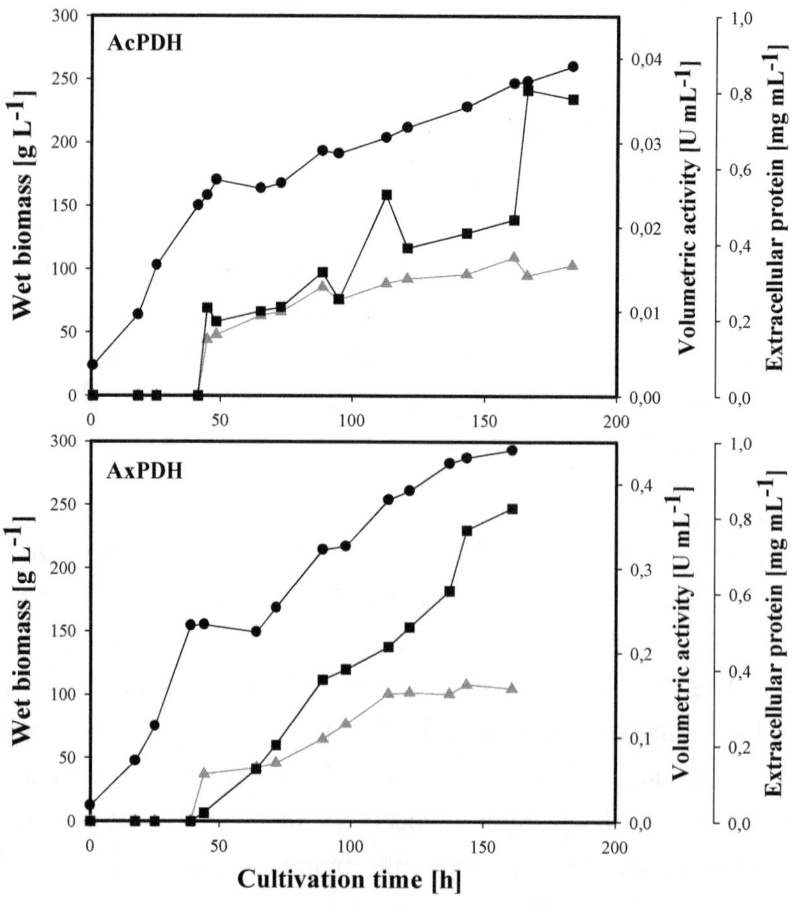

2.4. Purification of Recombinant PDHs

Recombinant AcPDH was purified from the cultivation broth in a four-step protocol (Table 1) including an additional hydrophobic interaction chromatography step (phenyl-source) compared to the three-step purification of AxPDH, which consisted of hydrophobic interaction chromatography, anion exchange chromatography and gel filtration as a polishing step. Before the AcPDH pool was loaded on the phenyl-source column, solid ammonium sulfate was added to a saturation of 40%, similar to the first purification step. Unexpectedly, the protein did not bind to the column in this case, and the whole PDH activity was found in the flow-through. This purification step indeed increased the specific activity from 0.4 U mg^{-1} to 2.8 U mg^{-1} but the flow-through still contained impurities due to the lack of restrictive pooling. Therefore, a subsequent gel filtration step was conducted, resulting in apparent homogeneity for both AcPDH and AxPDH (Figure 2), with final specific activities of 4.9 U mg^{-1} and 16.6 U mg^{-1}, respectively. For AcPDH, two pools with different specific activities were formed due to rather low overall yields. Pool 1 represented the center of the elution peak and all further analyses were performed with this enzyme preparation. The low purification yield for AcPDH corresponds to the fact that PDH accounted only for 0.1% of the total protein in the extracellular fraction, whereas AxPDH accounted for more than 2%. The purification step that decreased the yield most dramatically is the first for both enzymes. As the starting volume for purification after centrifugation was around 4 L in both cases, the calculation of the total activity based on the activity per mL could lead to imprecise results. This is especially true in the case of AcPDH, where the volumetric activity per mL was below the detection limit of the standard ferrocenium/glucose activity assay and only represents an estimate. A reduction of the volume and therefore concentration of the cultivation broth would have been useful for more precise measurements of the initial volumetric activity. The higher degree of glycosylation of AcPDH could also play a disadvantageous role for the purification (Figure 2, Table 2). All concentrated protein pools showed the typical light yellow color of flavoproteins and were stable over several months at 4 °C in 65 mM sodium phosphate buffer pH 7.5.

Table 1. Purification schemes of recombinant PDHs.

Purification step	Total protein [mg]	Total activity [U]	Specific activity [U mg^{-1}]	Purification [-fold]	Yield [%]
AcPDH					
Crude extract	1730	154.4	0.1	1	100
Phenyl sepharose	124.4	27.4	0.2	2.5	18
DEAE sepharose	28.2	21.5	0.4	4.1	14
Phenyl source	5.5	15.2	2.8	31.0	10
Gel filtration pool 1	1.2	5.6	4.9	54.5	4
Gel filtration pool 2	0.4	1.5	3.7	41.9	1
AxPDH					
Crude extract	1226.4	1298.9	1.1	1	100
Phenyl sepharose	109.9	538.6	4.9	4.6	41
DEAE sepharose	40.2	524.9	13.1	12.3	40
Gel filtration	25.8	428.8	16.6	15.7	33

Table 2. Molecular properties of recombinant PDHs.

PDH	Mass SDS-PAGE [kDa]	Mass SDS-PAGE deglyc. [kDa]	Theor. mass [kDa]	Glycan mass [%]	*N*-Glyc sites predicted	*O*-Glyc sites predicted
Ac	98	68	61.8	31	6	3
Ax	73	68	62.3	7	5	0

Figure 2. SDS-PAGE of purified PDHs. M, molecular marker; 1, AcPDH; 2, AcPDH deglycosylated; 3, Ax PDH; 4, AxPDH deglycosylated.

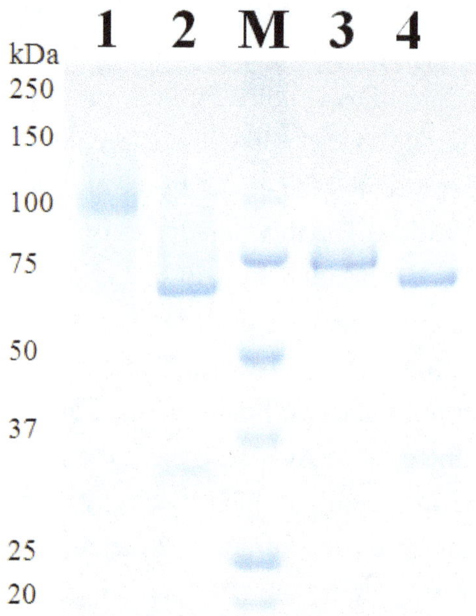

2.5. Molecular Properties

The molecular masses of AcPDH and AxPDH were determined by SDS-PAGE (Figure 2) and native PAGE (Figure 3). AcPDH formed a diffuse band around 98 kDa (Figure 2, lane 1), after deglycosylation with PNGase F under denaturing conditions, a sharp band at 68 kDa could be observed (lane 2). AxPDH showed a band at 73 kDa (lane 4); after deglycosylation, the band shifted to around 68 kDa (lane 5). The high degree of glycosylation of AcPDH compared to AxPDH (Table 2) is also observed in native PAGE (Figure 3). From the migration difference of the glycosylated and deglycosylated protein band in the gel, a glycan mass of 31% for AcPDH and 7% for AxPDH could be calculated (Table 2). The native AxPDH [4] showed a slightly smaller mass on the SDS-PAGE compared to the recombinant protein (around 65 kDa). This is most likely due to a difference in glycosylation in basidiomycete fungi compared to the yeast *P. pastoris*. Whereas proteins in (homo) basidiomycete fungi carry *N*-glycans of the oligomannosidic type (4-9 mannoses), *P. pastoris* tends to produce hyper-mannosylated glycans [20–22]. The NetNGlyc 1.0 server found nine glycosylation motifs in the sequence of AxPDH, five of them putatively glycosylated, and nine for AcPDH with six of them likely to carry a glycan structure. Concerning *O*-glycosylation, NetOGlyc 3.1 server predicted three sites for AcPDH with potential above the threshold and none for AxPDH. The potentially higher degree of *O*-glycosylation of AcPDH compared to AxPDH cannot account for the large difference in glycan mass between the two proteins. The deglycosylation was carried out using PNGase F, which exclusively removes *N*-glycans, and the deglycosylated proteins have a quite similar molecular mass. Furthermore, *O*-glycans in *P. pastoris* mostly consist of up to three, rarely four, mannose units in contrast to hyper-mannosylated *N*-glycans [23]. A substantial over-glycosylation with a comparable glycan content of approximately 30% of protein expressed in *P. pastoris* was also observed for AmPDH [9]. PDH is one of the rare flavoproteins carrying a covalently bound cofactor [24]. Covalent incorporation was proven by the method of Scrutton [25], the flavin associated with the protein gives a fluorescent signal when exposed to UV-light. The positive control (AmPDH), AcPDH and AxPDH showed a bright band under UV-light whereas the negative control glucose oxidase (GOx) from *A. niger* [26] did not give any signal (Figure S2).

Figure 3. Native PAGE of purified PDHs. M, molecular marker; 1, AcPDH; 2, AxPDH.

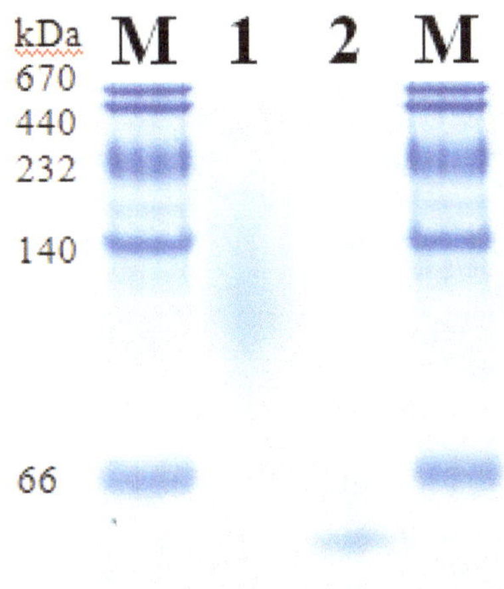

UV-Vis spectra of AcPDH and AxPDH in the oxidized state were recorded; typical flavoprotein absorbance maxima around 450 nm and 340 nm could be observed (Figure S3; AxPDH: data not shown).

2.6. Kinetic Properties

Catalytic constants for selected sugar substrates and electron acceptors were determined and are summarized in Tables 3 and 4. For both enzymes, D-glucose represents the preferred electron donor. This is mainly due to the K_m value, which displays a more than 10 times higher affinity for this substrate compared to e.g., D-galactose. The pentose sugar D-xylose is the second best substrate for Ac and AxPDH. This finding stands in contrast to the catalytic efficiencies of other so-far characterized PDHs like from *Agaricus meleagris* [5] and the native *A. xanthoderma* PDH [4], where L-arabinose and D-galactose are preferred over D-xylose. The catalytic efficiency of AxPDH with D-xylose is more than 75% of the catalytic efficiency for the main substrate D-glucose, whereas for AcPDH it is only 13%. The k_{cat}/K_m-values of AcPDH are in general lower for all sugar substrates. Remarkable here is that the catalytic efficiencies of Ac and AxPDH are similar for lactose but the K_m of AcPDH is 5.5-times lower than the K_m of AxPDH, whereas the k_{cat} behaves the opposite way.

Table 3. Apparent kinetic constants for selected electron donors determined at 30 °C with 0.2 mM ferrocenium hexafluorophosphate as the electron acceptor.

	AcPDH			AxPDH		
	K_m [mM]	k_{cat} [s^{-1}]	k_{cat}/K_m [mM^{-1} s^{-1}]	K_m [mM]	k_{cat} [s^{-1}]	k_{cat}/K_m [mM^{-1} s^{-1}]
D-glucose	0.35 ± 0.06	4.10 ± 0.19	11.7	0.49 ± 0.03	13.02 ± 0.38	26.6
D-galactose	7.13 ± 0.19	5.23 ± 0.53	0.7	4.99 ± 0.16	24.77 ± 2.23	5.0
D-xylose	4.19 ± 0.26	6.37 ± 0.05	1.5	1.44 ± 0.07	29.16 ± 0.25	20.3
L-arabinose	4.23 ± 0.02	3.05 ± 0.04	0.7	4.16 ± 0.58	22.90 ± 2.23	5.5
Lactose	53.16 ± 0.10	3.12 ± 0.17	0.1	293.84 ± 8.00	24.65 ± 0.37	0.1

Table 4. Apparent kinetic constants for selected electron acceptors determined at 30 °C with 25 mM D-glucose as the electron donor.

	AcPDH			AxPDH		
	K_m [mM]	k_{cat} [s^{-1}]	k_{cat}/K_m [mM^{-1} s^{-1}]	K_m [mM]	k_{cat} [s^{-1}]	k_{cat}/K_m [mM^{-1} s^{-1}]
Fc^+PF_6 (pH 8.5)	1.19 ± 0.17	19.92 ± 2.87	16.7	0.03 ± 0.00	22.07 ± 0.05	735.7
1,4-BQ (pH 4)	0.12 ± 0.01	34.82 ± 1.02	302.8	3.25 ± 0.51	12.89 ± 1.56	4.0
DCIP (pH 4)	0.11 ± 0.00	10.56 ± 0.57	96.0	0.09 ± 0.01	7.65 ± 0.66	85.0

Concerning the kinetic constants for the electron acceptors, the two enzymes have quite different preferences (Table 4). For AcPDH 1,4-benzoquinone is the favored substrate whereas AxPDH shows a clear preference for ferrocenium hexafluorophosphate. All PDHs that have been characterized to date show a higher catalytic activity with ferrocenium compared to 1,4-benzoquinone; AcPDH is the first PDH where a clear preference for 1,4-benzoquinone was observed [4,5]. The catalytic efficiencies for DCIP are nearly equal for both PDHs. In general, PDH shows activity only with a limited group of electron acceptors whereas it oxidizes a very broad range of sugar substrates. Many major mono- and oligosaccharide components of lignocellulose can be utilized as substrates, giving evidence for the putative biological function of PDH in lignin degradation [7]. Recently, the molecular mechanism of glucose oxidation by *A. meleagris* PDH was explored using MD simulation; the findings support the experimentally observed promiscuity of PDH concerning sugars [27].

The pH dependence of PDH activity was tested for the electron acceptors ferrocenium hexafluorophosphate and 1,4-benzoquinone with D-glucose as the electron donor (Figure 4). Using 1,4-benzoquinone as the electron acceptor, AcPDH displayed maximum activity at pH 7 (phosphate buffer) and already reached 95% activity at pH 5.5 (citrate buffer). AxPDH showed the highest activity at pH 5.5 (citrate buffer). With ferrocenium hexafluorophosphate, the optimum pH for AcPDH was 8.5 and 9 for AxPDH (borate buffer). Compared to the native AxPDH [4], the pH optimum for ferrocenium hexafluorophosphate is comparable but with 1,4-benzoquinone the native AxPDH exhibited highest activity at pH 2.5 and around 70% activity at pH 8. The recombinant AxPDH showed less than 2.5% activity at pH 2.5. In general, it can be observed that the pH optimum is highly dependent on the electron acceptor used. PDHs from *Agaricus sp.* showed pH optima in the basic region when using ferrocenium hexafluorophosphate whereas with 1,4-benzoquinone the enzymes were more active under acidic conditions and in some cases showed a second maximum in the alkaline region, which could be due to high blank readings caused by the formation of quinhydrone [2,4,5].

Figure 4. pH optima of AcPDH (black squares) and AxPDH (grey triangles) with the electron acceptors ferrocenium hexafluorophosphate and 1,4-benzoquinone; D-glucose as electron donor.

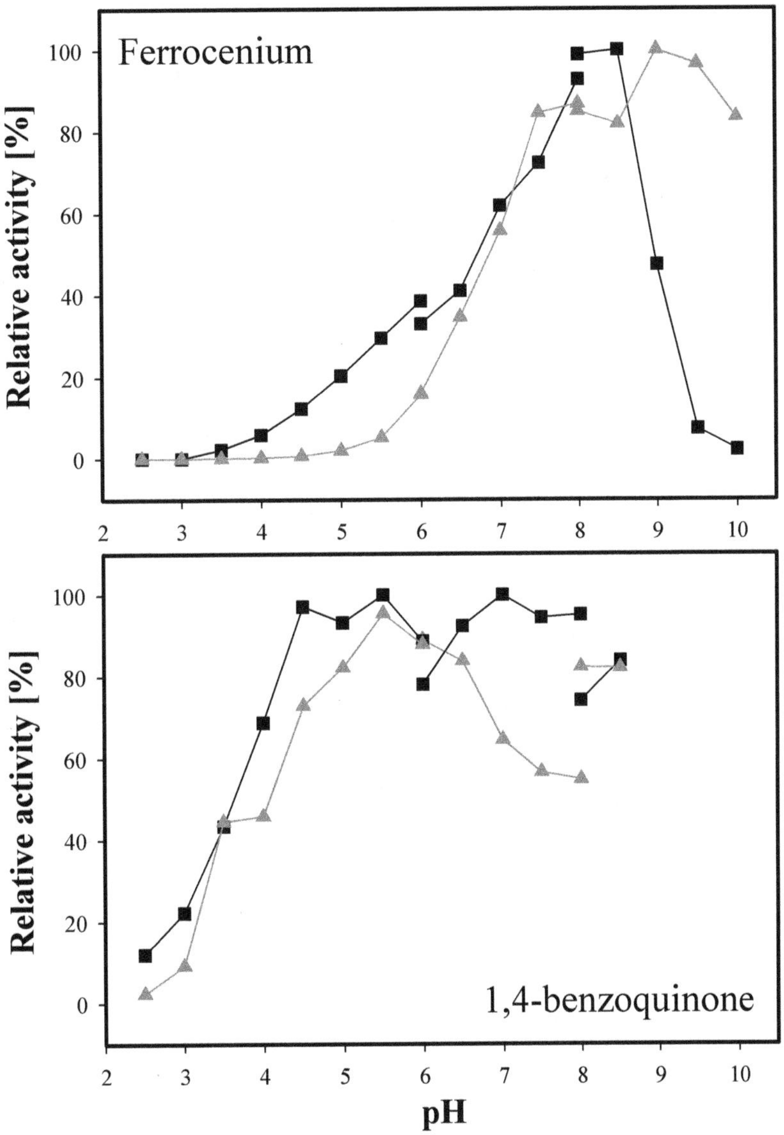

2.7. Batch Carbohydrate Conversion Experiments

Both enzymes were used for the conversion of 25 mM lactose together with equimolar amounts of the electron acceptor 1,4-benzoquinone in 1 mL batch experiments and the reaction products were analyzed by HPLC (Figure 5). Conversions were carried out in water due to interference of buffer salts with the HPLC analysis. 10 U of AxPDH converted lactose to nearly 100% (97%), producing 69% (67%) lactobionic acid and 31% (30%) 2-dehydrolactose (Figure 5a). Due to the fact that the cultivation yield of AcPDH was quite low, only one unit of the purified protein was used for the conversion experiment. Therefore, only 48% of the lactose was converted and yielded 88% (42%)

lactobionic acid and only 12% (6%) 2-dehydrolactose (Figure 5b). These results confirm the preference for C-1 oxidation in AcPDH-catalyzed conversions of lactose (the ratio of C-1 to C-2 oxidation is 7:1), compared to the AxPDH-catalyzed reaction where the ratio of C-1 to C-2 oxidation is only 2:1. C-2,3 oxidation, which can occur when oxidation at C-2 is complete under excess of 1,4-benzoquinone, could not be observed [12]. For industrial use, there is still need for improvement of the conversion yield. Acidification due to spontaneous hydrolysis of lactobionolactone to lactobionic acid slowed down the enzyme activity at the end of the conversion but could be avoided by buffering.

Figure 5. (**a**) HPLC analysis of lactose conversion by AcPDH; (**b**) and AxPDH at 0 h (A), 1 h (B), 3 h (C) and 7 h (D) incubation. Peaks: I, residual salt from enzyme preparation; II, lactobionic acid; III, lactose; IV, 2-dehydrolactose.

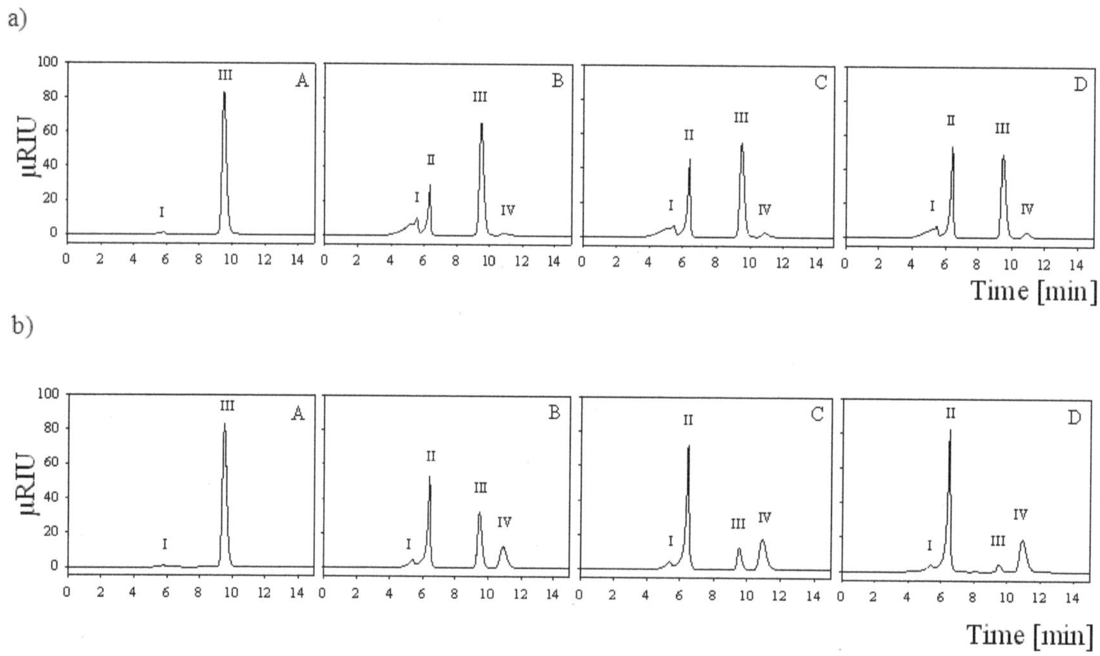

3. Experimental Section

3.1. Chemicals and Microorganisms

All chemicals, sugar standards for HPLC and media components were purchased from Sigma (Steinheim, Germany) unless otherwise stated and were of the highest purity available. Restriction endonucleases, T4 DNA ligase and Phusion High-Fidelity DNA Polymerase were obtained from Fermentas (St. Leon-Rot, Germany) unless otherwise stated and were used according to the manufacturer's instructions. *Agaricus campestris* (strain CCBAS 20649) and *Agaricus xanthoderma* (strain CCBAS 225) were obtained from the Culture Collection of Basidiomycetes of the Academy of Sciences (Prague, Czech Republic). *Escherichia coli* strain NEB5α (New England Biolabs, Ipswich, MA, USA) was used for subcloning, *Pichia pastoris* strain X-33 (Invitrogen, Carlsbad, CA, USA) was used for expression. Ferrocenium hexafluorophosphate and the various substituted quinones were obtained from Aldrich (Steinheim, Germany). Phenyl-Sepharose Fast Flow resin was purchased from Amersham Pharmacia Biotech (Uppsala, Sweden), DEAE Sepharose Fast Flow resin and Sephacryl

S300-HR resin were from GE Healthcare (Chalfont St. Giles, UK). GOx from *Aspergillus niger* was from Sigma.

3.2. Isolation of Genomic DNA and RNA

For DNA- and RNA-isolation, approximately 10 mL liquid Sabouraud medium was inoculated with mycelial fragments from freshly grown malt extract agar plates. The cultivations were performed in petri dishes at 25 °C without shaking over 3 weeks. Mycelia were harvested, squeezed dry between filter paper and shock-frozen in liquid nitrogen. Portions of approximately 100 mg of mycelium were used for DNA- and RNA-isolation. Genomic DNA extraction was performed according to Liu *et al.* [28]. Total RNA was extracted using Trizol reagent (Invitrogen) according to the manufacturer's instructions. To remove genomic DNA, the samples were incubated with *DNAseI* as recommended by the manufacturer. The isolated mRNA was reverse-transcribed using RevertAid First Strand cDNA Synthesis kit (Thermo Fisher Scientific, Waltham, MA, USA) and the anchor primer (Table S1).

3.3. Cloning and Sequencing of AcPDH Encoding Gene

Nucleic acid amplifications were done using Phusion High-Fidelity Polymerase, GC-Buffer, dNTP mix, oligonucleotide primers (VBC Biotech, Vienna, Austria) and a BioRad C-1000 thermocycler (BioRad, Vienna, Austria). Nucleotide sequences of all primers used in this work are shown in Table S1. To obtain the PDH-encoding gene from *A. campestris* degenerate primers (AcPDHfwd1-3) were designed according to a sequence analysis of conserved regions in *pdh* genes from *A. meleagris* and *A. bisporus* and used for the amplification of cDNA fragments of various lengths. 35 PCR cycles at temperatures of 98 °C (10 s), 57 °C (20 s) and 72 °C (1 min) and an initial denaturation at 98 °C (2 min) and a final extension at 72 °C (7 min), were employed with cDNA as template and "universal" as reverse primer. The amplicons were purified and sequenced by a commercial sequencing service (LGC Genomics, Berlin, Germany). For identification of the 5‘ flanking region, including the native signal sequence, the DNA Walking SpeedUp Premix Kit (Seegene, Seoul, South Korea) was used. Target-specific reverse primers (AcTSP1-3) were designed, genomic DNA was used as template and the PCRs were done according to the manufacturer's guidelines. The resulting PCR products were purified by using the illustra GFX PCR DNA and Gel Band Purification Kit (GE Healthcare) and sequenced. To obtain full-length cDNA clones a PCR was performed with the primer pair AcPDHfwd and AcPDHrev and cDNA as template. The resulting fragment was cloned into the pJET1.2 vector (Fermentas) according to the instructions in the manual and sequenced.

3.4. Cloning and Sequencing of AxPDH Encoding Gene

To obtain the 5‘- flanking region, including the native signal sequence, the DNA Walking SpeedUp Premix Kit (Seegene) was used. Target-specific reverse primers (AxTSP1-3) were designed according to *A. xanthoderma* PDH gene sequence, genomic DNA was used as template and the PCRs were done according to the manufacturer's guidelines. The resulting PCR products were purified and sequenced. To obtain full-length cDNA clones a PCR was performed with the primer pair AxPDHfwd and

AxPDHrev and cDNA as template. The amplified sequence was temporarily cloned into the pJET1.2 vector (Fermentas) according to the instructions in the manual and sequenced.

3.5. Construction of Expression Vectors for P. pastoris

The *acpdh* gene (in the pJET1.2 vector) was re-amplified with the primers AcPDHKpnIfwd and AcPDHNotIrev. The re-amplification of the *axpdh* gene was performed using the primer pair AxKpnIfwd and AxNotIrev. The resulting PCR products were digested with *KpnI* and *NotI* and ligated into the equally treated vector pPICZb (Invitrogen). After transformation into chemically competent *E. coli* NEB5α, according to the manufacturer's instructions, the plasmids were proliferated, linearized with *SacI* and transformed into electrocompetent *P. pastoris*, which were prepared according to Lin-Cereghino *et al.* [29]. For selection the Luria Bertani (LB) medium contained 25 μg mL^{-1} Zeocin in case of *E. coli*, while the YPD-medium contained 100 μg mL^{-1}. The resulting colonies were picked and grown in 96-well deep well plates.

3.6. Microscale Screening for High-Producing PDH Transformants

Microscale cultivation and expression in 96-well deep well plates was done according to Weis *et al.* [30] with some modifications. Cells were grown in 250 μL BMD1 (13.4 $g \cdot L^{-1}$ yeast nitrogen base, 0.4 $mg \cdot L^{-1}$ biotin, 10 $g \cdot L^{-1}$ D-glucose, 200 mM potassium phosphate [pH 6.0]) at 25 °C, 385 rpm and 60% humidity for approximately 60 h to reach the stationary growth phase. Induction was started by the addition of 250 μL of BMM2 medium (13.4 $g \cdot L^{-1}$ yeast nitrogen base, 0.4 $mg \cdot L^{-1}$ biotin, 1% methanol, 200 mM potassium phosphate [pH 6.0]) to reach a final concentration of 0.5% methanol. After 70, 82, and 108 h of incubation, 50 μL BMM10 (BMM2 with 5% methanol) were added to maintain inducing conditions. The cultivation was stopped after 130 h by centrifugation of the deep-well plates at 3000 rpm at room temperature for 20 min. PDH activity was measured using 2,6-dichloroindophenol (DCIP, ε520 = 6.8 mM^{-1} cm^{-1}) as electron acceptor and D-glucose as donor. Fifty μL of the supernatant were transferred to the 96-well screening plates and the time-dependent reduction of 300 μM DCIP in 100 mM sodium acetate buffer pH 4 containing 50 mM D-glucose was followed at 520 nm with a PerkinElmer EnSpire plate reader. The reaction was started by addition of 150 μL of the DCIP assay mixture to the screening plate and end-point measurements were carried out after incubation at 30 °C for 2 and 4 h.

3.7. Sequence Analysis

The translated amino acid sequences of the obtained cDNAs were analyzed using the programs Translate, Compute pI/MW and SignalP at http://www.expasy.org/ [31]. A multiple sequence alignment of AmPDH1, AbPDH, AcPDH and AxPDH was created using the MUSCLE algorithm (EMBL-EBI, Cambridgeshire, UK). Sequence identities were determined by BLAST search [32]. Predictions for *N*- and *O*-glycosylation sites were performed on the NetNGlyc 1.0 Server and the NetOGlyc 3.1 Server [19] of the Center for Biological Sequence Analysis (CBS) at the Technical University of Denmark (http://www.cbs.dtu.dk/services/). All predicted *N*- and *O*-glycosylation sites with a threshold above 0.5 except for Asn-Pro-sites were displayed.

3.8. Nucleotide and Protein Sequence Accession Numbers

The NCBI accession numbers for the sequences in this work are: AY53306, AY753308, DQ117577, AAW82996, AAW82997, AAW82998, AAW82999, AAZ94874, AAZ94875 (*A. meleagris* PDHs); AY764148, AAW92124, EKV41672 (*A. bisporus* PDH); AY764147, AAW92123, KF534751 (*A. xanthoderma* PDH); KF534750 (*A. campestris* PDH).

3.9. Recombinant Protein Production in P. pastoris

AxPDH and AcPDH were produced in a 7-L bioreactor (MBR, Wetzikon, Switzerland) with an initial volume of 4 L basal salts cultivation medium, according to the "*Pichia* Fermentation Process Guidelines" (Invitrogen) with slight modifications. After autoclaving the bioreactor, the temperature was set to 30 °C, the pH was adjusted to 5 and maintained by addition of 28% ammonium hydroxide solution during the cultivation. Dissolved oxygen was regulated to 4% by supplying filtered air and adjusting the stirrer velocity (around 800 rpm). Two shaking flasks with 20 mL YPD-Zeocin medium each were inoculated with the colony-PCR-verified *P. pastoris* clones and grown overnight at 30 °C and 120 rpm. The cultures were transferred to two shaking flasks with 200 mL YPD medium each and again grown overnight at 30 °C and 120 rpm. This culture (400 mL) was used to inoculate the bioreactor. After consumption of the glycerol in the batch medium (indicated by an O_2 spike), a feed of 50% glycerol containing 12 mL L^{-1} PTM1 trace salts was initiated with around 20 mL h^{-1} over night. Protein production was induced by changing to a feed of 100% methanol containing 12 mL PTM1 trace elements per liter. The feed rate was adjusted to maintain a dissolved oxygen concentration of around 4%. Samples were taken at least twice a day and biomass wet weight, PDH activity and total protein concentration was determined. The formation of foam was avoided by daily manual addition of approximately 10 mL a 10% antifoam 204 solution (Sigma). When no further increase in the specific activity could be observed, the bioreactor was harvested and the cultivation broth was centrifuged at 4 °C and 6000 rpm in a Sorvall Evolution RC centrifuge (Thermo Fisher Scientific).

3.10. Protein Purification

Solid ammonium sulfate was added to the cultivation supernatants to a saturation of 40%. The crude extracts (approximately 4 L) were applied to a 750 mL phenyl-sepharose FF column (GE Healthcare), and washed with binding buffer (50 mM potassium phosphate, pH 6.5 containing 1.5 M ammonium sulfate). The protein was eluted using a linear gradient from 0–100% elution buffer (50 mM potassium phosphate, pH 6.5) in 1 column volume (CV). Prior to the next purification step, the pooled fractions were desalted using cross-flow filtration (SpectrumLabs, Houston, TX, USA) to a conductivity equal to or less than 2 mS/cm. The pools of desalted fractions were loaded to a 60 mL DEAE sepharose column (GE Healthcare), washed with binding buffer (50 mM BisTris, pH 6) and eluted with a linear gradient from 0–100% elution buffer (50 mM BisTris, 1 M NaCl, pH 6) in 4 column volumes. For AcPDH, the concentrated pool from the anion exchange chromatography was subjected to a second hydrophobic interaction chromatography step using a 70 mL phenyl-source column (GE Healthcare). The purification was conducted similar to the first hydrophobic interaction chromatography step except for the elution of the protein, which was carried out in 5 CV.

The fractions with the highest Ax- and AcPDH activities were pooled and concentrated to a volume of around 2 mL using an Amicon Ultra Centrifugal Filter Unit (EMD Millipore, Billerica, MA, USA). The concentrated pools were applied to a 190 mL Sephacryl S300 gel filtration column (GE healthcare) equilibrated with 50 mM potassium phosphate buffer (pH 7.5) containing 150 mM NaCl. Fractions with the highest purity were pooled, concentrated for buffer exchange (65 mM sodium phosphate buffer, pH 7.5) and stored at 4 °C.

3.11. Enzyme Assay, Molecular Properties

Standard PDH activity was measured by following spectrophotometrically the D-glucose dependent reduction of the ferrocenium ion (Fc+) to ferrocene at 300 nm and 30 °C as described before [4] with modifications: The standard reaction mixture (1 mL) contained 50 μmol sodium phosphate buffer pH 7.5, 0.2 μmol of ferrocenium hexafluorophosphate and 25 μmol D-glucose. Protein concentration was determined using the method of Bradford using a BSA standard curve and a prefabricated assay solution (BioRad). Enzymatic deglycosylation and SDS-PAGE were carried out as described in Sygmund *et al.*, using the Precision Plus Protein Unstained Standard (BioRad) [9]. 1.5–2 μg of the protein samples were loaded in each lane. Native PAGE was performed using 10% and 5% polyacrylamide as the separation and stacking gels, respectively, and Tris-glycine buffer (pH 8.3) as the electrode buffer [33]. 5–10 μg of the protein samples were loaded in each lane. Staining procedure was carried out using Bio-Safe Coomassie (BioRad) according to the manufacturer's instructions. For determination of molecular weight, HMW Native Marker Kit (GE Healthcare) was used. To proof the covalent linkage of the FAD cofactor an additional SDS-PAGE was performed according to Scrutton [25]. 10 μg of the protein samples were loaded in each lane, the covalently linked FAD was visualized by exposure of the gel to UV-light (λ 302 nm, GelDoc2000, BioRad). As a positive control, recombinant *A. meleagris* PDH was loaded [1]; glucose oxidase from *A. niger* was used as a negative control.

Molecular masses of the proteins were calculated from their migration distances on the SDS-PAGE; the theoretical mass was derived from ExPASy ProtParam tool (http://web.expasy.org/protparam/) [31]. The glycan mass in% was calculated from the difference of the masses of the glycosylated and deglycosylated proteins on the SDS-PAGE.

UV-Vis absorbance spectra of 13 μM AcPDH and AxPDH were recorded in 65 mM sodium phosphate buffer pH 7.5 at room temperature using a U-3000 spectrophotometer (Hitachi, Tokyo, Japan) from 300–700 nm.

3.12. Kinetic Properties

Apparent kinetic constants for electron donors were measured using the standard activity assay with ferrocenium hexafluorophosphate as described above. Kinetic constants for ferrocenium hexafluorophosphate, 1,4-benzoquinone and 2,6-dichloroindophenol were determined using 25 mM D-glucose as electron donor. The observed data were fitted to the Michaelis-Menten equation and kinetic constants were calculated by nonlinear least-squares regression. Using the molecular mass, turnover numbers (k_{cat}) and catalytic efficiencies (k_{cat}/K_m) were calculated. The pH optima with the electron acceptors ferrocenium hexafluorophosphate (0.2 mM) and 1,4-benzoquinone (2 mM) were determined with the following buffers: 100 mM citrate (pH 2.5–6), 100 mM potassium phosphate (pH 6–8) and 100 mM borate (pH 8–10) and 25 mM D-glucose as the electron donor. Activities with 1,4-benzoquinone were not determined above pH 8.5 due to the formation of quinhydrone under basic conditions.

3.13. Batch Conversion Experiments

Small-scale lactose conversions were carried out in 1.5 mL Eppendorf vials containing 25 mM 1,4-benzoquinone, 25 mM lactose monohydrate and 1 U of purified AcPDH or 10 U of purified AxPDH in 1 mL deionized water. The vials were incubated at 30 °C and 400 rpm in a thermomixer, samples were taken in regular time intervals (50 μL). Immediately after sampling, PDH activity was stopped by heating the sample to 99 °C for 3 min. The samples were centrifuged, diluted 1:2 and subjected to HPLC analysis.

3.14. HPLC Analysis of Batch Conversion Products

HPLC analysis of the batch conversion products was performed on a Dionex Summit HPLC system (Thermo Fisher Scientific) fitted with a Shodex RI-101 refractive index detector (Shoko Scientific, Yokohama, Japan) using an Aminex HPX 87-K column (BioRad) with a guard column. Samples and standards were eluted at 80 °C with deionized water (0.5 mL min^{-1}). For the calculation of lactose and lactobionic acid concentrations, standards were included in the run. As there was no standard available for 2-dehydrolactose, the ratio of 2-dehydrolactose to lactobionic acid was estimated by comparing the peak areas.

4. Conclusions

This study demonstrates the successful expression of the PDH-encoding genes from the litter-degrading basidiomycetes *A. campestris* and *A. xanthoderma*, in the eukaryotic host organism *P. pastoris*. Small-scale conversion experiments with lactose as substrate revealed that AcPDH has a strong preference for C-1 oxidation, resulting in the production of lactobionic acid. Compared to AxPDH, which produces mixtures of C-2/C-1 oxidation products in a 1:2 ratio, AcPDH is a very attractive biocatalyst for the production of lactobionic acid. Further research towards a better expression yield is required for industrial applications/purposes.

Acknowledgments

The authors thank Cindy Lorenz for technical assistance and performing the HPLC analysis. This work was supported by the Austrian Science Fund FWF (grant P22094 to CKP), IK is a member of the doctoral program BioToP (Biomolecular Technology of Proteins) of the Austrian Science Fund (W1224).

References

1. Tan, T.C.; Spadiut, O.; Wongnate, T.; Sucharitakul, J.; Krondorfer, I.; Sygmund, C.; Haltrich, D.; Chaiyen, P.; Peterbauer, C.K.; Divne, C. The 1.6 å crystal structure of pyranose dehydrogenase from *Agaricus meleagris* rationalizes substrate specificity and reveals a flavin intermediate. *PLoS One* **2013**, *8*, e53567.
2. Volc, J.; Kubátová, E.; Wood, D.A.; Daniel, G. Pyranose 2-dehydrogenase, a novel sugar oxidoreductase from the basidiomycete fungus *Agaricus bisporus*. *Arch. Microbiol.* **1997**, *167*, 119–125.
3. Volc, J.; Kubátová, E.; Daniel, G.; Sedmera, P.; Haltrich, D. Screening of basidiomycete fungi for the quinone-dependent sugar C-2/C-3 oxidoreductase, pyranose dehydrogenase, and properties of the enzyme from *Macrolepiota rhacodes*. *Arch. Microbiol.* **2001**, *176*, 178–186.
4. Kujawa, M.; Volc, J.; Halada, P.; Sedmera, P.; Divne, C.; Sygmund, C.; Leitner, C.; Peterbauer, C.; Haltrich, D. Properties of pyranose dehydrogenase purified from the litter-degrading fungus *Agaricus xanthoderma*. *FEBS J.* **2007**, *274*, 879–894.
5. Sygmund, C.; Kittl, R.; Volc, J.; Halada, P.; Kubátová, E.; Haltrich, D.; Peterbauer, C.K. Characterization of pyranose dehydrogenase from *Agaricus meleagris* and its application in the C-2 specific conversion of D-galactose. *J. Biotechnol.* **2008**, *133*, 334–342.
6. Kittl, R.; Sygmund, C.; Halada, P.; Volc, J.; Divne, C.; Haltrich, D.; Peterbauer, C.K. Molecular cloning of three pyranose dehydrogenase-encoding genes from *Agaricus meleagris* and analysis of their expression by real-time RT-PCR. *Curr. Genet.* **2008**, *53*, 117–127.
7. Peterbauer, C.K.; Volc, J. Pyranose dehydrogenases: Biochemical features and perspectives of technological applications. *Appl. Microbiol. Biotechnol.* **2010**, *85*, 837–848.
8. Pisanelli, I.; Kujawa, M.; Gschnitzer, D.; Spadiut, O.; Seiboth, B.; Peterbauer, C. Heterologous expression of an *Agaricus meleagris* pyranose dehydrogenase-encoding gene in *Aspergillus* spp. and characterization of the recombinant enzyme. *Appl. Microbiol. Biotechnol.* **2010**, *86*, 599–606.
9. Sygmund, C.; Gutmann, A.; Krondorfer, I.; Kujawa, M.; Glieder, A.; Pscheidt, B.; Haltrich, D.; Peterbauer, C.; Kittl, R. Simple and efficient expression of *Agaricus meleagris* pyranose dehydrogenase in *Pichia pastoris*. *Appl. Microbiol. Biotechnol.* **2012**, *94*, 695–704.
10. Volc, J.; Sedmera, P.; Halada, P.; Přikrylová, V.; Daniel, G. C-2 and C-3 oxidation of D-glc, and C-2 oxidation of D-gal by pyranose dehydrogenase from *Agaricus bisporus*. *Carbohydr. Res.* **1998**, *310*, 151–156.
11. Volc, J.; Sedmera, P.; Halada, P.; Prikrylová, V.; Haltrich, D. Double oxidation of D-xylose to D-glycero-pentos-2,3-diulose (2,3-diketo-D-xylose) by pyranose dehydrogenase from the mushroom *Agaricus bisporus*. *Carbohydr. Res.* **2000**, *329*, 219–225.
12. Volc, J.; Sedmera, P.; Kujawa, M.; Halada, P.; Kubátová, E.; Haltrich, D. Conversion of lactose to β-D-galactopyranosyl-(1→4)-D-arabino-hexos-2-ulose-(2-dehydrolactose) and lactobiono-1,5-lactone by fungal pyranose dehydrogenase. *J. Mol. Catal. B Enzym.* **2004**, *30*, 177–184.
13. Sedmera, P.; Halada, P.; Peterbauer, C.; Volc, J. A new enzyme catalysis: 3,4-dioxidation of some aryl β-D-glycopyranosides by fungal pyranose dehydrogenase. *Tetrahedron. Lett.* **2004**, *45*, 8677–8680.

14. Sedmera, P.; Halada, P.; Kubátová, E.; Haltrich, D.; Přikrylová, V.; Volc, J. New biotransformations of some reducing sugars to the corresponding (di)dehydro(glycosyl) aldoses or aldonic acids using fungal pyranose dehydrogenase. *J. Mol. Catal. B Enzym.* **2006**, *41*, 32–42.
15. Haltrich, D.; Leitner, C.; Neuhauser, W.; Nidetzky, B.; Kulbe, K.D.; Volc, J. A convenient enzymatic procedure for the production of aldose-free D-tagatose. *Ann. N. Y. Acad. Sci.* **1998**, *864*, 295–299.
16. Leitner, C.; Neuhauser, W.; Volc, J.; Kulbe, K.D.; Nidetzky, B.; Haltrich, D. The cetus process revisited: A novel enzymatic alternative for the production of aldose-free D-fructose. *Biocatal. Biotransform.* **1998**, *16*, 365–382.
17. Gutiérrez, L.F.; Hamoudi, S.; Belkacemi, K. Lactobionic acid: A high value-added lactose derivative for food and pharmaceutical applications. *Int. Dairy J.* **2012**, *26*, 103–111.
18. Schuster-Wolff-Bühring, R.; Fischer, L.; Hinrichs, J. Production and physiological action of the disaccharide lactulose. *Int. Dairy J.* **2010**, *20*, 731–741.
19. Julenius, K.; Mølgaard, A.; Gupta, R.; Brunak, S. Prediction, conservation analysis, and structural characterization of mammalian mucin-type o-glycosylation sites. *Glycobiology* **2005**, *15*, 153–164.
20. Berends, E.; Ohm, R.A.; de Jong, J.F.; Rouwendal, G.; Wösten, H.A.B.; Lugones, L.G.; Bosch, D. Genomic and biochemical analysis of N glycosylation in the mushroom-forming basidiomycete *Schizophyllum commune*. *Appl. Environ. Microbiol.* **2009**, *75*, 4648–4652.
21. Gemmill, T.R.; Trimble, R.B. Overview of N- and O-linked oligosaccharide structures found in various yeast species. *Biochim. Biophys. Acta* **1999**, *1426*, 227–237.
22. Wilson, I.B.; Zeleny, R.; Kolarich, D.; Staudacher, E.; Stroop, C.J.; Kamerling, J.P.; Altmann, F. Analysis of asn-linked glycans from vegetable foodstuffs: Widespread occurrence of lewis a, core alpha1,3-linked fucose and xylose substitutions. *Glycobiology* **2001**, *11*, 261–274.
23. Duman, J.G.; Miele, R.G.; Liang, H.; Grella, D.K.; Sim, K.L.; Castellino, F.J.; Bretthauer, R.K. O-mannosylation of *Pichia pastoris* cellular and recombinant proteins. *Biotechnol. Appl. Biochem.* **1998**, *28*, 39–45.
24. Heuts, D.P.H.M.; Scrutton, N.S.; McIntire, W.S.; Fraaije, M.W. What's in a covalent bond? On the role and formation of covalently bound flavin cofactors. *FEBS J.* **2009**, *276*, 3405–3427.
25. Scrutton, S. Identification of Covalent Flavoproteins and Analysis of the Covalent Link. In *Methods in Molecular Biology: Flavoprotein Protocols*; Chapman, K., Reid, A., Eds.; Humana Press Inc.: Totowa, NJ, USA, 1999; Volume 131, pp. 181–193.
26. Hecht, H.J.; Kalisz, H.M.; Hendle, J.; Schmid, R.D.; Schomburg, D. Crystal structure of glucose oxidase from *Aspergillus niger* refined at 2.3 å resolution. *J. Mol. Biol.* **1993**, *229*, 153–172.
27. Graf, M.M.; Bren, U.; Haltrich, D.; Oostenbrink, C. Molecular dynamics simulations give insight into D-glucose dioxidation at C2 and C3 by *Agaricus meleagris* pyranose dehydrogenase. *J. Comput. Aided Mol. Des.* **2013**, *27*, 295–304.
28. Liu, D.; Coloe, S.; Baird, R.; Pedersen, J. Rapid mini-preparation of fungal DNA for PCR. *J. Clin. Microbiol.* **2000**, *38*, 471.
29. Lin-Cereghino, J.; Wong, W.W.; Xiong, S.; Giang, W.; Luong, L.T.; Vu, J.; Johnson, S.D.; Lin-Cereghino, G.P. Condensed protocol for competent cell preparation and transformation of the methylotrophic yeast *Pichia pastoris*. *BioTechniques* **2005**, *38*, 44–48.

30. Weis, R.; Luiten, R.; Skranc, W.; Schwab, H.; Wubbolts, M.; Glieder, A. Reliable high-throughput screening with *Pichia pastoris* by limiting yeast cell death phenomena. *FEMS Yeast Res.* **2004**, *5*, 179–189.
31. Artimo, P.; Jonnalagedda, M.; Arnold, K.; Baratin, D.; Csardi, G.; de Castro, E.; Duvaud, S.; Flegel, V.; Fortier, A.; Gasteiger, E.; *et al.* Expasy: Sib bioinformatics resource portal. *Nucleic Acids Res.* **2012**, *40*, W597–W603.
32. Altschul, S.F.; Madden, T.L.; Schäffer, A.A.; Zhang, J.; Zhang, Z.; Miller, W.; Lipman, D.J. Gapped blast and psi-blast: A new generation of protein database search programs. *Nucleic Acids Res.* **1997**, *25*, 3389–3402.
33. Laemmli, U.K. Cleavage of structural proteins during the assembly of the head of bacteriophage t4. *Nature* **1970**, *227*, 680–685.

12

Research Applications of Proteolytic Enzymes in Molecular Biology

János András Mótyán, Ferenc Tóth and József Tőzsér *

Department of Biochemistry and Molecular Biology, Faculty of Medicine,
Medical and Health Science Center, University of Debrecen, POB 6, Debrecen H-4012, Hungary;
E-Mails: motyan.janos@med.unideb.hu (J.A.M.); tothfree@gmail.com (F.T.)

* Author to whom correspondence should be addressed; E-Mail: tozser@med.unideb.hu

---

Abstract: Proteolytic enzymes (also termed peptidases, proteases and proteinases) are capable of hydrolyzing peptide bonds in proteins. They can be found in all living organisms, from viruses to animals and humans. Proteolytic enzymes have great medical and pharmaceutical importance due to their key role in biological processes and in the life-cycle of many pathogens. Proteases are extensively applied enzymes in several sectors of industry and biotechnology, furthermore, numerous research applications require their use, including production of Klenow fragments, peptide synthesis, digestion of unwanted proteins during nucleic acid purification, cell culturing and tissue dissociation, preparation of recombinant antibody fragments for research, diagnostics and therapy, exploration of the structure-function relationships by structural studies, removal of affinity tags from fusion proteins in recombinant protein techniques, peptide sequencing and proteolytic digestion of proteins in proteomics. The aim of this paper is to review the molecular biological aspects of proteolytic enzymes and summarize their applications in the life sciences.

Keywords: proteolytic enzymes; proteases; molecular biology research applications

---

1. Scope of the Review

Proteolytic enzymes are capable of hydrolyzing peptide bonds and are also referred to as peptidases, proteases or proteinases [1].

The physiological function of proteases is necessary for all living organisms, from viruses to humans, and proteolytic enzymes can be classified based on their origin: microbial (bacterial, fungal and viral), plant, animal and human enzymes can be distinguished.

Proteolytic enzymes belong to the hydrolase class of enzymes (EC 3) and are grouped into the subclass of the peptide hydrolases or peptidases (EC 3.4). Depending on the site of enzyme action the proteases can also be subdivided into exopeptidases or endopeptidases. Exopeptidases catalyze the hydrolysis of the peptide bonds near the *N*- or *C*-terminal ends of the substrate. Aminopeptidases (Figure 1) can liberate single amino acids (EC 3.4.11), dipeptides (dipeptidyl peptidases, EC 3.4.14) or tripeptides (tripeptidyl peptidases EC 3.4.14) from the N-terminal end of their substrates. Single amino acids can be released from dipeptide substrates by dipeptidases (EC 3.4.13) or from polypeptides by carboxypeptidases (EC 3.4.16-3.4.18) (Figure 1), while peptidyl dipeptidases (EC 3.4.15) liberate dipeptides from the C-terminal end of a polypeptide chain. Endopeptidases (Figure 1) cleave peptide bonds within and distant from the ends of a polypeptide chain [2].

Figure 1. Action of aminopeptidases and carboxypeptidases removing the terminal amino acid residues as well as endopeptidases on a polypeptide substrate (having n residues). Red arrows show the peptide bonds to be cleaved.

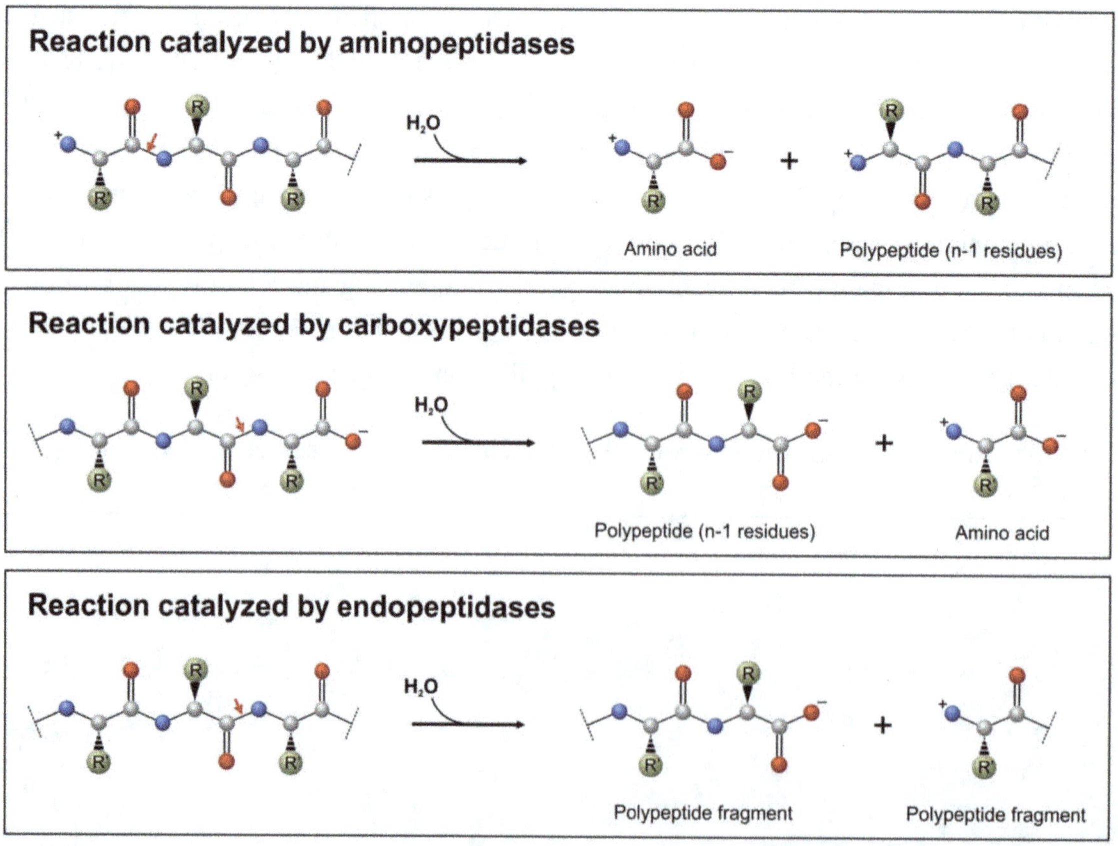

Based on the catalytic mechanism and the presence of amino acid residue(s) at the active site the proteases can be grouped as aspartic proteases, cysteine proteases, glutamic proteases, metalloproteases, asparagine proteases, serine proteases, threonine proteases, and proteases with mixed or unknown catalytic mechanism [3].

The current classification system further classifies the proteases into families based on sequence similarities, furthermore, homologous families are grouped into clans using a structure-based classification [3,4]. Classification and nomenclature of proteolytic enzymes as well as a detailed description of individual proteases is available in the MEROPS database [3].

Action of the proteolytic enzymes is essential in several physiological processes, e.g., in digestion of food proteins, protein turnover, cell division, blood-clotting cascade, signal transduction, processing of polypeptide hormones, apoptosis and the life-cycle of several disease-causing organisms including the replication of retroviruses [5,6]. Due to their key role in the life-cycle of many hosts and pathogens they have great medical, pharmaceutical, and academic importance [7–9].

It was estimated previously that about 2% of the human genes encode proteolytic enzymes [8] and due to their necessity in many biological processes proteases have become important therapeutic targets [8]. They are intensively studied to explore their structure-function relationships, to investigate their interactions with the substrates and inhibitors, to develop therapeutic agents for antiviral therapies [9] or to improve their thermostability, efficiency and to change their specificity by protein engineering for industrial or therapeutic purposes [7]. Studying proteolytic enzymes is highly justified by their key role in several fields of industry [2,10–12], as well. The worldwide market of industrial enzymes was estimated to reach $3.3 billion value in 2010 and the largest segment of this market is related to proteases [13].

Proteases are extensively applied enzymes in several sectors of industry and biotechnology, furthermore, numerous research applications require the use of them, including the production of Klenow fragments, peptide synthesis, digestion of unwanted proteins during nucleic acid purification, use of proteases in cell culture experiments and in tissue dissociation, preparation of recombinant antibody fragments for research, diagnostics and therapy, exploration of the structure-function relationships by structural studies, removal of affinity tags from fusion proteins in recombinant protein techniques, peptide sequencing, and proteolytic digestion of proteins in proteomics.

This review focuses on the application of proteolytic enzymes in life sciences, especially in the field of molecular biology. The summary table of proteases discussed in this review (Table 1) contains the substrate specificities of the enzymes which are grouped based on their catalytic mechanisms.

2. Molecular Biology Research Applications

2.1. Klenow Fragment Production

The Klenow fragment is the large fragment of the *E. coli* DNA polymerase I enzyme. While the holoenzyme has 5'→3' polymerase, 3'→5' and 5'→3' exonuclease activities, the Klenow fragment has only the polymerase and the 3'→5' exonuclease activities. The Klenow fragment has several applications in the recombinant DNA technology, including the labeling, sequencing, and site-specific mutagenesis of DNA.

Table 1. Substrate specificity of some proteolytic enzymes used in molecular biology research. Proteases are classified based on their catalytic mechanisms, furthermore, the main sources and enzyme specificities are indicated. The arrows indicate the sites of cleavages.

Enzyme	Main source	Cleavage site
Endopeptidases		
Serine proteases		
Trypsin	bovine	-Arg or Lys↓nonspecific-
Chymotrypsin	bovine	-Trp (or Phe, Leu, Tyr)↓nonspecific-
Enterokinase	bovine	Asp-Asp-Asp-Lys↓nonspecific-
Endoproteinase Arg-C	microbial	-Arg↓nonspecific-
Endoproteinase Glu-C	microbial	-Glu (or Asp)↓nonspecific-
Endoproteinase Lys-C	microbial	-Lys↓nonspecific-
Elastase	porcine	-Ala (or Gly or Val)↓nonspecific-
Subtilisin	microbial	-Trp (or Tyr, Phe, Leu)↓nonspecific-
Proteinase K	fungal	-aromatic, aliphatic or hydrophobic ↓nonspecific-
Thrombin	bovine	-Arg (or Lys)↓nonspecific- specific for -Leu-Val-Pro-Arg-↓Gly-Ser-
Factor Xa	bovine	-Arg (or Lys)↓nonspecific- specific for -Leu-Val-Pro-Arg-↓Gly-Ser-
WNV protease	*E. coli*	-Lys (or Arg)-Arg↓Gly-Ser-
Cysteine proteases		
Bromelain	plant	-nonspecific↓nonspecific-
Papain	plant	-Arg (or Lys)↓nonspecific-
Ficin (ficain)	plant	-nonspecific↓nonspecific-
Rhinovirus 3C	*E. coli*	Gly-Pro dipeptide after the scissile bond highly specific for -Leu-Glu-Val-Leu-Phe-Gln↓Gly-Pro-
TEV protease	*E. coli*	specific for -Gln-Asn-Leu-Tyr-Phe-Gln↓Gly-
TVMV protease	*E. coli*	specific for -Glu-Thr-Val-Arg-Phe-Gln↓Ser-
Metalloproteases		
Endoproteinase Asp-N	microbial	-nonspecific↓Asp-
Thermolysin	microbial	-Leu (or Phe)↓Leu (or Phe, Val, Met, Ala, Ile)-
Collagenase	microbial	-Pro-neutral↓Gly-Pro-
Dispase	microbial	-nonspecific↓non-polar-
Aspartic proteases		
Pepsin	porcine	-Phe (or Tyr, Leu, Trp)↓Trp (or Phe, Tyr, Leu)-
Cathepsin D	bovine	-Phe (or Leu)↓nonspecific (not Val, Ala)-
Exopeptidases		
Serine proteases		
Carboxypeptidase Y	yeast	-nonspecific↓nonspecific
Cysteine proteases		
Cathepsin C	bovine	removes N-terminal dipeptide
DAPase	porcine	removes N-terminal dipeptide
Metalloproteases		
Carboxypeptidase A	bovine	-nonspecific↓aromatic or branched preferred
Carboxypeptidase B	porcine	specific for C-terminal Arg or Lys

The enzymatic method to release the large protein fragment from the DNA polymerase I holoenzyme by proteolysis was published in 1970 [14]. Subtilisin-catalyzed proteolytic cleavage was used to produce Klenow fragment leading to the retention of the polymerase and the 3'→5' exonuclease activities and to the loss of 5'→3' exonuclease activity of the intact polymerase [6].

Nowadays, commercially available Klenow fragment is produced in recombinant ways in *E. coli* strains which carry the gene of large fragment of DNA polymerase I, therefore, the proteolytic production of Klenow fragment has mainly historical significance.

2.2. Enzymatic Peptide Synthesis

While enzymatic peptide synthesis has been frequently used to synthesize peptides for pharmaceutical and nutritional purposes (Table 2), this method is also essential for several research applications. The enzymatic method has several advantages compared to chemical methods, such as stereo specificity with side-chain protection, and the non-toxic nature of solvents coupled with the possibility of recycling the reagents used for synthesis. Enzymes have been selected considering their specificity for amino acid residues (Table 2), but this type of application is limited by the possibility of the hydrolysis of the peptide bond. The types of the enzymatic synthesis and its requirements have been reviewed [15–17]. Enzymatic peptide synthesis can be made by equilibrium- or kinetically-controlled methods.

Table 2. Examples of peptides synthesized by proteases.

Peptide	Sequence	Enzyme(s)	Reference
Aspartame	Asp-Phe	Thermolysin	[18]
Nutritional peptide	Tyr-Trp-Val	α-Chymotrypsin, papain	[19]
Somatostatin	Ala-Gly-Cys-Lys-Phe-Phe-Trp-Lys-Thr-Phe-Thr-Ser-Cys	Thermolysin, chymotrypsin	[20]
Vasopressin	Tyr-Phe-Phe-Gln	Thermolysin, chymotrypsin	[21]
Oxytocin	Cys-Tyr Tyr-Ile Pro-Leu Leu-Gly	Papain, thermolysin, chymotrypsin	[21]
mouse EGF (21–31)	His-Ile-Glu-Ser-Leu-Asp-SerTyr-Thr-Cys	Papain, trypsin	[22]

2.2.1. Kinetically Controlled Peptide Synthesis

The scheme for chymotrypsin-catalyzed kinetically-controlled Z-D-Leu-L-Leu-NH_2 synthesis [23] is illustrated in Figure 2. The acyl donor Z-D-Leu that is activated by carbamoylmethyl (Cam) ester and chymotrypsin (E) form the enzyme-substrate complex first and after that the covalently linked Z-D-Leu-E intermediate with the loss of the carbamoylmethyl ester. If this intermediate is attacked by water, hydrolysis occurs, which results in the Z-D-Leu-OH fragment. However if a more powerful nucleophile (e.g., alcohol or thiol) is present in the media, the enzyme produces a peptide bond instead of the cleavage [15] and the Z-D-Leu-L-Leu-NH_2 dipeptide may be formed in the presence of H-L-Leu-NH_2 nucleophile (Figure 2). The product yield depends on the kinetics of the two nucleophilic reactions, however, the reaction is faster and requires lower substrate enzyme ratios compared to the equilibrium-controlled synthesis, due to the activated acyl donor.

Figure 2. Kinetically-controlled synthesis of Z-D-Leu-L-Leu-NH_2 dipeptide. After the formation of enzyme-substrate complex (K_1) a covalent enzyme-substrate intermediate is formed (K_2). The intermediate is subjected to the attack from H_2O or other nucleophiles (Nu). K_H is the equilibrium constant of hydrolysis, K_T is the equilibrium constant of the transferase reaction.

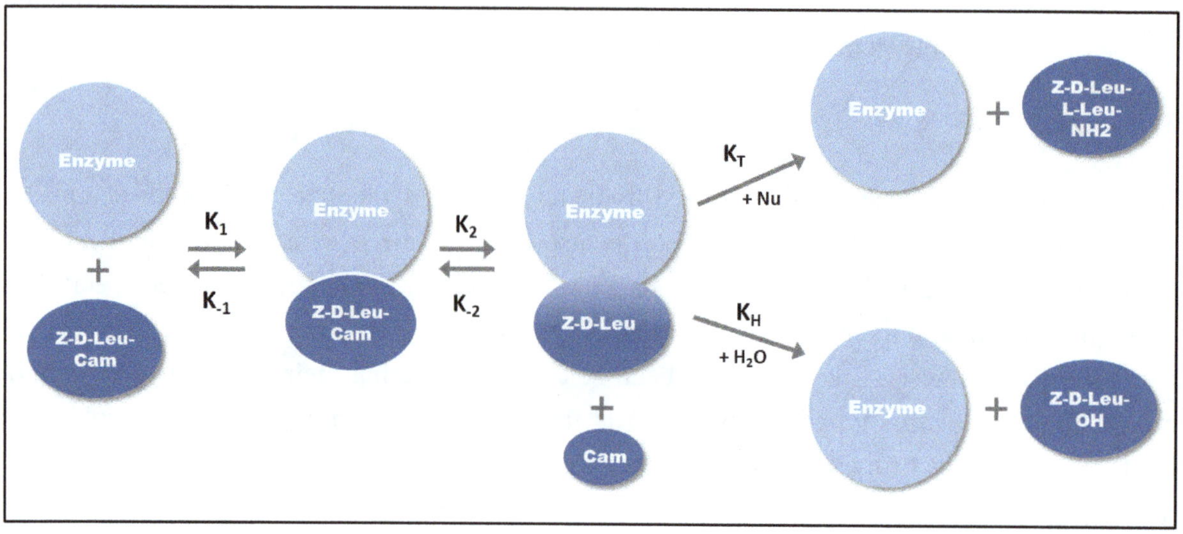

The acyl donor activating agent must not only be an ester, but an amide or a nitrile as well. Only serine or cysteine proteases can be used to perform the kinetically controlled peptide synthesis, as these enzymes can act as transferase and hence are able to catalyze the transfer of an acyl group from the acyl donor to the nucleophile through the formation of a covalent acyl enzyme intermediate. Papain, thermolysin, trypsin and α-chymotrypsin are mostly used for kinetically-controlled peptide synthesis [24]. The yield of peptide product will depend on the apparent ratio of transferase to hydrolase rate constants $(K_T/K_H)_{app}$ and the rate at which the peptide product is hydrolyzed. The $(K_T/K_H)_{app}$ values of proteases used for kinetically-controlled synthesis are in the range of 10^2–10^4 [17].

2.2.2. Equilibrium-Controlled Synthesis

In the case of the equilibrium-controlled synthesis the process is the reverse of hydrolysis. Important problems of this enzymatic method of peptide synthesis are the low reaction rates and the need of increased yield because proteases do not alter the equilibrium of the reaction. A high amount of the enzyme is often required together with the precise reaction conditions to drive the equilibrium towards synthesis [15]. A higher rate of peptide bond formation can be reached by the appropriate pH of the reaction medium (changing the equilibrium of ionization) but there are several other ways to increase the yield of the reaction:

(a) Precipitation of the product is the classical method. When certain soluble carboxyl amine components are used as reactants the products will precipitate and the reduced concentration of soluble products will drive the equilibrium towards synthesis. In this case the yield of the product depends on the concentration of the starting materials and can be determined by the solubility of the product [17].

(b) In a biphasic system the enzyme acts in an aqueous environment that is surrounded by a water immiscible medium; the water content of the system is around 2–5%. The reactants (dissolved in high concentrations) can diffuse from the organic phase into the water until the equilibrium is reached; the enzyme-catalyzed synthesis is followed by the diffusion of the products back into the organic phase. The organic phase reduces the dielectric constant of the medium and thus the acidity of the carboxyl group of the acyl donor as well, which in turn promotes the synthesis of the peptide bond instead of hydrolysis [24]. This method is applicable only for the synthesis of water insoluble products.

(c) The dissolved state system can be used for the synthesis of water-soluble products (short peptides, high molecular weight peptides and proteins). In this environment forcing the reaction towards peptide synthesis requires the mass action, the addition of a water-miscible organic co-solvent in high concentration or the excess of one reactant. In serine protease-catalyzed reaction in water, the rate determining feature is the acylation of the enzyme while the product yield at equilibrium depends on the partition coefficient and the ratio of the aqueous and organic volumes [17]. The chymotrypsin- and subtilisin-catalyzed synthesis of N-Bz-L-Tyr-L-Leu-NH_2 is more efficient in hydrophobic organic solvents; adding water in sub-saturating concentration increases the yield of the chymotrypsin-catalyzed peptide synthesis [25].

2.2.3. Strategies Used in Enzymatic Synthesis

The *use of enzymes in organic solvents* have several advantages compared to aqueous solvents which have led to their widespread application: the thermodynamic equilibrium can be shifted towards synthesis, the undesirable side reactions can be reduced, the nonpolar substrates are more soluble in organic solvents, the separation process and enzyme recovery is more effective in a low water-containing environment [26–29]. Many proteases, such as thermolysin, subtilisin and α-chymotrypsin [26,29] can maintain their active conformation in organic solvents and show good functionality in the synthesis of aspartame and demorphin derivatives.

However, the use of enzymes in organic solvent has also disadvantages such as the unfavorable effects of the organic solvents on enzyme activity and stability. The *modification of biocatalysts* by protein engineering [30,31] and/or chemical modification or the use of naturally solvent-tolerant proteases [32] for peptide synthesis is a developing field. The driving force of this field is the aim of making biocatalysts with proper features to suit them for the reactions under specific synthesis conditions. Site-directed mutagenesis is a very effective tool and can be used by protein engineers to screen mutants with enhanced stability, activity or specificity, furthermore, this method can be used to explore structure-function relationships (rational design).

Subtilisin has been extensively studied and engineered via site-directed mutagenesis to make it more capable of peptide bond formation in aqueous solution [33]. Single and multiple mutations have been introduced into subtilisin to increase its stability and make it more resistant against oxidizing agents, thermal denaturations and inactivation effects of polar solvents [28].

Thermolysin has a higher synthesis rate compared to the solvent stable PST-01 protease from *Pseudomonas aeruginosa.* Considering the high structural similarity of these enzymes, the synthetic activity of PST-01 protease was increased by the Y114F mutation [31], moreover, the Y114R and Y114S mutations resulted in better activity enhancement.

Chemical modification is also an efficient method to modify the properties of enzymes used for peptide synthesis. Thiol-subtilisin, in which the serine residue has been chemically changed to cysteine at the active site, shows an enhanced aminolysis to hydrolysis ratio in aqueous solution and in dimethyl sulphoxide. The stability of proteases can also be increased by chemical modification e.g., a hydrophilic carbohydrate-polyacrylate polymer coat can make the enzymes highly active and stable in polar solvents and more resistant against thermal inactivation [28].

Immobilization of proteases is the most frequently applied method for the recovery of products without great loss of the catalysts, which greatly decreases the cost of the synthesis. This approach also ensures better operational stability of biocatalysts and control of the reaction. Enzyme immobilization techniques can be divided into five groups: (a) covalent attachment to solid support; (b) absorption on solid support; (c) entrapment in polymeric gel; (d) crosslinking with bifunctional reagents and (e) encapsulation [17,34].

Substrate engineering means the manipulation of the leaving group and is a powerful tool to increase the specificity of the proper enzyme and/or increase the rate and the yield of the reaction [24].

Protease-catalyzed synthesis of stereochemically modified peptides is also a preferable application compared to chemical synthesis due to stereospecificity of the proteases.

Proteases can bind not only natural substrates, but also specifically designed substrate mimetics, which are also very useful tools to increase the yield of peptide synthesis. Substrate mimetics can bind to the active site of the enzyme and in this way proteases can be used for the synthesis of products containing non-specific amino acids. The undesired cleavage of the newly synthesized peptide bonds can be avoided using this method and it is not required to change the properties of the medium or the enzyme [17].

The production of peptides with amides at their C-termini is a great challenge for enzymatic peptide synthesis, but amidation may be required to retain biological activity.

2.3. Nucleic Acid Isolation

Generally, the first step of nucleic acid isolation protocols is the lysis of the biological material containing the DNA or RNA of interest. Before the purification and concentration of nucleic acids the contaminating proteins and other macromolecules have to be removed from the sample. Undamaged nucleic acids can be isolated when the degradation of the DNA and RNA present in the sample is avoided by the inhibition and removal of DNases and RNases. The nucleases can be inhibited by the addition of chelators (e.g., EDTA) which bind the ions essential for their action. Besides the inactivation of nucleases, proteolytic enzymes are applied during the nucleic acid isolation to remove total protein content of the sample.

The most widely used proteolytic enzyme in nucleic acid purification is the Proteinase K, which was described in 1974 [35]. Proteinase K is a non-specific serine endopeptidase which can catalyze the cleavage of peptide bonds at the carboxylic side of aromatic, aliphatic, or hydrophobic amino acid residues. Besides the digestion of unwanted proteins, Proteinase K also quickly inactivates the nucleases which might degrade the nucleic acids present in the sample [36]. This proteolytic digestion decreases the level of contaminants in the nucleic acid extract and prevents nucleic acids from degradation leading to a higher yield of the DNA or RNA to be isolated.

2.4. Cell Isolation and Tissue Dissociation

Cell biology studies frequently require the dissociation of primary tissues and the isolation of viable cells for tissue culturing. The most common method for cell isolation is the enzymatic digestion of the junctions connecting the cells and the components of the surrounding extracellular matrix, by which the cells can be released from a wide variety of tissues. Several enzymes are available in the market for the detachment of cultured cells, cell dissociation and cell component or membrane-associated protein isolation [37,38]. Besides the polysaccharidases, nucleases and lipases, the proteases are the most important enzymes used widely to dissociate cells from tissues.

The experimental conditions of cell isolation are functions of several parameters, including the type of the tissue and the source of its origin. Cells with high viability can be isolated in high yield using a suitable enzyme or the optimal combination of enzymes. As proteases differ in their specificities, different enzymes are recommended to be used use for most effective tissue disruption, depending on the origin and type of the tissue. We describe below the enzymes most commonly used for cell isolation.

The matrix metalloproteinase collagenase was first isolated in 1953 [39]. This endopeptidase can digest the collagenous extracellular matrix in a zinc-dependent manner. Collagenase cleaves the peptide bonds within the triple helices of native collagen, between a neutral amino acid and Gly within the Pro-X-Gly-Pro sequence. This sequence can be found most frequently in the collagen; therefore collagenases digest other proteins less efficiently. A commercially available collagenase (clostridiopeptidase A) is produced by *Clostridium histolyticum*, and it is capable of digesting collagen fibers very effectively. Solutions supplied for tissue dissociation contain collagenase and other additional proteinases which can digest the components of the extracellular matrix [38,40]. The serine protease elastase is a unique enzyme which can cleave the peptide bonds in elastin, therefore, it is generally used to dissociate tissues containing a high amount of elastin connective fibers. Elastase cleaves peptide bonds next to smaller neutral amino acids and besides its protease activity it also has esterase and amidase activities. Papain is a cysteine peptidase of *Carica papaya* latex. Papain, similarly to elastase, also has amidase and esterase activities and has a broad specificity. Papain has less damaging effects on tissues and therefore it is typically applied for cell dissociation of neuronal tissues. Besides cell dissociation, papain is also widely used for integral membrane protein solubilization and for digestion of proteoglycans. The serine protease trypsin is a very specific proteinase cleaving the peptide bonds at the C-terminal end of positively charged Lys and Arg side chains. Due to the high specificity of trypsin the digestion of tissue proteins is less effective and it is generally used for tissue dissociation together with other proteolytic enzymes. Serine protease chymotrypsin cleaves peptide bonds preferentially at the carboxyl side of aromatic Tyr, Trp and Phe residues. Chymotrypsin is less widely used for tissue dissociation; the use of other additional proteases is required for efficient digestion. The Zn-metalloprotease dispase is also a neutral protease. This non-specific protease cleaves the peptide bonds of proteins at the amino side of non-polar amino acid residues.

2.5. Cell Culturing

Cells isolated from a tissue can be cultured separately from the organism in cell culture flasks using appropriate growth medium. Adherent cells grown in a cell culture flask are attached to the surface by protein bridges which have to be disrupted during passaging. The cells can be released from the cell flask surface mechanically using a cell scraper or can be detached by a protease treatment using trypsin solution.

Trypsinization means the process used for the detachment of adherent cells using trypsin solution to digest the adhesion molecules by which the cells are attached to the surface of the culture flask. Trypsin solutions generally contain EDTA to reduce the concentration of metal ions that might inhibit trypsin.

Although trypsinization is the most commonly used method to detach adherent cells from cell culture flasks, the effects of this protease treatment were only recently studied in detail. It was found that trypsinization can affect the extracellular matrix surrounding the cells [41] and has physiological effects on cells grown in cell cultures [42]. Trypsin treatment can lead to cleavage of membrane proteins and receptors, which can cause significant changes in the expression level of different proteins: level of growth- and metabolism-related protein expressions were found to be down-regulated after trypsinization, while up-regulation of apoptosis-related protein expressions was seen after the protease treatment [42,43]. This effect should be taken into account when trypsinization is involved in experimental design.

2.6. Antibody Fragment Production

Antibody molecules are produced by the immune system against foreign substances and are classified into the immunoglobulin superfamily of the proteins (Figure 3A). They consist of four polypeptide chains, two identical heavy chains (H) and two identical light chains (L) which are connected by disulfide bridges. Both the H and the L chains contain variable (V_H and V_L) and constant (C_H1, C_H2, C_H3 and C_L) regions, respectively. The V_H and V_L chains, containing hypervariable regions, are responsible for the antigen-antibody interactions and determine the antigen specificity [6].

Fragments of the monoclonal antibodies are widely used in diagnostics, therapeutics and in biopharmaceutical research [44–46] having beneficial properties compared to the whole immunoglobulin molecules due to their smaller size and lower immunogenicity [44]. Fragments of whole immunoglobulin molecules can be produced using recombinant DNA technology or can be generated by enzymatic digestion. Here we discuss the proteolytic antibody fragmentation method.

Generally, the papain, pepsin and ficin proteases are used for the specific digestion of IgG molecules. Digestion of an antibody by the cysteine protease papain results in three fragments due to the cleavage of peptide bonds in the hinge region between C_H1 and C_H2 domains: one Fc (crystallizable) and two identical Fab (antigen binding) fragments are released (Figure 3B). While both released Fab fragments carry one antigen-binding site, the Fc fragment does not have antigen-binding ability. The aspartic acid protease pepsin cleaves the peptide bonds of the antibody near the disulfide bonds connecting the H chains (Figure 3C). This digestion results in the release of the peptides of the Fc region and one $F(ab')_2$ fragment containing both antigen binding sites. The cysteine protease ficin can release both $F(ab')_2$ or Fab fragments (Figure 3D), depending on the cysteine concentration [44].

Figure 3. Structure of IgG antibody molecules (**A**) and fragments released after proteolytic digestion using papain (**B**), pepsin (**C**) or ficin (**D**).

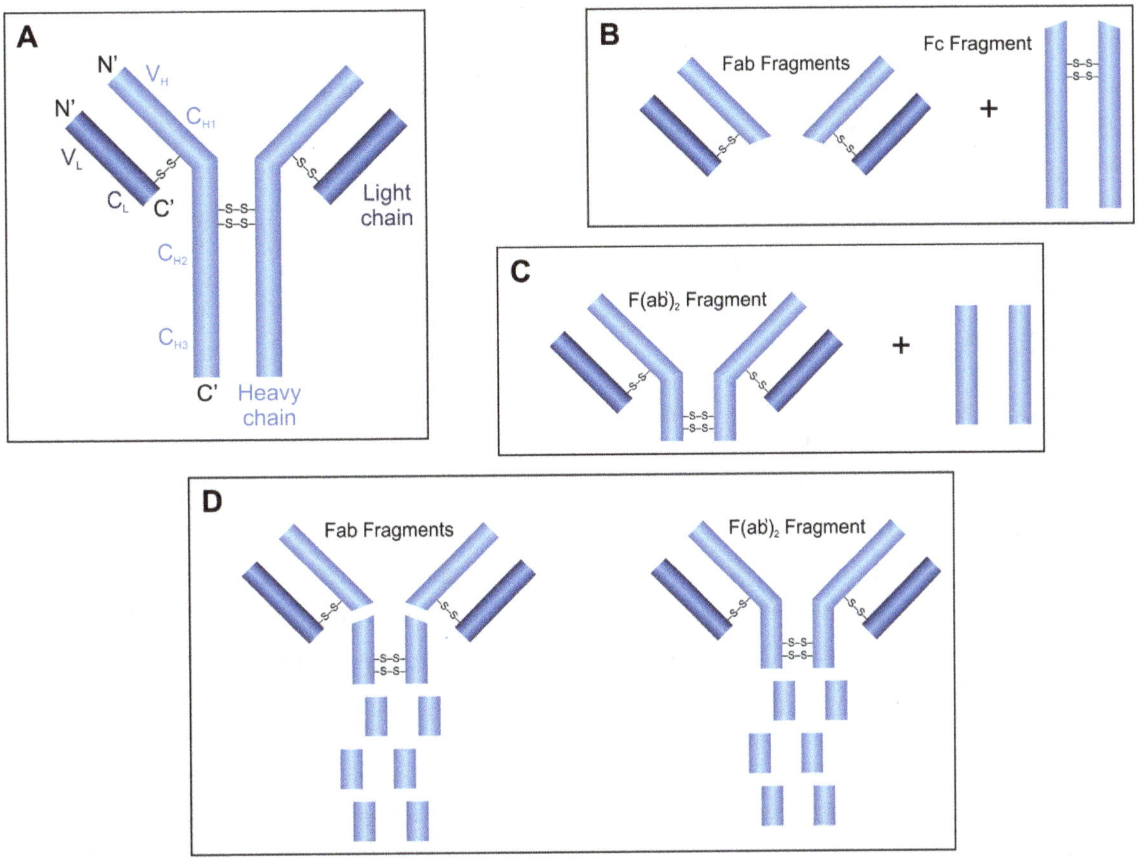

The *in vivo* applications of the large, pentameric IgM immunoglobulines could be restricted because of their large size. Proteolytic fragmentation of IgM moleucles is also a useful method to produce smaller, active Fab, Fab', $F(ab')_2$ or "IgG-like" fragments which have beneficial properties compared to the whole IgM molecules. Trypsin and pepsin are useful enzymes for IgM fragmentation, pepsin can generate Fab and $F(ab')_2$ fragments, while trypsin can generate "IgG-like" fragments, as well. Papain cannot be used for IgM fragmentation because of the production of heterogeneous fragments.

Digestion of monoclonal antibodies by papain or pepsin is still used to produce Fab or $F(ab')_2$ fragments [44]. There are several products available on the market obtained by proteolytic cleavage of antibodies, furthermore, several companies supply fragmentation kits for antibody fragment preparation. In some cases it could be preferable to produce the Fab fragments in high quality by recombinant expression in cell lines and obtain the fragment in sufficient quantity. For example, the Fab fragments produced by recombinant expression were found to be more preferable for the crystallization experiments because of the higher quality of the protein sample [47].

Antibody fragments have several beneficial properties for *in vivo* applications compared to whole antibody molecules. Due to their smaller size they have higher mobility and can penetrate tissues or permeate cell membranes more easily. As antigen-binding fragments lack the Fc fragments, they have lower immunogenicity, as they contain only their antigen-binding domain sites (Fab, Fab' or $F(ab')_2$ fragments) and do not carry the regions responsible for antibody effector functions, furthermore, they do not form large immunocomplexes.

The antigen-binding fragments of antibodies have great importance from the viewpoint of clinical and therapeutic applications. Antibody fragments could be administered to prevent the development of a disease (e.g., restenosis), could be applied during the diagnosis (e.g., metastatic breast and colon cancer) or for the treatment of some diseases (e.g., macular degeneration) or to detect toxins or neutralize snake venoms [44,45]. The current number of antibody-based therapeutics approved by the FDA is 35, while several other antibodies are in clinical trials. The relevance of antibody fragments in structural studies is discussed in detail in the following paragraph. Further important biotechnological applications of antibody fragments, e.g., as immunodetection, immunopurification and detoxification, have been reviewed in the recent past [46].

2.7. Structural Studies

Crystallographers have great challenges in the structure determination of transmembrane proteins. It is difficult to crystallize these proteins due to their high molecular flexibility, hydrophobic surfaces and low solubility. Antibody fragments are very useful tools for solving these problems. Specific binding of antibody fragments can increase the overall hydrophilicity of proteins and the solubility of the transmembrane proteins, furthermore, they can decrease the flexibility and stabilize the conformation of the molecule [48]. The structures of the membrane proteins co-crystallized with antibody fragments can be determined at higher resolution because these crystals have a higher diffraction quality [49–51].

The proteases have great indirect significance in structure determination, because the antibody fragments produced by proteolytic cleavage of whole immunoglobulins can be used to improve the crystallization properties [52–54].

The proteases are not only tools for crystallographers but are also important target molecules for structural biologists and have great relevance in antiviral therapies, drug and therapeutics development from the viewpoint of structural biology. Besides interest in proteases with unknown structure, the results of structural studies can help researchers evaluate the structure-function relationships more efficiently; increasing knowledge on the structural organization of viral proteases can help to explore their action, perform comparative studies by which we can better understand the structure-function and evolutionary relationships and recognize general or specific features [55–57]. One of the main driving forces of structure determination of proteases is the need for the development of efficient drugs for antiviral therapies [7,9,58–60]. Both structural and enzymatic inhibition studies are required for the structure-based drug development of protease inhibitors [61]. Structural data can also help protein engineers to alter the specificity and to improve the enzymatic properties of proteases by structure-guided mutagenesis [62–64] for several purposes.

2.8. Fusion Tag Removal

The proteins produced by recombinant techniques are typically linked with a fusion partner termed a fusion tag. The introduction of a fusion tag means the fusion of an additional protein or peptide to the recombinant protein. These fusion tags are extensively used from basic research to high-throughput structural biology owing to the several advantages they provide in the expression of different recombinant proteins [65,66]. The tags largely aid the detection and purification of proteins; moreover they also could have a favorable effect on protein yield and/or solubility. Tags can prevent proteins from proteolytic digestion, can protect antigenicity or facilitate the folding of the fusion protein. On the other hand, they can also negatively alter solubility, structural integrity and biological activity [47,67,68] or may cause a disadvantage for further application of the protein, so the removal of a tag can be crucial [66].

Commonly used affinity tags and fusion protein partners are the hexahistidine-tag (His_6), FLAG-tag, maltose binding protein (MBP), glutathione S-transferase (GST), thioredoxin (TRX), small ubiquitin-like modifier (SUMO), ubiquitin (Ub) and green fluorescent protein (GFP) [69].

In some cases the tag can be removed by a chemical treatment but those methods are rather unspecific compared to enzymatic cleavages and may lead to protein denaturation and/or side chain modifications of amino acids in the target protein. The specificity and detergent sensitivity of common proteases used for tag removal have been examined and reviewed previously [70–73]. Both endo- and exo-proteases could be suitable for fusion tag removal.

2.8.1. Endoproteases

Serine proteases such as enterokinase (also referred to as enteropeptidase), factor Xa and thrombin have been widely used for many years to remove N-terminal tags, but several cases have been reported in which they cleaved not only at the desired cleavage site but also in the protein of interest. These incidents led to the extensive application of viral proteases like human rhinovirus (HRV) 3C protease and tobacco etch virus (TEV) protease. While sequences recognized by a cellular serine protease and the viral proteases could be similar, the viral proteases cleave the protein substrates at the undesired sites less efficiently due to their high specificity and low catalytic rate, moreover, recombinant viral proteases can be produced in high quantities in *E. coli* [73]. These findings and the limited activity of the generally used serine proteases in some detergents, which are needed to study the membrane proteins, inspired the search for other viral proteases for tag removal, such as Tobacco Vein Mottling Virus (TVMV) protease [74], West Nile Virus (WNV) protease [75] and some alphaviral proteases: Venezuelan Equine Encephalitis Virus (VEEV) protease, Semliki Forest Virus (SFV) and Sindbis Virus (SIN) protease [76].

2.8.2. Exopeptidases

Aminopeptidases and carboxypeptidases are not as widely used as endopeptidases, as they frequently leave amino acid residues on the target protein, while it is easier to design cleavage sites with endopeptidases not to leave extra residues after the cleavage. However, if is still desired to add a tag onto the C-terminal of the target protein a carboxypeptidase may be used for its removal. Among metallocarboxypeptidases, type A carboxypeptidases remove mostly aromatic or branched aliphatic side chain containing amino acids, while type B carboxypeptidases prefer basic amino acids. Carboxypeptidases can be utilized to remove a His_6 tag from the C-terminal end of a protein [77]. Dipeptidyl aminopeptidase (DAPase) is a useful enzyme for the removal of N-terminal dipeptides.

2.9. Proteomic Applications

Proteomic studies are made with the aim to identify, characterize, and quantify the required samples and typically involve mass spectrometric (MS) analysis. Besides determination of the composition of protein complexes, chemical properties, post-translational modifications and structural properties of proteins can also be revealed by MS.

MS is a powerful analytical method to measure the mass of proteins or peptides by the analysis of mass-to-charge ratio (*m*/*z*) ratio. Generally, the samples to be analyzed contain a mixture of various proteins and/or polypeptides that have to be separated and digested into smaller fragments before the MS analysis. Protein separation can be performed efficiently by polyacrylamide gel electrophoresis. The separated proteins can be digested after the electrophoresis by chemical cleavage or by enzyme-catalyzed digestion of peptide bonds. The process in which the bands or spots are cut out from the gel followed by the addition of protease(s) to the gel containing the protein(s) of interest is called in-gel digestion [78]. Proteolysis of whole proteins leads to the release of smaller peptides with different molecular masses which are suitable for MS analysis (Figure 4).

Figure 4. Steps of proteomic analysis using mass-spectrometry after separation and in-gel digestion of proteins of interest.

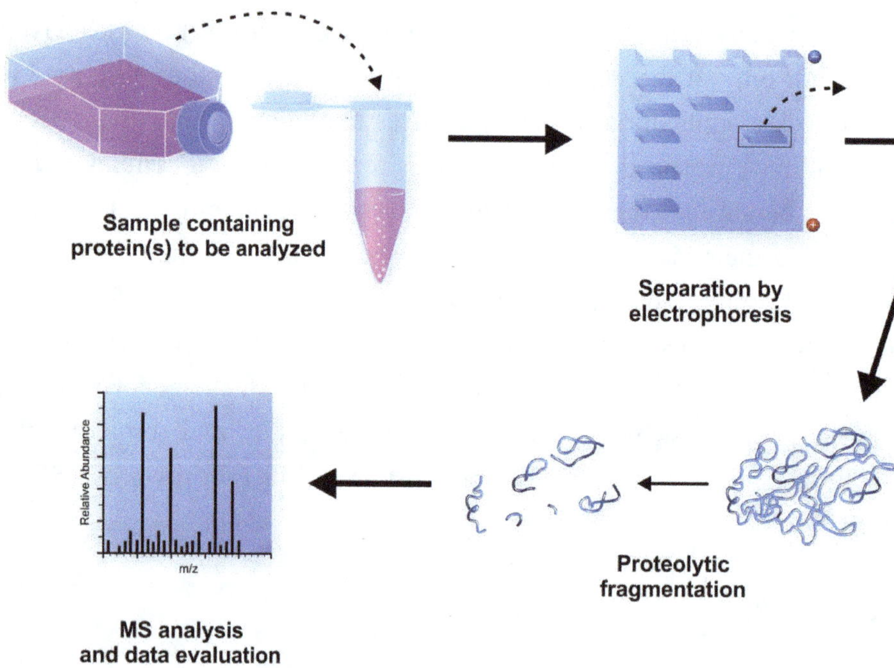

Trypsin is the most widely used proteolytic enzyme for protein digestion in MS analysis. Due to its high specificity it is easy to predict the cleavage sites and to compare the results of experimental enzymatic and theoretical *in-silico* digestion. The Asp-N, Lys-C, Arg-C and Glu-C enyzmes are also highly sequence-specific endoproteases but they are less active [78]. Chymotrypsin, pepsin, papain, elastase, subtilisin, proteinase K, thrombin, factor Xa and some other proteases are also suitable enzymes for fragmentation in MS analysis [79].

Several proteases are available for this fragmentation of which specificities are well established [79,80]. The number and length of released peptide fragments depends on the protease(s) applied for selective proteolysis of the targeted protein. Applications developed for *in silico* protein fragmentation are useful to predict the proteolytic fragments and to choose the most proper enzyme for the most effective digestion based on the enzyme specificities (http://prospector.ucsf.edu).

Generally, highly sequence-specific proteases are preferred for protein fragmentation instead of less specific enzymes, as the latter ones produce a very complex mixture of fragments. In the case of the efficient fragmentation the peptide, the fragments have a proper length and are released in high yield, and the complete sequence of the whole protein can be covered by the analysis of the proteolytic fragments.

3. Summary

Besides extended application for nutritional and pharmaceutical purposes, proteases from natural sources are also widely used tools in molecular biology practice. Their degradative properties make them useful for general protein digestion in tissue dissociation, cell isolation, and cell culturing. The specificity and the predictability of cleavages by proteases enables their use for more specific tasks such as antibody fragment production, the removal of affinity tags from recombinant proteins and specific protein digestion in the proteomics field mainly for protein sequencing. Moreover, the already mentioned specificity makes proteases—in a water restricted environment—able to synthesize the peptide bonds instead of hydrolyzing them. This property combined with their enantioselectivity has also promoted their use in peptide synthesis.

The expansion of knowledge has assisted the increase of applications of proteolytic enzymes for several purposes, and the application fields are widening with the help of protein engineering techniques and by chemical modification of the enzymes [7,62]. Studies made with the aim to better understand the structure and function of existing proteolytic enzymes and to obtain new, engineered proteases with altered properties for therapeutic, industrial or research fields require the use of the applications discussed in this paper.

Acknowledgments

This work was supported by the TÁMOP 4.2.2.A-11/1/KONV-2012-0023 "VÉD-ELEM" and by the Hungarian Science and Research Fund (OTKA 101591) implemented through the New Hungary Development Plan co-financed by the European Social Fund and the European Regional Development Fund.

References

1. Barrett, A.J.; McDonald, J.K. Nomenclature: Protease, proteinase and peptidase. *Biochem. J.* **1986**, *237*, 935.
2. Rao, M.B.; Tanksale, A.M.; Ghatge, M.S.; Deshpande, V.V. Molecular and biotechnological aspects of microbial proteases. *Microbiol. Mol. Biol. Rev.* **1998**, *62*, 597–635.

3. Rawlings, N.D.; Barrett, A.J.; Bateman, A. MEROPS: The database of proteolytic enzymes, their substrates and inhibitors. *Nucleic Acids Res.* **2012**, *40*, D343–D350.
4. Rawlings, N.D.; Barrett, A.J. Evolutionary families of peptidases. *Biochem. J.* **1993**, *290*, 205–218.
5. Neurath, H.; Walsh, K.A. Role of proteolytic enzymes in biological regulation (a review). *Proc. Natl. Acad. Sci. USA* **1976**, *73*, 3825–3832.
6. Devlin, T.M. *Textbook of Biochemistry with Clinical Correlations*, 5th ed.; Wiley & Sons: New York, NY, USA, 2002.
7. Li, Q.; Yi, L.; Marek, P.; Iverson, B.L. Commercial proteases: Present and future. *FEBS Lett.* **2013**, *587*, 1155–1163.
8. Craik, C.S.; Page, M.J.; Madison, E.L. Proteases as therapeutics. *Biochem. J.* **2011**, *435*, 1–16.
9. Antonelli, G.; Turriziani, O. Antiviral therapy: Old and current issues. *Int. J. Antimicrob. Agents* **2012**, *40*, 95–102.
10. Kirk, O.; Borchert, T.V.; Fuglsang, C.C. Industrial enzyme applications. *Curr. Opin. Biotechnol.* **2002**, *13*, 345–351.
11. Rani, K.; Rana, R.; Datt, S. Review on latest overview of proteases. *Int. J. Curr. Life Sci.* **2012**, *2*, 12–18.
12. Ray, A. Protease enzyme- potential industrial scope. *Int. J. Technol.* **2012**, *2*, 1–4.
13. Sarrouh, B.; Santos, T.M.; Miyoshi, A.; Dias, R.; Azevedo, V. Up-to-date insight on industrial enzymes applications and global market. *J. Bioprocess. Biotech.* **2012**, S4:002
14. Klenow, H.; Henningsen, I. Selective elimination of the exonuclease activity of the deoxyribonucleic acid polymerase from Escherichia coli B by limited proteolysis. *Proc. Natl. Acad. Sci. USA* **1970**, *65*, 168–175.
15. Morihara, K. Using proteases in peptide synthesis. *Trends Biotechnol.* **1987**, *5*, 164–170.
16. Bhalla, T.C.; Kumar, D.; Gajju, H.; Agrawal, H.O. Thermophilic bacterial proteases. *J. Punjab Acad. Sci.* **1999**, *1*, 77–91
17. Kumar, D.; Bhalla, T.C. Microbial proteases in peptide synthesis: Approaches and applications. *Appl. Microbiol. Biotechnol.* **2005**, *68*, 726–736.
18. Kühn, D.; Dürrschmidt, P.; Mansfeld, J.; Ulbrich-Hofmann, R. Biolysin and thermolysin in dipeptide synthesis: A comparative study. *Biotechnol. Appl. Biochem.* **2002**, *36*, 71–76.
19. Kimura, Y.; Muraya, K.; Araki, Y.; Matsuoka, H.; Nakanishi, K.; Matsuno, R. Synthesis peptides consisting of essential amino acids by a reactor system using three proteinases and an organic solvent. *Agric. Biol. Chem.* **1990**, *54*, 3331–3333.
20. Bille, V.; Ripak, C.; van Aasche, I.; Forni, I.; Degelaen, L.; Searso, A. A Semi-Enzymatic Synthesis of Somatostatin. In Proceedings of 21st European Peptide Symposium. ESCOM, Leiden, The Netherlands, 1991; Giralt, E., Andreu, D., Eds.; pp. 253–254.
21. Rizo, J.; Gierarch, L.M. Constrained peptides: Models of bioactive peptides and protein structures. *Ann. Rev. Biochem.* **1992**, *61*, 387–418.
22. Widmer, F.; Bayne, S.; Houen, G.; Rigby, R.B.; Whittaker, R.G.; Johansen, J.T. Use of Proteolytic Enzymes for the Synthesis of Fragments of Mouse Epidermal Growth Factor. In Proceedings of Peptides 1984, Almquist, Stockholm, 1985; Ragnurrson, U., Ed.; pp. 193–196.

23. Salam, S.M.; Kagawa, K.; Kawashiro, K. Alpha-chymotrypsin-catalyzed peptide synthesis using N-protected D-amino acid carbamoylmethyl esters as acyl donors. *Biotechnol. Lett.* **2005**, *27*, 1199–1203.
24. Guzmán, F.; Barberis, S.; Illanes, A. Peptide synthesis: Chemical or enzymatic. *Electron. J. Biotechnol.* **2007**, *10*, 279–314.
25. Sergeeva, M.V.; Paradkar, V.M.; Dordick, J.S. Peptide synthesis using proteases dissolved in organic solvents. *Enzyme Microb. Technol.* **1997**, *20*, 623–628.
26. Dordick, J.S. Enzymatic catalysis in monophasic organic solvents. *Enzyme Microb. Technol.* **1989**, *11*, 194–211.
27. Khemlnitski, Y.L.; Levashov, A.V.; Klyachko, N.L.; Martinek, K. Engineering biocatalytic systems in organic media with low water content. *Enzyme Microb. Technol.* **1988**, *10*, 710–724.
28. Gill, I.; Fandino, R.L.; Jobra, X.; Vulfson, E.N. Biologically active peptides and enzymatic approaches to their production. *Enzyme Microb. Technol.* **1996**, *18*, 162–183.
29. Shen, H.Y.; Tian, G.L.; Ye, Y.H. Synthesis of demorphin (1–4) derivatives catalysed by proteases in organic solvents. *J. Pept. Res.* **2004**, *65*, 143–148.
30. Wells, J.A.; Estell, D.A. Subtilisin: An enzyme designed to be engineered. *Trends Biochem. Sci.* **1988**, *13*, 291–297.
31. Ogino, H.; Tsuchiyama, S.; Yasuda, M.; Doukyu, N. Enhancement of the aspartame precursor synthetic activity of an organic solvent-stable protease. *Protein Eng. Des. Sel.* **2010**, *23*, 147–152.
32. Xu, J.X.; Jiang, M.; Sun, H.L.; He, B.F. An organic solvent-stable protease from organic solvent-tolerant Bacillus cereus WQ9–2: Purification, biochemical properties, and potential application in peptide synthesis. *Bioresour. Technol.* **2010**, *101*, 7991–7994.
33. Abrahmsén, L.; Tom, J.; Burnier, J.; Butcher, K.A.; Kossiakoff, A.; Wells, J.A. Engineering subtilisin and its substrates for efficient ligation of peptide bonds in aqueous solution. *Biochemistry* **1991**, *30*, 4151–4159.
34. Adamczak, M.; Krishna, S.H. Enzyme for efficient biocatalysis. *Food Technol. Biotechnol.* **2004**, *42*, 251–264.
35. Ebeling, W.; Hennrich, N.; Klockow, M.; Metz, H.; Orth, H.D.; Lang, H. Proteinase K from Tritirachium album Limber. *Eur. J. Biochem.* **1974**, *47*, 91–97.
36. Carpi, F.M.; di Pietro, F.; Vincenzetti, S.; Mignini, F.; Napolioni, V. Human DNA extraction methods: Patents and applications. *Recent Pat. DNA Gene Seq.* **2011**, *5*, 1–7.
37. Sigma-Aldrich Corporation. Enzymes for cell dissociation and lysis. *Biofiles Life Sci. Res.* **2006**, Avaiable online: http://www.sigmaaldrich.com/ etc/medialib/docs/Sigma/ General_Information/2/biofiles_issue2.Par.0001.File.tmp/biofiles_issue2.pdf (accessed on 20 July 2013).
38. Santangelo, C. Worthington biochemical online tissue dissociation guide. *Worthington Biochem. Corp.* **2011**. Avaiable online: http://www.worthington-biochem.com/ tissuedissociation/ default.html (accessed on 20 July 2013).
39. Mandl, I.; Maclennan, J.D.; Howes, E.L. Isolation and characterization of proteinase and collagenase from Cl. histolyticum. *J. Clin. Investig.* **1953**, *32*, 1323–1329.
40. Kin, T.; Johnson, P.R.; Shapiro, A.M.; Lakey, J.R. Factors influencing the collagenase digestion phase of human islet isolation. *Transplantation* **2007**, *83*, 7–12.

41. Canavan, H.E.; Cheng, X.; Graham, D.J.; Ratner, B.D.; Castner, D.G. Cell sheet detachment affects the extracellular matrix: A surface science study comparing thermal liftoff, enzymatic, and mechanical methods. *J. Biomed. Mater. Res. A* **2005**, *75*, 1–13.
42. Huang, H.L.; Hsing, H.W.; Lai, T.C.; Chen, Y.W.; Lee, T.R.; Chan, H.T.; Lyu, P.C.; Wu, C.L.; Lu, Y.C.; Lin, S.T.; *et al.* Trypsin-induced proteome alteration during cell subculture in mammalian cells. *J. Biomed. Sci.* **2012**, *17*, doi:10.1186/1423-0127-17-36.
43. Danhier, P.; Copetti, T.; de Preter, G.; Leveque, P.; Feron, O.; Jordan, B.F.; Sonveaux, P.; Gallez, B. Influence of cell detachment on the respiration rate of tumor and endothelial cells. *PLoS One* **2013**, *8*, e53324.
44. Rader, C. Overview on concepts and applications of Fab antibody fragments. *Curr. Protoc. Protein Sci.* **2009**, *55*, 6.9.1–6.9.14.
45. Flanagan, R.J.; Jones, A.L. Fab antibody fragments: Some applications in clinical toxicology. *Drug Saf.* **2004**, *27*, 1115–1133.
46. De Marco, A. Biotechnological applications of recombinant single-domain antibody fragments. *Microb. Cell Fact.* **2011**, *10*, doi:10.1186/1475-2859-10-44.
47. Zhao, Y.; Gutshall, L.; Jiang, H.; Baker, A.; Beil, E.; Obmolova, G.; Carton, J.; Taudte, S.; Amegadzie, B. Two routes for production and purification of Fab fragments in biopharmaceutical discovery research: Papain digestion of mAb and transient expression in mammalian cells. *Protein Expr. Purif.* **2009**, *67*, 182–189.
48. Hunte, C.; Michel, H. Crystallisation of membrane proteins mediated by antibody fragments. *Curr. Opin. Struct. Biol.* **2002**, *12*, 503–508.
49. Kovari, L.C.; Momany, C.; Rossmann, M.G. The use of antibody fragments for crystallization and structure determinations. *Structure* **1995**, *3*, 1291–1293.
50. Caffrey, M. Membrane protein crystallization. *J. Struct. Biol.* **2003**, *42*, 108–132.
51. Bill, R.M.; Henderson, P.J.; Iwata, S.; Kunji, E.R.; Michel, H.; Neutze, R.; Newstead, S.; Poolman, B.; Tate, C.G.; Vogel, H. Overcoming barriers to membrane protein structure determination. *Nat. Biotechnol.* **2011**, *29*, 335–340.
52. Zhou, Y.; Morais-Cabral, J.H.; Kaufman, A.; MacKinnon, R. Chemistry of ion coordination and hydration revealed by a K+ channel-Fab complex at 2.0 A resolution. *Nature* **2001**, *414*, 43–48.
53. Rasmussen, S.G.F.; Choi, H.J.; Rosenbaum, D.M.; Kobilka, T.S.; Thian, F.S.; Edwards, P.C.; Burghammer, M.; Ratnala, V.R.; Sanishvili, R.; Fischetti, R.F.; *et al.* Crystal structure of the human beta2 adrenergic G-protein-coupled receptor. *Nature* **2007**, *450*, 383–387.
54. Kwong, P.D.; Wyatt, R.; Robinson, J.; Sweet, R.W.; Sodroski, J.; Hendrickson, W.A. Structure of an HIV gp120 envelope glycoprotein in complex with the CD4 receptor and a neutralizing human antibody. *Nature* **1998**, *393*, 648–659.
55. Bagossi, P.; Sperka, T.; Fehér, A.; Kádas, J.; Zahuczky, G.; Miklóssy, G.; Boross, P.; Tőzsér, J. Amino acid preferences for a critical substrate binding subsite of retroviral proteases in type 1 cleavage sites. *J. Virol.* **2005**, *79*, 4213–4218.
56. Eizert, H.; Bander, P.; Bagossi, P.; Sperka, T.; Miklóssy, G.; Boross, P.; Weber, I.T.; Tőzsér, J. Amino acid preferences of retroviral proteases for amino-terminal positions in a type 1 cleavage site. *J. Virol.* **2008**, *82*, 10111–10117.

57. Tőzsér, J. Comparative studies on retroviral proteases: Substrate specificity. *Viruses* **2010**, *2*, 147–165.
58. Patick, A.K.; Potts, K.E. Protease inhibitors as antiviral agents. *Clin. Microbiol. Rev.* **1998**, *11*, 614–627.
59. Wlodawer, A.; Gustchina, A. Structural and biochemical studies of retroviral proteases. *Biochim. Biophys. Acta* **2000**, *1477*, 16–34.
60. De Clercq, E. Strategies in the design of antiviral drugs. *Nat. Rev. Drug Discov.* **2000**, *1*, 13–25.
61. Wlodawer, A. Structure-based design of AIDS drugs and the development of resistance. *Vox Sang.* **2002**, *83 (Suppl. 1)*, 23–26.
62. Pogson, M.; Georgiou, G.; Iverson, B.L. Engineering next generation proteases. *Curr. Opin. Biotechnol.* **2009**, *20*, 390–397.
63. Lim, E.J.; Sampath, S.; Coll-Rodriguez, J.; Schmidt, J.; Ray, K.; Rodgers, D.W. Swapping the substrate specificities of the neuropeptidases neurolysin and thimet oligopeptidase. *J. Biol. Chem.* **2007**, *282*, 9722–9732.
64. Villa, J.P.; Bertenshaw, G.P.; Bond, J.S. Critical amino acids in the active site of meprin metalloproteinases for substrate and peptide bond specificity. *J. Biol. Chem.* **2003**, *278*, 42545–42550.
65. Derewenda, Z.S. The use of recombinant methods and molecular engineering in protein crystallization. *Methods* **2004**, *34*, 354–363.
66. Waugh, D.S. Making the most of affinity tags. *Trends Biotechnol.* **2005**, *23*, 316–320.
67. Renzi, F.; Panetta, G.; Vallone, B.; Brunori, M.; Arceci, M.; Bozzoni, I.; Laneve, P.; Caffarelli, E. Large-scale purification and crystallization of the endoribonuclease XendoU: Troubleshooting with His-tagged proteins. *Acta Crystallogr. Sect. F Struct. Biol. Cryst. Commun.* **2006**, *62*, 298–301.
68. Horchani, H.; Ouertani, S.; Gargouri, Y.; Sayari, A. The N-terminal His-tag and the recombination process affect the biochemical properties of Staphylococcus aureus lipase produced in Escherichia coli. *J. Mol. Catal. B Enzym.* **2009**, *61*, 194–201.
69. Terpe, K. Overview of tag protein fusions: From molecular and biochemical fundamentals to commercial systems. *Appl. Microbiol. Biotechnol.* **2003**, *60*, 523–533.
70. Jenny, R.J.; Mann, K.G.; Lundblad, R.L. A critical review of the methods for cleavage of fusion proteins with thrombin and factor Xa. *Protein Expr. Purif.* **2003**, *31*, 1–11.
71. Arnau, J.; Lauritzen, C.; Petersen, G.E.; Pedersen, J. Current strategies for the use of affinity tags and tag removal for the purification of recombinant proteins. *Protein Expr. Purif.* **2005**, *48*, 1–13.
72. Vergis, J.M.; Wiener, M.C. The variable detergent sensitivity of proteases that are utilized for recombinant protein affinity tag removal. *Protein Expr. Purif.* **2011**, *78*, 139–142.
73. Waugh, D.S. An overview of enzymatic reagents for the removal of affinity tags. *Protein Expr. Purif.* **2011**, *80*, 283–293.
74. Nallamsetty, S.; Kapust, R.B.; Tőzsér, J.; Cherry, S.; Tropea, J.E.; Copeland, T.D.; Waugh, D.S. Efficient site-specific processing of fusion proteins by tobacco vein mottling virus protease *in vivo* and *in vitro*. *Protein Expr. Purif.* **2004**, *38*, 108–115.
75. Huang, Q.; Li, Q.; Chen, A.S.; Kang, C. West Nile virus protease activity in detergent solutions and application for affinity tag removal. *Anal. Biochem.* **2013**, *435*, 44–46.

76. Zhang, D.; Tőzsér, J.; Waugh, D.S. Molecular cloning, overproduction, purification and biochemical characterization of the p39 nsp2 protease domains encoded by three alphaviruses. *Protein Expr. Purif.* **2009**, *64*, 89–97.
77. Austin, B.P.; Tőzsér, J.; Bagossi, P.; Tropea, J.E.; Waugh, D.S. The substrate specificity of Metarhizium anisopliae and Bos taurus carboxypeptidases A: Insights into their use as tools for the removal of affinity tags. *Protein Expr. Purif.* **2011**, *77*, 53–61.
78. Steen, H.; Mann, M. The ABC's (and XYZ's) of peptide sequencing. *Nat. Rev. Mol. Cell Biol.* **2004**, *5*, 699–711.
79. Granvogl, B.; Plöscher, M.; Eichacker, L.A. Sample preparation by in-gel digestion for mass spectrometry-based proteomics. *Anal. Bioanal. Chem.* **2007**, *389*, 991–1002.
80. Meyers, A.; Trauger, S.; Webb, W.; Reisdorph, N.; Wranik, C.; Peters, E.; Siuzdak, G. Protein identification and profiling with mass spectrometry. *Spectroscopy* **2003**, *17*, 1–15.

Permissions

All chapters in this book were first published in MDPI; hereby published with permission under the Creative Commons Attribution License or equivalent. Every chapter published in this book has been scrutinized by our experts. Their significance has been extensively debated. The topics covered herein carry significant findings which will fuel the growth of the discipline. They may even be implemented as practical applications or may be referred to as a beginning point for another development.

The contributors of this book come from diverse backgrounds, making this book a truly international effort. This book will bring forth new frontiers with its revolutionizing research information and detailed analysis of the nascent developments around the world.

We would like to thank all the contributing authors for lending their expertise to make the book truly unique. They have played a crucial role in the development of this book. Without their invaluable contributions this book wouldn't have been possible. They have made vital efforts to compile up to date information on the varied aspects of this subject to make this book a valuable addition to the collection of many professionals and students.

This book was conceptualized with the vision of imparting up-to-date information and advanced data in this field. To ensure the same, a matchless editorial board was set up. Every individual on the board went through rigorous rounds of assessment to prove their worth. After which they invested a large part of their time researching and compiling the most relevant data for our readers.

The editorial board has been involved in producing this book since its inception. They have spent rigorous hours researching and exploring the diverse topics which have resulted in the successful publishing of this book. They have passed on their knowledge of decades through this book. To expedite this challenging task, the publisher supported the team at every step. A small team of assistant editors was also appointed to further simplify the editing procedure and attain best results for the readers.

Apart from the editorial board, the designing team has also invested a significant amount of their time in understanding the subject and creating the most relevant covers. They scrutinized every image to scout for the most suitable representation of the subject and create an appropriate cover for the book.

The publishing team has been an ardent support to the editorial, designing and production team. Their endless efforts to recruit the best for this project, has resulted in the accomplishment of this book. They are a veteran in the field of academics and their pool of knowledge is as vast as their experience in printing. Their expertise and guidance has proved useful at every step. Their uncompromising quality standards have made this book an exceptional effort. Their encouragement from time to time has been an inspiration for everyone.

The publisher and the editorial board hope that this book will prove to be a valuable piece of knowledge for researchers, students, practitioners and scholars across the globe.

List of Contributors

Jaepil Jeong, Hyejin Kim and Jong Bum Lee
Department of Chemical Engineering, University of Seoul, Seoul 130-743, Korea

Ryan M. Bouldin
Department of Natural and Applied Sciences, Bentley University, Waltham, MA 02452, USA

Michael Magaletta and Sean Connor
Department of Plastics Engineering, University of Massachusetts, Lowell, MA 01854, USA

Ankita Singh and Ramaswamy Nagarajan
Department of Plastics Engineering, University of Massachusetts, Lowell, MA 01854, USA
Center for Advanced Materials, University of Massachusetts, Lowell, MA 01854, USA

Jayant Kumar
Center for Advanced Materials, University of Massachusetts, Lowell, MA 01854, USA
Department of Physics, University of Massachusetts, Lowell, MA 01854, USA

Fengjie Cui and Bin Zou
School of Food and Biological Engineering, Jiangsu University, Zhenjiang 212013, Jiangsu, China

Zhongming Wang and Zhenhong Yuan
Key Laboratory of Renewable Energy, Chinese Academy of Sciences, Guangzhou 510640, Guangdong, China

Shuhao Huo
School of Food and Biological Engineering, Jiangsu University, Zhenjiang 212013, Jiangsu, China
Key Laboratory of Renewable Energy, Chinese Academy of Sciences, Guangzhou 510640, Guangdong, China

Pengxiang Zhao
Beijing Biomass Energy Technology Center, State Grid Energy Conservation Service Ltd., Beijing 100053, China

Gábor Náray-Szabó
Laboratory of Structural Chemistry and Biology and HAS-ELTE Protein Modeling Group, Eötvös Loránd University, Pázmány Péter St. 1A, Budapest H-1117, Hungary

Julianna Oláh and Balázs Krámos
Department of Inorganic and Analytical Chemistry, Budapest University of Technology and Economics, Gellért tér 4, Budapest H-1111, Hungary

Sergio D. Aguirre and M. Monsur Ali
Department of Biochemistry and Biomedical Sciences, McMaster University, 1280 Main St. W., Hamilton, ON L8S 4K1, Canada

Bruno J. Salena
Dvision of Gastroenterology, Department of Medicine, McMaster University, 1280 Main St. W., Hamilton, ON L8S 4K1, Canada

Yingfu Li
Department of Biochemistry and Biomedical Sciences, McMaster University, 1280 Main St. W., Hamilton, ON L8S 4K1, Canada
Department of Chemistry and Chemical Biology, McMaster University, 1280 Main St. W., Hamilton, ON L8S 4K1, Canada
Michael G. DeGroote Institute for Infectious Disease Research, McMaster University, 1280 Main St. W., Hamilton, ON L8S 4K1, Canada

Jose L. Adrio
Neol Biosolutions SA, BIC Granada, Granada 18016, Spain

Arnold L. Demain
Research Institute for Scientists Emeriti (R.I.S.E.), Drew University, Madison, NJ 07940, USA

Poonam Singh Nigam
Biomedical Science Research Institute, University of Ulster, Coleraine BT52 1SA, UK

Kouichi Kuroda and Mitsuyoshi Ueda
Division of Applied Life Sciences, Graduate School of Agriculture, Kyoto University, Sakyo-ku, Kyoto 606-8502, Japan

Jun-ichi Kadokawa
Graduate School of Science and Engineering, Kagoshima University, 1-21-40 Korimoto, Kagoshima 890-0065, Japan
Research Center for Environmentally Friendly Materials Engineering, Muroran Institute of Technology, 27-1 Mizumoto-cho, Muroran, Hokkaido 050-8585, Japan

Francesca Valetti and Gianfranco Gilardi
Department of Life Sciences and Systems Biology, University of Torino, via Accademia Albertina 13, Torino 10123, Italy

Petra Staudigl, Iris Krondorfer, Dietmar Haltrich and Clemens K. Peterbauer
Food Biotechnology Laboratory, BOKU-University of Natural Resources and Life Sciences Vienna, Muthgasse 11, Vienna 1190, Austria

János András Mótyán, Ferenc Tóth and József Tőzsér
Department of Biochemistry and Molecular Biology, Faculty of Medicine, Medical and Health Science Center, University of Debrecen, Debrecen H-4012, Hungary

Index

www.ingramcontent.com/pod-product-compliance
Lightning Source LLC
LaVergne TN
LVHW061514220926
848485LV00018B/1037
9781641167963